Smart Cyber-Physical Systems

Smart cyber-physical systems (SCPS) are revolutionizing the interaction between the physical and digital realms, driving innovation across diverse fields. This book provides an extensive overview of cutting-edge SCPS technologies, highlighting their applications, challenges, and transformative potential. Designed for researchers, professionals, and students, it offers foundational insights and advanced applications in urban planning, healthcare, transportation, finance, and cybersecurity. It also bridges the gap between theory and practice through real-world case studies and in-depth analysis, showcasing how SCPS address critical global challenges.

This book:

- Provides a comprehensive introduction to SCPS, their components, and their applications in various domains and explores their role in enabling sustainable smart cities and urban planning.
- Provides advanced SCPS-based solutions for optimizing supply chain logistics and route planning and real-world applications of SCPS in healthcare, including wearable technology and remote patient monitoring, and financial modeling, comparing investment indices using advanced techniques.
- Discusses the integration of artificial intelligence and machine learning in autonomous vehicle systems and transportation.
- Analyzes cybersecurity challenges in cryptocurrency and blockchain ecosystems.
- Focuses on innovations in energy management, including smart grids and sustainable practices powered by SCPS, and explores adaptive human–machine interaction frameworks for enhanced decision-making.

This book serves as a vital resource for understanding the transformative power of SCPS and inspires further research and development in this dynamic and rapidly evolving field.

Computational and Intelligent Systems

Series Editor:
Ramesh Chandra Poonia
CHRIST (Deemed to be University), Bangalore, Karnataka, INDIA - 560029

About the Series:

In today's world, the systems that integrate intelligence into machine-based applications are known as intelligent systems. In order to simplify the human–machine interaction, intelligent systems play an important role. The books under the proposed series will explain the fundamentals of intelligent systems, review the computational techniques, and offer step-by-step solutions to practical problems. Aimed at senior undergraduate students, graduate students, academic researchers, and professionals, the proposed series will focus on broad topics including artificial intelligence, deep learning, image processing, cyber-physical systems, wireless security, mechatronics, cognitive computing, and Industry 4.0.

Application of Soft Computing Techniques in Mechanical Engineering
Amar Patnaik, Vikas Kukshal, Pankaj Agarwal, Ankush Sharma, and Mahavir Choudhary

Computational Intelligence based Optimization of Manufacturing Process for Sustainable Materials
Deepak Sinwar, Kamalakanta Muduli, Vijaypal Singh Dhaka, and Vijander Singh

Advanced Mathematical Techniques in Computational and Intelligent Systems
Sandeep Singh, Aliakbar Montazer Haghighi and Sandeep Dalal

Natural Language Processing and Information Retrieval: Principles and Applications
Edited By Muskan Garg, Sandeep Kumar, Abdul Khader Jilani Saudagar

Soft Robotics for Medical and Healthcare Applications
Edited by Shaik Himam Saheb, Tharakeshwar Appala, and Mohammad S Khan

Smart Cyber-Physical Systems: Innovations and Real-World Implications
Edited by Ramesh Chandra Poonia, Kamal Upreti, and Mohammad S Khan

For more information on this series page, please visit our website link: https://www.routledge.com/Computational-and-Intelligent-Systems/book-series/CRCCIS

Smart Cyber-Physical Systems

Innovations and Real-World Implications

Edited by

Ramesh Chandra Poonia, Kamal Upreti, and Mohammad S. Khan

CRC Press
Taylor & Francis Group
Boca Raton London New York

CRC Press is an imprint of the
Taylor & Francis Group, an **informa** business

Designed cover image: Shutterstock

First edition published 2026
by CRC Press
2385 NW Executive Center Drive, Suite 320, Boca Raton FL 33431

and by CRC Press
4 Park Square, Milton Park, Abingdon, Oxon, OX14 4RN

CRC Press is an imprint of Taylor & Francis Group, LLC

ISBN: 9781032892931 (hbk)
ISBN: 9781032892948 (pbk)
ISBN: 9781003542513 (ebk)

DOI: 10.1201/9781003542513

Typeset in Sabon
by codeMantra

Contents

Preface

The convergence of the physical and digital worlds has ushered in a new era of technological innovation, with smart cyber-physical systems (SCPS) at the forefront. SCPS integrate computational capabilities with physical processes, enabling real-time monitoring, dynamic decision-making, and seamless interconnectivity across various domains. These systems are no longer theoretical constructs; they are transforming industries, redefining urban landscapes, enhancing healthcare, and optimizing critical infrastructures worldwide.

This book, *Smart Cyber-Physical Systems: Innovations and Real-World Implications*, comprehensively explores how SCPS reshape modern life. By addressing their potential, challenges, and practical applications, this book serves as both an academic and useful resource for researchers, professionals, and students in engineering, urban planning, healthcare, finance, transportation, and beyond. It bridges the gap between theory and application, shedding light on how SCPS can drive sustainable development, improve operational efficiency, and enhance the quality of life in unprecedented ways.

This book's structure reflects the multifaceted nature of SCPS, with each chapter delving into a unique aspect or application. A detailed overview of the content is given as follows:

Chapter 1 introduces advancements in space object detection using computer vision. The authors present cutting-edge methodologies that blend SCPS with machine vision to tackle increasing space exploration and object-tracking challenges.

Chapter 2 examines the role of SCPS in creating sustainable smart cities, showcasing case studies where these systems have revolutionized urban planning, resource management, and environmental sustainability.

Chapter 3 discusses the integration of SCPS in digital payment ecosystems. It provides an in-depth analysis of security risks and presents evidence-based solutions for mitigating vulnerabilities in the era of Digital Payments 5.0.

Chapter 4 focuses on how AI-driven SCPS optimize supply chain logistics and route planning. This chapter highlights dynamic models for minimizing costs, improving efficiency, and adapting to real-time changes in supply networks.

Chapter 5 explores the integration of SCPS into urban planning frameworks, envisioning the future of not only connected but also resilient and adaptive cities.

Chapter 6 delves into the innovations in intelligent transportation and autonomous systems. It discusses the potential of SCPS in reducing congestion, enhancing safety, and fostering seamless mobility.

Chapters 7 and 8 present transformative applications of SCPS in the healthcare sector. They emphasize how SCPS are revolutionizing remote patient monitoring, wearable health technologies, and health security, thus contributing to modern medicine's rapid evolution.

Chapter 9 examines SCPS applications in finance, comparing the Nifty 50 and Nifty Midcap 100 indices using advanced machine learning models. This chapter highlights how SCPS enhance investment potential and financial decision-making.

Chapter 10 provides insights into collaborative machine learning techniques for improving the performance of autonomous SCPS. The authors propose methodologies for achieving efficiency and reliability in real-world systems.

Chapter 11 investigates the integration of artificial intelligence and machine learning in advancing autonomous vehicle technology within SCPS. The discussion spans innovations, challenges, and future opportunities in this critical domain.

Chapter 12 applies SCPS and AI to sports analytics, focusing on real-time player performance monitoring and decision-making for enhancing team strategies and individual training.

Chapter 13 addresses cybersecurity challenges in cryptocurrency and blockchain solutions. This chapter highlights SCPS's role in safeguarding decentralized financial systems against emerging threats.

Chapter 14 presents innovations in smart grids and energy management systems powered by SCPS. The authors explore how SCPS drive efficiency, reduce wastage, and enable sustainable energy practices.

Chapter 15 investigates system vulnerabilities in digital environments. The chapter provides actionable strategies for identifying, mitigating, and preventing risks in SCPS deployments.

Chapter 16 concludes with an exploration of adaptive multimodal human–machine interaction frameworks. By focusing on decision-making enhancements, this chapter highlights the role of SCPS in fostering intuitive and effective communication between humans and machines.

Through its detailed exploration of SCPS, this book aims to provide readers with a nuanced understanding of how these systems address contemporary challenges and create opportunities for innovation. Each chapter integrates cutting-edge research with real-world applications, making this book a valuable guide for those seeking to harness the power of SCPS in their respective fields.

We hope this book catalyzes further research and development, inspiring solutions that will shape the future of intelligent, interconnected systems and their implications for society.

About the Editors

Ramesh Chandra Poonia completed a Postdoctoral Fellowship from the Cyber-Physical Systems Laboratory (CPS Lab), Department of ICT and Natural Sciences, Norwegian University of Science and Technology (NTNU), Alesund, Norway. He also completed a Part-Time Postdoctoral Fellowship on the International Collaborative Project of a Prediction Model for Pandemic Disease using Machine Learning at Oakland University, USA, and Imam University, Saudi Arabia. He received his Ph.D. in Computer Science from Apaji Institute of Mathematics & Applied Computer Technology, Banasthali University, Banasthali, India, in July 2013. He has also completed Master of Technology (M.Tech.) in Data Science & Engineering, Work Integrated Learning Programme (WILP) from Birla Institute of Technology and Science, Pilani (BITS Pilani), India. He is a Professor in the Department of Computer Science, CHRIST (Deemed to be University), Bangalore, India. His research interests are sustainable technologies, cyber-physical systems, computational intelligence, and network protocol evaluation. He is Chief Editor of the *TARU Journal of Sustainable Technologies and Computing (TJSTC)* and Associate Editor of the *Sustainable Computing: Informatics and Systems*, Elsevier. He also serves on the editorial boards of a few international journals. He is the principal author and co-author of six books and an editor of more than 25 special issues of journals, books, and conference proceedings. He has also authored/co-authored over 75 research publications in peer-reviewed reputed journals, book chapters, and conference proceedings. Moreover, he is a regular reviewer for international journals and conferences. He also serves as Conference Chair of Conference Series, International Conference on Sustainable Computing in Science, Technology and Management (SUSCOM-20XX). He was endorsed with the prestigious "Faculty Appreciation Award 2013" for commendable services. H-index as of July 2022 is 23, according to Google Scholar.

Kamal Upreti is currently working as Associate Professor in the Department of Computer Science, CHRIST (Deemed to be University), Delhi NCR, Ghaziabad, India. He completed his B.Tech (Hons) degree from UPTU, M.Tech (Gold Medalist) and PGDM(Executive) from IMT Ghaziabad, and Ph.D. in the Department of Computer Science & Engineering. He has completed his postdoctorate degree at the National Taipei University of Business, TAIWAN, funded by MHRD. He has published 50+ patents, 32+ magazine issues, and 165+ research papers in various reputed journals and international conferences. His areas of interest include modern physics, data analytics, cyber security, machine learning, healthcare, embedded systems, and cloud computing. He has published more than 45+ authored and edited books in reputed publications. He has years of enriched experience in corporate and teaching fields in engineering colleges. He has worked with HCL, NECHCL, Hindustan Times, Dehradun Institute of Technology, and Delhi Institute of Advanced Studies, with more than 15+ years of enriching experience in research, academics, and corporate fields. He has also worked at NECHCL in Japan, where he had a project called "Hydrastore" funded by a collaboration between HCL and NECHCL Company. He worked on a government project: "Integrated Power Development Scheme (IPDS)," launched by the Ministry of Power, Government of India, to strengthen the sub-transmission and distribution network in the urban areas. Currently, he has completed work in collaboration with GB PANT & AIIMS Delhi, the funded project of the ICMR Scheme on cardiovascular disease prediction strokes using machine learning techniques from 2017 to 2020, which has a fund of 80 lakhs. He got 5 lakhs funding from DST SERB for conducting the International Conference, ICSCPS-2024, 13–14 Sept 2024. Recently, he received a fund of 10 lakhs from AICTE – Inter-Institutional Biomedical Innovations and Entrepreneurship Program (AICTE-IBIP) for 2024–2026. He has attended as a session chairperson at national and international conferences and was a keynote speaker on various platforms such as skill-based training, corporate training, guest faculty, and faculty development programs. He was awarded as the best teacher, researcher, extra academic performer, and gold medalist in the M.Tech program.

Mohammad S. Khan (SM' 19) is Associate Professor of Computing at East Tennessee State University and Director of the Network Science and Analysis Lab (NSAL). He received his M.Sc. and Ph.D. in Computer Science and Computer Engineering from the University of Louisville, Kentucky, USA, in 2011 and 2013, respectively. His primary area of research is in *ad hoc* networks, wireless sensor networks, network tomography, connected vehicles, vehicular social networks, and cyber-physical systems. He is an associate editor of IEEE Access, IET ITS, IET WSS, Telecommunication Systems, and Neural Computing and Applications. He has been on technical program committees of various international conferences and is a technical reviewer of various international journals in his field. He is a senior member of IEEE.

Contributors

Adanoi Joseph Abraham
CHRIST (Deemed to be University)
Delhi NCR, India

Srinivasa Rao Adapa
OpenLogix LLC
Bloomfield Hills, Michigan

Gaurav Agarwal
Invertis University
Bareilly, India

Faraz Ahmad
CHRIST (Deemed to be University)
Delhi NCR, India

Bader Eddin Alasali
Civil and Environmental Engineering Department
College of Engineering
National University of Science and Technology
Muscat, Oman

Khuloud Nasser Al Balushi
Civil and Environmental Engineering Department
College of Engineering
National University of Science and Technology
Muscat, Oman

Yashmita Awasthi
School of Commerce, Finance and Accountancy
Christ University, Bengaluru

Aishwarya Balan
CHRIST (Deemed to be University)
Delhi NCR, India

Bhavana Bharati
CHRIST (Deemed to be University)
Delhi NCR, India

Maligi Bindu
CGI
Hyderabad, India

Rajat Dang
SEAS, GWU
Washington, District of Columbia

Prapti D. Deshmukh
G.Y. Pathrikar College of Computer Science and Information Technology
MGM University
Aurangabad, India

Shubham Rajendra Ekatpure
Kulicke and Soffa
Fort Washington, Pennsylvania

Parth Gharat
Regional Ayurveda Research Institute Pune CCRAS
Ministry of AYUSH Govt. of India
Pune
Pune, India

Ashwini Dattatraya Gopwad
Deogiri College
Aurangabad, India

Anuj Gupta
Chandigarh University
Gharun, India

Rituraj Jain
Marwadi University
Rajkot, Gujarat, India

Rasika Kolhe
Regional Ayurveda Research Institute Pune CCRAS
Ministry of AYUSH Govt. of India
Pune
Pune, Maharashtra, India

Puja Kumari
Gautam Buddha University
Greater Noida, Uttar Pradesh, India

Virendra Singh Kushwah
VIT Bhopal University
Sehore, Madhya Pradesh, India

Rishit Maheshwari
Pandit Deendayal Energy University
Gandhinagar, Gujarat, India

Bhavya Mathur
VIT Bhopal University
Sehore, India

Pratham Nijhawan
CHRIST (Deemed to be University)
Delhi NCR, India

Shivani Odedra
Pandit Deendayal Energy University
Gandhinagar, India

Damodharan Palaniappan
Marwadi University
Rajkot, India

Brijal Pandey
CHRIST (Deemed to be University)
Delhi NCR, India

Kumar Parmar
Marwadi University
Rajkot, India

Ramesh Chandra Poonia
CHRIST (Deemed to be University)
Delhi NCR, India

T. Premavathi
Marwadi University
Rajkot, India

Shashank Rai
School of Commerce, Finance and Accountancy
Christ University, Bengaluru

Vineet Kumar Ramnani
CHRIST (Deemed to be University)
Delhi NCR, India

Shreya Saha
CHRIST (Deemed to be University)
Delhi NCR, India

Sonali Samdarshini
Barclays
Pune, India

Ansu Samuel
Go Daddy
Tempe, Arizona

Akash Sanghi
Invertis University
Bareilly, India

Asmita Sarkar
VIT Bhopal University
Sehore, India

Ayesha Anam Irshad Siddiqui
G.Y. Pathrikar College of Computer Science and Information Technology
MGM University
Aurangabad, India

Aman Singh
State Street Corporation
Bengaluru, India

Aromma Singh
CHRIST (Deemed to be University)
Delhi NCR, India

Shanu Singh
School of Commerce, Finance and Accountancy
Christ University, Bengaluru

Sitesh Kumar Singh
Civil and Environmental Engineering Department
College of Engineering
National University of Science and Technology
Muscat, Oman

S. Sridhar Raj
Mepco Schlenk Engineering College
Sivakasi, India

Om Sutariya
Pandit Deendayal Energy University
Gandhinagar, Gujarat, India

Sumegh Tharewal
School of Advance Computing
DBS Global University
Dehradun, India

Reeta Tomar
CHRIST (Deemed to be University)
Delhi NCR, India

Kamal Upreti
Department of Computer Science
School of Sciences
CHRIST (Deemed to be University)
Delhi NCR, India

Shitiz Upreti
Lloyd Business School
Lloyd Group of Institutions
Greater Noida
Uttar Pradesh, India

Acknowledgments

The journey of creating this book, *Smart Cyber-Physical Systems: Innovations and Real-World Implications*, has been an intellectually enriching and collaborative endeavor, and it would not have been possible without the support, guidance, and contributions of numerous individuals and institutions.

First and foremost, we extend our deepest gratitude to all the authors and researchers who contributed their groundbreaking work to this volume. Their dedication and innovative ideas have shaped this book into a valuable resource for understanding and exploring the potential of smart cyber-physical systems.

We are profoundly thankful to the academic and professional communities who provided critical insights and feedback throughout development. Their expertise and encouragement have been instrumental in refining the content and ensuring the book's relevance and rigor.

We acknowledge the unwavering support of our institutions and colleagues, whose resources and encouragement helped create an environment conducive to academic pursuit. We also thank the technical editors and publishing team, whose meticulous attention to detail and professionalism ensured the seamless completion of this work.

Special recognition is reserved for our Research Scholar, Pratham Nijhawan (CHRIST Deemed to be University, Delhi NCR, India), whose contributions have been pivotal in compiling this book. His relentless pursuit of excellence, deep understanding of smart cyber-physical systems, and commitment to advancing research in this field have enriched the book significantly. Beyond his intellectual input, Pratham's dedication and collaborative spirit have inspired us throughout this journey. We sincerely appreciate his endeavors and wish him continued success in his academic and professional pursuits. We would also like to thank Mrs. Anju Nijhawan, whose contributions and suggestions have been significant in compiling this book.

Finally, we thank our families and loved ones for their enduring patience and support during the long hours dedicated to this project. Their belief in us has been our cornerstone.

This book is a testament to the collective efforts of everyone involved, and we hope it serves as a valuable resource for researchers, practitioners, and readers exploring the transformative world of smart cyber-physical systems.

Chapter 1

Exploring advancements in space object detection through computer vision

Ayesha Anam Irshad Siddiqui, Prapti D. Deshmukh, Kamal Upreti, Puja Kumari, and Pratham Nijhawan

1.1 INTRODUCTION

The significance of space object detection cannot be undervalued due to its application in controlling satellites and exploring space. Due to the deployment of computer vision in this space, computer vision as a subfield of machine learning and imaging technologies is also included in space object detection [1]. This review discusses recent progress, prominent approaches, and new discussions concerning space object detection focused on computer vision.

Space object detection through the use of computer vision applications is a vital technology that combines space imaging capabilities with artificial intelligence for the purposes of target identification and tracking. On this front, the combination of disciplines such as astronomy, satellite platforms, and information technology is used to realistically resolve the problems of space surveillance and exploration [2].

The aspects that are common among the available literature and are identified as key features include the following:

- **Image Acquisition:** The use of telescopes and cameras for taking pictures of space objects [3].
- **Image Processing:** Applying algorithms on captured images or videos to enhance an object and effectively eliminate noise [4].
- **Object Orientation:** Utilization of machine learning to automatically locate and organize different types of orbiting features like satellites, space junk, or planets [5].
- **Tracking Algorithms:** Research on the methods developed for maintaining the position and direction of the detected object in motion over a span of time [6].
- **Multi-sensor Fusion:** Integrating the outputs of different sensors [7].

The detection of space objects is significant due to its numerous applications. They include the following:

DOI: 10.1201/9781003542513-1

a. **Space Situational Awareness Program (SSAP):** This enables the characterization and tracking of various objects in Earth's orbit, such as satellites and other risky objects [8].
b. **Collision Prevention:** This involves the accurate identification of orbital debris and accurate tracking of their motion, which helps space vehicle operators avoid collisions with other objects, such as satellites [9].
c. **Space Debris Mitigation:** Space orbital debris can be hidden, and a proper system of identification and cataloging can assist in cleaning up the orbit and safeguarding future satellites [10].
d. **Astronomy:** When integrated with computer vision systems, they assist with the detection and observation of space objects, which include asteroids and comets.
e. **Security:** Space object detection plays a fundamental role in the space-based components essential to national defense and intelligence/surveillance [11].
f. **Satellites:** These satellite systems are made more efficient by integrating accurate orbital detection systems and greatly enhance the management of satellite constellations [12].
g. **Space Science:** Accurate orbital measuring systems are useful for investigating other planetary bodies along with their moons and satellites. 8. Impact studies: Possible asteroid detection and tracking systems can protect the Earth from future impacts [13].
h. **Space Traffic Control:** The mapping out and tracking of a congested space is made possible through the use of artificial intelligence and satellite-based surveillance technologies for future space traffic management [13].

1.2 LITERATURE REVIEW ON OBJECT DETECTION TECHNIQUES

This literature review explores various approaches and methodologies in the field of object detection within computer vision. It encompasses a wide range of scholarly works, each contributing unique insights and advancements to the domain.

Object Detection Techniques and Applications: A significant number of related works have been reported in the object detection domain. A few them are mentioned as follows:

a. **Tools, Techniques, Datasets, and Application Areas for Object Detection** [14]: This paper gives a detailed overview of object detection techniques, datasets, and evaluation metrics. It also presents a comparative analysis among different methods along with the challenges and future aspects, but it did not specifically account about the detection of space objects.

b. **Current and Future Trends in Object Detection** [15]: In this paper, both traditional methods as well as deep learning-based methods have been used in order to analyze current trends with future ones. Traditional-based methods use visual features such as shape and texture, whereas evolutionary-based methods use deep learning techniques such as convolutional neural networks.
c. **Evolutionary Learning Approach for Object Detection using an Analytical Review** [16]: This study performs an analytical review with research papers up to the last 4 years. Here, researchers classify machine learning-based object detection approaches discussed related on different research challenges/gaps considering fast R-CNN and gradient-based algorithms.
d. **Improving Classification Accuracy and Robustness** [17]: Presents an overall survey of object detection methods for improving the efficiency, accuracy, and robustness, but without any specific detailed methodology.
e. **Computer Vision for Explosive Ordnance Detection** [18]: Analyze the state of the art of computer vision techniques and their applications in explosive ordnance detection, including a comprehensive review of imaging sensors and object detection methods designed to work on mobile platforms which integrate unmanned ground vehicles and handheld devices.
f. **Development of Object Detection Techniques** [19]: Provides an overview on object detection using traditional machine learning and deep learning methods like R-CNN and YOLO, but not specific to space object detection.
g. **Object Detection Models in Various Industries** [20]: Compares two-stage and single-stage object detection models in different industries and does not address space objects explicitly.
h. **Recent Evolutions in Object Detection Algorithms** [21]: Gives a brief about the traditional approaches like support vector machine (SVM) and recent deep learning, which is a review without showing the methodology.
i. **Deep Neural Networks for General Object Detection** [22]: This paper reviews state of the art and open challenges of deep learning-based object detection (DetCo-like solutions and DETR-like modifications).
j. **Twenty Years of Object Detection Research** [23]: A comprehensive survey on object detection advances from the 1990s to 2022 for nearly 25 years.
k. **Deep Learning-based Object Detection Methods** [24]: Review and comparison of deep learning-based methods explaining trade-offs between accuracy and speed for different applications and datasets.

l. **Object Detection and Tracking Algorithms** [25]: A literature review of state-of-the-art detection and tracking algorithms to reduce the tracking drift in computer vision.
m. **Object Detection in UAV Surveillance** [26]: Reviews different object detection techniques applicable to UAV images (see Table 1.1).

Table 1.1 Tabular summary of object detection techniques and applications

Paper/authors	*Summary*	*Key techniques & applications*	*Challenges identified*
Tools, Techniques, Datasets, and Application Areas for Object Detection [14]	Provides a detailed overview of object detection techniques, datasets, and evaluation metrics. Comparative analysis among different methods, challenges, and future aspects discussed.	Object detection techniques, datasets, and evaluation metrics	Does not address detection of space objects
Current and Future Trends in Object Detection [15]	Analyzes current and future trends using both traditional and deep learning methods.	Traditional (shape and texture) and deep learning (CNNs)	Limited focus on applications beyond general object detection
Evolutionary Learning Approach for Object Detection using Analytical Review [16]	Reviews machine learning-based object detection methods from the last 4 years. Classifies methods based on challenges and gaps.	Fast R-CNN and gradient-based methods	Limited to methods from the last 4 years
Improving Classification Accuracy and Robustness [17]	Surveys methods for improving efficiency, accuracy, and robustness in object detection.	General object detection methods	No specific methodology provided
Computer Vision for Explosive Ordnance Detection [18]	A comprehensive review of computer vision techniques in explosive ordnance detection, particularly for mobile platforms.	Imaging sensors, mobile platforms, UGVs, and handheld devices	Limited focus on general object detection
Development of Object Detection Techniques [19]	Overview of traditional and deep learning object detection methods, like R-CNN and YOLO.	R-CNN, YOLO, traditional machine learning, and deep learning methods	Does not specifically address space object detection

(*Continued*)

Table 1.1 (Continued) Tabular summary of object detection techniques and applications

Paper/authors	*Summary*	*Key techniques & applications*	*Challenges identified*
Object Detection Models in Various Industries [20]	Compares two-stage and single-stage models across various industries.	Two-stage vs. single-stage models	Lacks focus on specific industry challenges like space object detection
Recent Evolutions in Object Detection Algorithms [21]	Briefly reviews traditional and recent deep learning object detection methods.	Traditional (SVM) and deep learning approaches	No specific methodology discussed
Deep Neural Networks for General Object Detection [22]	Reviews deep learning-based object detection methods and challenges.	DetCo-like and DETR-like modifications	Generalized approach without detailed applications
Twenty Years of Object Detection Research [23]	Comprehensive survey on object detection advances from the 1990s to 2022.	Review of techniques and approaches over 25 years	Limited focus on current or niche applications
Deep Learning-based Object Detection Methods [24]	Comparison of deep learning-based methods, discussing trade-offs between the accuracy and speed.	Deep learning-based methods	Trade-offs between accuracy and processing speed
Object Detection and Tracking Algorithms [25]	Reviews state-of-the-art detection and tracking algorithms to reduce tracking drift.	Object detection and tracking algorithms	Challenges in minimizing tracking drift
Object Detection in UAV Surveillance [26]	Reviews object detection techniques for UAV imagery analysis.	UAV-specific object detection methods	Limited generalizability outside UAV imagery applications

1.3 SPACE OBJECT DETECTION METHODOLOGIES

1.3.1 Traditional methods of image processing in the past

In the past, the detection of objects in space focused on finding images by primarily utilizing the methods of image processing that were available at that time. Most of these techniques involved the use of feature extraction methods in the images, such as using thresholding or edge detection. Sobel and Canny edge detectors have, for example, been employed to ascertain the edges of celestial objects [27]. In any case, these approaches tend to be limited by their inability to cope with noise and variable illumination, which is bound to be the case in any space image.

1.3.2 Machine learning techniques

Another fairly recent trend has been the application of machine learning techniques in order to improve the detection rate. Random forests and SVM have been employed to recognize space objects based on the thus obtained features. Methods of Machine Learning: The use of machine learning methods to increase the detection accuracy is a recent development. To identify space objects based on extracted features, Random forests and SVMs have been used [28]. So machine learning with SVMs is one of the useful machine learning techniques. With it, the target feature classes can easily learn from the training dataset.

1.3.3 Deep learning

Neural Networks and Deep Learning: Deep learning's better feature extraction capabilities have completely changed the way that space objects are detected. Because they can learn hierarchical feature representations, convolutional neural networks (CNNs) are currently widely employed for recognition tasks [29]. Real-time space object identification has seen the adaptation of models such as You Only Look Once (YOLO) and fast R-CNN, which have demonstrated considerable promise in terms of accuracy and speed [30] (see Figure 1.1).

1.4 EMERGING DEBATES AND THEMES

1.4.1 Precision vs. real-time data processing

The main topic of discussion in this discipline is the trade-off between accurate detection and real-time processing. Despite the CNN's excellent

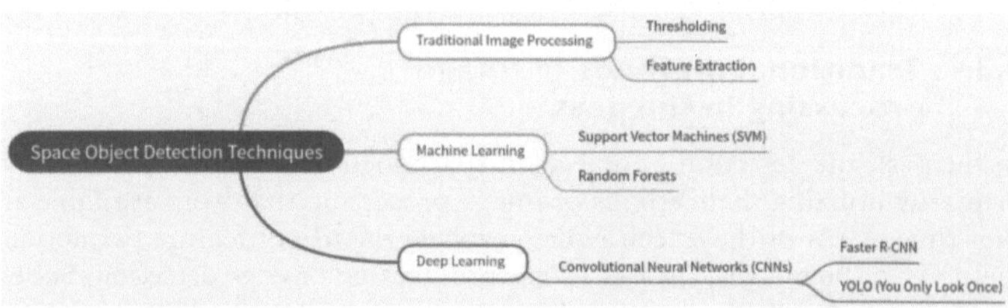

Figure 1.1 Space object detection methodologies (Author's interpretation).

performance, real-time implementation is often constrained due to the high computational requirements of neural networks in shallow systems such as satellites [31]. Availability of data and labeling: Several challenges remain, including attaining access to massive, tagged and labeled datasets. Despite the use of synthetic datasets together with real datasets, issues of porting models trained from fake data to real-world scenarios subsist (Zhuang, Tang, & Guo, 2021). This again underlines the need for collaboration in order to develop large training databases.

1.4.2 Real-time processing vs. accuracy

One of the primary debates in this field concerns the trade-off between real-time processing capabilities and detection accuracy. While CNN-based models provide high accuracy, their computational demand poses challenges for real-time processing, especially on resource-constrained platforms like satellites [31].

1.4.3 Data availability and annotation

The availability of large, annotated datasets remains a significant challenge. While synthetic datasets are being used to augment real-world data, questions remain about the transferability of models trained on synthetic data to real-world scenarios [32]. This gap highlights the need for collaborative initiatives to develop comprehensive datasets for training.

1.4.4 Robustness and adaptability

There's ongoing research into making models more robust and adaptable to varying conditions, such as lighting changes and object occlusions. Transfer learning and domain adaptation are actively explored to minimize these issues and improve model generalization across different tasks [33].

1.5 GAPS AND FUTURE DIRECTIONS

1.5.1 Addressing dataset limitations

A significant gap is the lack of expansive, labeled datasets representative of real-world space conditions. Future research should focus on collaborative datasets that incorporate data from multiple sources and conditions.

1.5.2 Improving computational efficiency

Further work is needed to enhance the computational efficiency of deep learning models, particularly for deployment in space where processing power and battery life are limited.

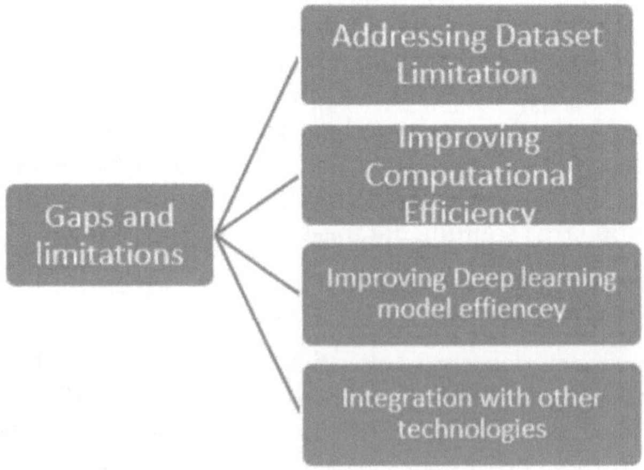

Figure 1.2 Gaps and future direction (Author's interpretation).

1.5.3 Integration with other technologies

There is potential for integrating space object detection systems with other technologies, such as automated decision-making systems or communication networks, to enable performing more autonomous space operations (see Figure 1.2).

1.6 SPACE OBJECT DETECTION AND TRACKING: A LITERATURE REVIEW

This literature review investigates the advancement of computer vision techniques applied to the detection, tracking, identification, and characterization of space objects. Attention is given strictly to the methods designed to tackle those unique challenges posed by space environments and objects, most notably dim and eccentric objects.

1.6.1 Key papers and contributions

a. **Computer Vision Techniques for Space Object Detection:** This paper investigates the application of computer vision techniques such as the Harris Corner method and phase congruency for detecting and tracking space objects. These techniques extract static and streaking point sources from images faster and serve to address the challenges posed with dim and dynamic objects. Other discussions include geometric inference based upon detections and methods for assessing localization uncertainties [34].

b. **Detection of Asteroids/Comets in Space Satellite Images:** The authors present a computer vision-based approach to detecting moving objects such as asteroids and comets in NEOSSat satellite images. Their method can detect objects that differ in brightness by less than 2% from the background, including nonlinearly moving objects. This is efficient and user-friendly, winning the Canadian Space Agency's Space Apps Challenge 2019 [35].
c. **EfficientDet for Space Object Localization and Classification:** This publication reports on using the EfficientDet detector to enhance space situational awareness by localizing and classifying space objects. It describes the main features [36].
d. This AstroVision dataset represents a very useful asset for training as well as the evaluation of feature detection networks, being compiled from 115,970 images of 16 celestial bodies. This dataset is essential for benchmarking models; however, the esthetic lack of training data poses a challenge in building robust models that can aid autonomous space exploration. On occasion, deep learning has been proven beneficial for facilitating autonomous decision-making and characterizing celestial bodies on its own. Applications of deep learning in space missions indicate that this work requires resilient fully autonomous systems as a challenging obstacle on the road to space operations [37].
e. The SpaceYOLO model is proposed by Mahendrakar [38], which addresses several challenges of real-time detection of space objects with noncooperative tumbling. The proposed model fits better for onboard feature detection on spacecraft, enabling autonomous navigation and debris removal. Such capabilities have recently become very important for servicing and maintaining satellite constellations deployed in the Earth's orbit.
f. To further reduce collision dangers, San Martín [39] introduced an All-sky image detection pipeline to locally detect sunlit objects in low-light conditions through processing images from a synoptic telescope. Other notable improvements comprised noise reduction as well as successful tracking in fluctuating lighting situations, all of which remain pivotal for safe and effective space operations.
g. Multiscale detection in satellite imagery has been investigated by Tiwari et al. [40], Etten [41], and Tahir et al. [42], all using some version of YOLO to scan images of on-orbit satellite deployments, which offers real-time object detection with a fair amount of accuracy given the small amounts of data. All of these models, permitted by implementations such as YOLOv4, YOLT, and CNNs, could provide insights into the chains of satellite data, potentially offering high-speed object detection; however, many issues still remain, especially concerning the operational complexity and data scalability in a practical environment.

1.7 APPROACHES AND DIFFICULTIES

Harris corner: Among the basic methods for locating important areas of interest in pictures are the corner and phase congruency methods. They are especially appropriate for detecting both stationary and mobile point sources in space.

Shift-and-Add: This technique aligns incoming images on each other and then superimposes them to bring forth the visibility of moving objects. This is computationally intensive but effective for linear trajectory detection.

Streak Detection: This is a method that finds patterns of streaks in frames, which can then be utilized to track the motion of objects across a number of frames. It is particularly capable of handling nonlinear trajectories.

EfficientDet Detector: This modern machine learning method presents a vigorous solution for object localization and classification in space, addressing variation in object appearances and environmental conditions (see Table 1.2).

Table 1.2 Tabular summary of findings and methods: space object detection and tracking

Category	*Paper*	*Methodologies used*	*Key findings/contributions*	*Challenges addressed*
Computer Vision Techniques for Space Object Detection	[34]	Harris corner method and phase congruency	Extracts static and streaking sources from images to detect dim and dynamic objects; includes geometric inference and localization uncertainty analysis.	Detection of dim and dynamic objects
Detection of Asteroids/ Comets in Space Satellite Images	[35]	NEOSSat Image Analysis	Detects objects with brightness differences <2%; efficient for nonlinear trajectories; won Canadian Space Agency's Space Apps Challenge 2019.	Detecting faint, nonlinearly moving objects
EfficientDet for Space Object Localization and Classification	[36]	EfficientDet Detector	Enhances space situational awareness by localizing and classifying space objects.	Complexity in detecting diverse space objects

(Continued)

Table 1.2 (Continued) Tabular summary of findings and methods: space object detection and tracking

Category	*Paper*	*Methodologies used*	*Key findings/contributions*	*Challenges addressed*
AstroVision Dataset for Autonomous Space Missions	[37]	Deep learning for feature detection	Contains 115,970 images across 16 celestial bodies, essential for model benchmarking. Highlights the need for annotated data.	Scarcity of annotated data; need for autonomous systems
Deep Learning Applications in Space Missions	[37]	Optical feature tracking with deep learning	Facilitates autonomous operations and supports celestial body characterization.	Need for resilient autonomous systems
Real-Time Space Object Detection with SpaceYOLO	[38]	SpaceYOLO model	Enables real-time detection on spacecraft for autonomous navigation and debris removal, particularly for non-cooperative, tumbling objects.	Handling non-cooperative, tumbling objects
All-Sky Image Detection Pipeline	[39]	Synoptic telescope image processing	Detects sunlit objects, reduces noise, and tracks in low-light conditions, crucial for collision risk mitigation.	Tracking in variable lighting conditions
Multi-Scale Detection in Satellite Imagery	[40–42]	YOLOv4, YOLT, CNNs	Provides real-time object detection with limited data; potential for high-speed, multiscale detection in satellite imagery.	Complexity and scalability in operational data processing

1.8 REVIEW OF THE LITERATURE ON THE DETECTION AND CLASSIFICATION OF SPACE OBJECTS

1.8.1 Secure space navigation and operations management

Space situational awareness (SSA) is crucial for space navigation since spacecraft and space debris occurrences continue to increase. One of the significant components of SSA is the identification and categorization of space objects that enable the increase in the potential for space collisions and vehicle access for servicing. The strengths of SSA systems could be advanced by recent advancements in machine learning, particularly object recognition systems such as EfficientDet and SpaceYOLO.

An EfficientDet Application to Detect Space Objects Scientific Reports: Aldahoul et al. [36] performed the detection and classification of space objects using models of EfficientDet and EfficientNet-B4. Pointing out the difficulty of recognition of objects in complicated backgrounds, this study intends to implement a decision fusion method that greatly refines the detection metrics-A at last: M.P. 78.45%, M.R. 92.00%, and C.A. 94%. Practical SSA applications of this method were illustrated by tests conducted on the SPARK dataset.

Automated Identification through All-Sky Imaging: San Martín [39] provided a customizable pipeline for automatic space object detection based on all-sky images from a synoptic survey telescope.

The pipeline addresses image distortion and noise reduction to assist in collision anticipation and avoidance with debris surrounding near-Earth orbits, thus making illuminated objects in space much easier to spot.

Detection in Real-Time with SpaceYOLO: Mahendrakar et al. [38] in the IEEE Aerospace Conference presented SpaceYOLO, an onboard spacecraft real-time feature detection model built on the cognitive processes of human beings. The model aims to deliver accurate identification of features on tumbling and less-than-cooperative space objects to assist trajectory planning and autonomous navigation. This capability will be crucial for safe robotic interventions in the removal of space debris and on-orbit servicing (see Table 1.3).

1.8.2 Review of literature on object detection based on satellite imagery

Although the reasons behind the rise of object detection have been paved by the challenging development of satellite images and machine learning

Table 1.3 Tabular summary review of the literature on the detection and classification of space objects

Author(s)	*Study*	*Findings*	*Method*
Aldahoul et al. [36]	EfficientDet Application to Detect Space Objects	Detection metrics: M.P. 78.45%, M.R. 92.00%, and C.A. 94%.	Uses EfficientDet and EfficientNet-B4 for detection.
San Martín [39]	Automated Identification Through All-Sky Imaging	Improved detection of illuminated space objects for collision avoidance.	Customizable pipeline for automatic detection from all-sky images.
Mahendrakar et al. [38]	Detection in Real-time with SpaceYOLO	Accurate identification of features on tumbling, less-cooperative objects for trajectory planning.	SpaceYOLO model for real-time feature detection.

techniques, this paper duly examined three relevant but splendid works related to this novel imagination implemented in such modern schemes of satellite image object detection.

a. Rishabh Tiwari, Ashwani Kumar Dubey, and Á. Rocha proposed a Peninsula of Modern Yolov4 Procedure, which demonstrated efficient object detection in satellite images. Using the TensorFlow framework, the Yolov4 architecture merges many object detection mechanisms. Their consolidated architecture has been shown to perform well in enhancing the overall detection accuracy in challenging environments [40].
b. YOLTT is an acronym for "You Only Look Twice" in a changing architecture. The YOLT pipeline, conceptualized by A.V. Etten [41], is meant for rapid multiscale object detection in satellite imagery; this method can detect various-sized objects with considerably less training time as its object detection is performed efficiently over satellite images on a scale greater than 0.5 km^2/s. The YOLT pipeline will be very helpful in analyzing large volumes of satellite data.
c. A study conducted by Arsalan Tahir et al. [4] comparing the performance of several deep-learning algorithms for open-area TDis detection in satellite imagery is very comparable in scope. According to the authors, the convolutional neural networks have been shown to operate very well as an attempt to improve their targeting at the said level using the advantages of satellite data. Urban planning and defence-type things are, therefore, highly important in real-world applications (see Table 1.4).

Table 1.4 Tabular summary of review of literature on object detection based on satellite imagery

Author(s)	*Study*	*Findings*	*Method*
Rishabh Tiwari, Ashwani Kumar Dubey, Á. Rocha [40]	Peninsula of Modern Yolov4 Procedure	Efficient object detection in satellite images, enhancing overall detection accuracy in challenging environments.	Yolov4 architecture using the TensorFlow framework.
A.V. Etten [41]	YOLTT (You Only Look Twice)	Fast multiscale object detection in satellite imagery, with less training time and efficiency for large-scale analysis.	YOLT pipeline for object detection in satellite imagery (>0.5 km^2/s).
Arsalan Tahir et al. [4]	Comparison of Deep-Learning Algorithms for TDis Detection	Convolutional neural networks perform well for open-area TDis detection, important for urban planning and defense applications.	Evaluation of multiple deep-learning algorithms in satellite imagery.

1.8.3 Summary of computer vision for space object classification, detection, tracking, and autonomous feature detection review

1.8.3.1 Space object detection and tracking

Focuses on tracking and acquiring space objects, such as satellites and other orbital debris. It guarantees adequate environmental awareness and prevents collisions while forecasting object movements using radar, tracking algorithms, and telescopes.

1.8.3.2 AstroVision – autonomous feature detection for space missions

The Successful Autonomous Feature Recognition System for Space Missions: AstroVision for AFM: Includes online tools to help a craft dock, land, or steer by identifying things such as a specific landmark or even a planet. The space missions become more autonomous as a result.

1.8.3.3 Space object detection and classification

Finding space objects and classifying them as satellites or space debris are two aspects of the detection and classification of space objects. Classification aids in the creation of an object database and facilitates STM.

1.8.3.4 Object detection in satellite imagery

Object Recognition in Satellite Images: This technique is used in Earth observation to analyze satellite imagery and concentrate on identifying particular objects, such as roads or buildings in satellite photos, to provide aid in areas affected by natural disasters, or satellite imagery used in environmental studies and city planning (see Figure 1.3 and Table 1.5).

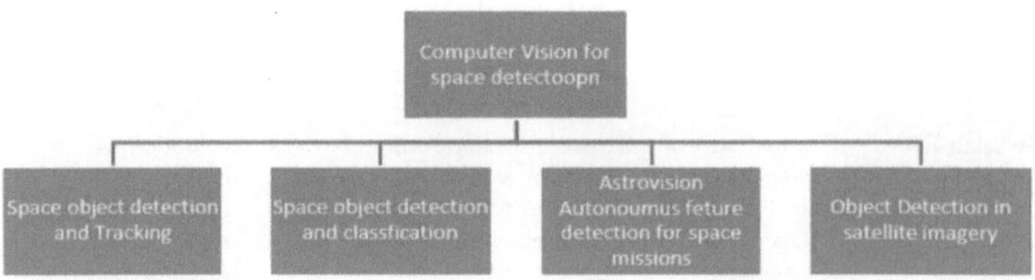

Figure 1.3 Computer vision for space detection categories (Author's interpretation).

Table 1.5 Tabular summary of computer vision for space detection categories

Aspect	*Space object detection and tracking*	*AstroVision – autonomous feature detection for space missions*	*Space object detection and classification*	*Object detection in satellite imagery*
Purpose	Identify and monitor the movement of space objects (e.g., debris and satellites)	Detect mission-critical features autonomously during space missions	Detect and classify objects in space (e.g., satellites and debris) by type	Identify objects (e.g., buildings and roads) on Earth's surface from satellite images
Domain	Space Situational Awareness (SSA)	Autonomous spacecraft navigation and operations	Space object cataloging and identification	Remote sensing and geospatial analysis
Focus	Real-time tracking of orbits and trajectories	Real-time feature detection for navigation or docking	Detection with classification (e.g., types of debris or spacecraft)	Localization of objects in high-resolution satellite images
Techniques Used	Radar, telescopes, Kalman filters, and machine learning for tracking	Computer vision algorithms, and SLAM (Simultaneous Localization and Mapping)	Neural networks (e.g., CNNs) and classifiers	Image segmentation and object detection models (e.g., YOLO and Mask R-CNN)
Applications	Collision avoidance and traffic management in space	Navigation, landing, rendezvous, and docking in autonomous missions	Space debris identification and satellite categorization	Disaster management, urban planning, and environmental monitoring

1.9 CONCLUSION

This literature review highlights the contribution provided by datasets like AstroVision in advancing computer vision for space applications. It thus promotes the growth of autonomous systems for future space missions by defining existing gaps and outlining specific new directions for research. The methodologies provide development for space object detection and classification. EfficientDet and Space YOLO are fruitful models that exhibit an upgrade in their performance scores for real-time detection settings. A huge problem exists, that is, to further refine these models. Future works will involve integrating such models with other technologies about SSA to fit into an integrated autonomous space navigation system. This literature review showcases recent studies and discusses their merits within the existing methodologies while providing synergy-oriented recommendations to bolster model robustness against varying space conditions and extend the dataset used to support modeling. This paper shows some noteworthy progress in object detection within satellite photography through novel approaches and frameworks. The major topics requiring attention include investigating multiscale detection methods and modifications to present deep learning models, allowing for greater efficiency and accuracy. Further research may consider evaluating the fits of these methods across varied environmental settings, together with their real-time data handling capabilities.

To conclude, progress in space object detection and tracking is critical for ensuring safety, reliability, and operational efficiency in space. A wide variety of applications ranging from space debris monitoring and satellite observations to autonomous feature detection are associated with mission-critical tasks. Each demonstrates a unique application of cutting-edge technologies – i.e., machine learning, computer vision, or real-time tracking systems. These improvements aid not just in navigation and collision avoidance in space but also contribute extensively to our knowledge of the Earth's surface through remote sensing. Ongoing developments will continue to prop up the application of various technological platforms in addressing the challenges of SSA and sustainable management of space resources.

REFERENCES

1. Z. Shangguan, L. Wang, J. Zhang, W. Dong, "Vision-based object recognition and precise localization for space body control," *International Journal of Aerospace Engineering*, 2019, p. 7050915, 2019.
2. A. Mazouz, C.P. Bridges, "Multi-sensory CNN models for close proximity satellite operations," in *2019 IEEE Aerospace Conference,* Big Sky, MT, USA, 2018.
3. V. Baranova, "Autonomous streaming space objects detection basedon a remote optical system," *Devices and Methods of Measurements*, vol. 12, no. 4, pp. 272–279, 2021.
4. A. Tahir, H.S. Munawar, J. Akram, M. Adil, S. Ali, A.Z. Kouzani, M.A.P. Mahmud, "Automatic target detection from satellite imagery using machine learning," *Sensors*, vol. 22, no. 3, p. 1147, 2022.
5. M.C. Müller, S. Swami, B. Haser, M. Bilal, A. Kinzel, R. Förstner, C. Mundt, "Data pipeline of a multi-spectral satellite experiment for object detection and artificial intelligence-based processing," in *SPIE Defense + Commercial Sensing*, p. 20. Vol. 12521, id. 125210N, Orlando, Florida, United States, May 2023.
6. T. Mahendrakar, R.T. White, M. Wilde, M. Tiwari, "SpaceYOLO: a human-inspired model for real-time, on-board spacecraft feature detection," in *IEEE Aerospace Conference 2023,* Big Sky, MT, USA, Feb. 2023.

7. B. Yekkehkhany, P. Shokri, A. Zadeh, “A computer vision approach for detection of asteroids/comets in space satellite images,” *The International Archives of the Photogrammetry, Remote Sensing and Spatial Information Sciences*, vol. XLIII, no. B3, pp. 1185–2020, 2020.
8. H.N. Do, T.J. Chin, N. Moretti, M.K. Jah, M. Tetlow, “Robust foreground segmentation and image registration for optical detection of GEO objects,” *Advances in Space Research*, vol. 64, no. 3, pp. 733–746, 2019.
9. N.W. Oumer, S. Kriegel, H. Ali, P. Reinartz, “Appearance learning for 3D pose detection of a satellite at close-range,” *ISPRS Journal of Photogrammetry and Remote Sensing*, vol. 125, pp. 1–15, 2017.
10. Y. Chen, J. Gao, K. Zhang, “R-CNN-based satellite components detection in optical images,” *International Journal of Aerospace Engineering*, vol. 2020, pp. 1–10, 2020.
11. Z. Li, Y. Wang, W. Zheng, “Space-based optical observations on space debris via multipoint of view,” *International Journal of Aerospace Engineering*, vol. 2020, 2020.
12. S. Lim, J.-H. Kim, H.D. Kim, “Strategy for on-orbit space object classification using deep learning,” *The Journal of Aerospace Engineering*, vol. 235, no. 15, pp. 2326–2341, 2023.
13. B. Wang, S. Li, J. Mu, X. Hao, W. Zhu, J. Hu, “Research advancements in key technologies for space-based situational awareness,” *Space: Science & Technology*, vol. 2020, pp. 1–31, 2022.
14. J. Kaur, W. Singh, “Tools, techniques, datasets and application areas for object detection in an image: a review,” *Multimedia Tools and Applications*, vol. 81, no. 27, pp. 38297–38351, 2022.
15. A. Kumar, A. Kalia, A. Sharma, “Object detection: a comprehensive review of the state-of-the-art methods,” *International Journal of Next-Generation Computing*, vol. 11, no. 1, pp. 52–75, 2020.
16. K.V. Sriram, R.H. Havaldar, “Analytical review and study on object detection techniques in the image,” *International Journal of Modeling, Simulation, and Scientific Computing*, vol. 12, no. 5, 2021.
17. D. Patel, P.K. Gautam, “A review paper on object detection for improve the classification accuracy and robustness using different techniques,” *International Journal of Computer Applications*, vol. 112, no. 11, pp. 5–7, 2015.
18. V. Mishchuk, H.V. Fesenko, “Analysis of computer vision methods and means for explosive ordnance detection mobile systems,” *Electronic Modeling*, vol. 46, pp. 90–111, 2024.
19. X. Zou, “A review of object detection techniques,” in *International Conference on Smart Grid and Electrical Automation ICSGEA*, IEEE, Xiangtan, China, 2019. doi: 10.1109/ICSGEA.2019.00065
20. A.K. Shetty, I. Saha, R.M. Sanghvi, S.A. Save, Y.J. Patel, “Deep Learning Models,” in *6th International Conference for Convergence in Technology (I2CT)*, IEEE, Maharashtra, India, 2021.
21. S. Nandhini, D. Easwaramoorthy, R. Abinands, “An extensive review on recent evolutions in object detection algorithms,” *International Journal of Emerging Trends in Engineering Research*, vol. 8, no. 7, pp. 3766–3776, 2020.

22. J. You, "Deep neural networks for object detection," in *4th International Conference on Material Science, Environment Science and Computer Science (MSESCS 2022)*, HSET, vol. 17, pp. 159–165, Guangzhou City, China, 2022. doi: 10.54097/hset.v17i.2576
23. Z. Zou, K. Chen, Z. Shi, Y. Guo, J. Ye, "Object detection in 20 years: a survey," *Computer Vision and Pattern Recognition IEEE*, vol. 111, pp. 257–276, 2023.
24. D. Mishra, "Deep learning based object detection methods: a review," *Medicon Engineering Themes*, vol. 2, no. 4, pp. 20–28, 2022.
25. R. Reddy, "Object detection and tracking – a survey," in *International Conference on Computational Intelligence and Communication Networks (CICN)*, Jabalpur, India, 2015.
26. A. Ramachandran, A.K. Sangaiah, "A review on object detection in unmanned aerial vehicle surveillance aerial vehicle surveillance," *International Journal of Cognitive Computing in Engineering*, vol. 2, pp. 215–228, 2021.
27. R.C. Gonza Lez, R.E. Woods, *Digital Image Processing* (4th ed.), Pearson, 2018.
28. T. Hastie, R. Tibshirani, J. Friedman, *The Elements of Statistical Learning: Data Mining, Inference, and Prediction* (2nd ed.), Springer, 2009.
29. Y. LeCun, Y. Bengio, G. Hinton, "Deep learning," *Nature*, vol. 28, no. 521, pp. 435–444, 2015.
30. J. Redmon, A. Farhadi, "*YOLOv3: An Incremental Improvement*" arXiv:1804.02767, 08-Apr-2018.
31. X. Zhang et al., "Remote sensing object detection meets deep learning: A metareview of challenges and advances," in *Remote Sensing Object Detection Meets Deep Learning: A metareview of challenges and advances*, vol. 11, no. 4, pp. 8–44, 2020. doi: 10.1109/MGRS.2023.3312347
32. G.Y. Zhuang, "Transfer learning for space object detection: challenges and solutions," *Neural Computing and Applications*, vol. 33, no. 9, pp. 5673–5684, 2021.
33. S.J. Pan, Q. Yang, "A survey on transfer learning," *IEEE Transactions on Knowledge and Data Engineering*, vol. 22, no. 10, pp. 1345–1359, 2010.
34. B. Flewelling, "Computer vision techniques applied to space object detect, track, ID, and characterize," in Report Documentation Page, 2014. https://apps.dtic.mil/sti/tr/pdf/ADA616753.pdf#page=12.46
35. B. Yekkehkhany, P. Shokri, "A computer vision approach for detection of asteroids/comets in space satellite images," in *The International Archives of the Photogrammetry, Remote Sensing and Spatial Information Sciences*, Volume XLIII-B3-2020, 2020. https://doi.org/10.5194/isprs-archives-XLIII-B3-2020-1185-2020.
36. N. AlDahoul, H.A. Karim, A. De Castro, M.J.T. Tan, Localization and classification of space objects using EfficientDet detector for space situational awareness," *Scientific Report*, vol. 12, no. 1, p. 21896, 2022.
37. T. Driver, et al., "AstroVision: Towards Autonomous Feature Detection and Description for Missions to Small Bodies Using Deep Learning." *Acta Astronautica*, vol. 210, Jan. 2023, pp. 393–410. https://doi.org/10.1016/j.actaastro.2023.01.009.

38. T. Mahendrakar, R. T. White, M. Wilde and M. Tiwari, “SpaceYOLO: A Human-Inspired Model for Real-time, On-board Spacecraft Feature Detection,” *2023 IEEE Aerospace Conference*, Big Sky, MT, USA, 2023, pp. 01–11, doi: 10.1109/AERO55745.2023.10115705.
39. F.I. San Martín, “A flexible pipeline for automatic detection of sunlit space objects using all-sky images,” *Advances in Space Research*, vol. 66, no. 1, pp. 123–135, 2020.
40. R. Tiwari, A.K. Dubey, A. Rocha, “A novel YOLOv4-modified approach for efficient object detection in satellite imagery,” *Expert Systems,* vol. 40, no. 5, p. e13180, 2023.
41. A.V. Etten, “*You Only Look Twice: Rapid Multi-Scale Object Detection In Satellite Imagery,*”arXiv:1805.09512, 2018.
42. A. Tahir, H.S. Munawar, J. Akram, M. Adil, S. Ali, A.Z. Kouzani, M.A.P. Mahmud, “Automatic target detection from satellite imagery using machine learning,” *Sensors (Basel, Switzerland),* vol. 22, no. 3, p. 1147, 2022.

Chapter 2

Smart cities and the role of smart cyber-physical systems toward sustainability

Case studies

Shreya Saha, Adanoi Joseph Abraham, Shivani Odedra, Om Sutariya, Gaurav Agarwal, and Akash Sanghi

2.1 INTRODUCTION

2.1.1 Smart cyber-physical systems

Smart cyber-physical systems (SCPS) are highly complex structures that follow a multidisciplinary approach from fields like cybernetics, mechatronics, and process science to create structures that integrate computer components like sensors, control centers, and machines alongside physical processes such as traffic control or fire detection to work together through application of real-time data [1]. This involves computers using real-world dynamic data to influence and manipulate tangible processes from machinery and vehicles, for instance, to ecosystems such as cities.

This is highly similar to process control, also known as embedded systems. However, it must be noted that in embedded systems, the computational components are more significant than the interaction between them and physical elements, as exhibited in SCPS. The advent of SCPS can also be traced to the rise of the Internet of Things (IoT), which helps connect machinery with limited computational functions to the Internet through the TCP/IP stack protocol, allowing for better communication and service exposure [2]. While the two share a similar architecture, SCPS has higher integration between the computer and physical components [3].

2.1.2 Role of SCPS in smart cities

The traditional understanding of a city can be related to a group of interconnected objects that individuals living within a specific area interact with regularly. A smart city is essentially an improved version of these in terms of their functional and structural advantages. These improvements refer to strategic implementation of smart factors, which has foundations in Information Communications Technology (ICT) for

 DOI: 10.1201/9781003542513-2

sustainable growth, improved health, and better operations, all leading to increased quality of life for the residents of the city [4]. CPS was first integrated into physical infrastructure as the Smart Building (SB), which was named by the United Technology Building Systems Corporation and was based at Hartford, Connecticut [5]. As such, smart cities are an example of a cyber-physical system (CPS) wherein cyber systems like processors, sensors, and IoT devices are integrated into the physical infrastructure of a city and are able to regulate the conditions continuously based on current needs. This integration can be of two types: integrated into functional attributes such as data collection and predictive analytics on energy, transport, healthcare, manufacturing, and so on and integration into external attributes that relate to culture as well as tourism of the place [4].

SCPS applications in functional attributes can be diverse and can essentially improve services, economic growth, and increase the overall quality of life for the citizens of the cities and communities in which it is implemented [6]. For instance, in the transportation sector, CPS optimizes the flow of traffic, addressing congestion, and the efficiency of public transport through the collection and simultaneous analysis of real-time data. This entails to the ability of dynamic control of traffic signals and the formation of multimodal transportation networks on a public level as well as development of autonomous vehicles [7]. Moreover, CPS in the energy sector is used to create smart grids that manage all aspects of electricity from production, distribution, to consumption, leading to an increment in energy efficiency through reduction in wastages, and increased use of renewable energy sources [8,9]. Likewise, for healthcare, CPS functions can be multidimensional, such as monitoring vital signs of patients, automated medication, and better remote patient care, ultimately reducing the burden on healthcare systems [10,11]. Such integration of CPS technologies across urban systems and infrastructure makes cities not just sustainable but also resilient and responsive to citizens' needs.

2.1.3 Importance of SCPS in modern urbanization

Contemporary problems of rapid urbanization lead to problems in efficient construction of infrastructure, management of finite resources, and most importantly, public services [12]. Governments' responsibilities range into services involving not just education, transportation, public safety, water supply and other utilities but also the sustainable use of natural resources, regulating land use, maintaining the urban landscape, and so forth [13]. Climate change and resource depletion have added dimensions to these issues, making it an increasingly difficult challenge to address. Cities today also contend with tremendous socioeconomic crises while competing for businesses, tourists, and, most importantly, human talent [14]. Thereby, local governments can improve the urban environment, gain a competitive edge, manage environmental risks, and meet modern urban governance's difficulties with innovation, especially technological innovation. Cities need to become integrated with technology and innovative to manage themselves. SCPS offer the potential to address these issues and have led to the formation of Smart cities, which are core enablers of sustainable urban transformation.

Puliafito et al. [4] highlight the structure of a smart city as being divided into four layers of cointegrating elements to build up a whole system, which can be seen in Figure 2.1. These layers begin with the Infrastructure layer, followed by the Management and Application layers. The Stakeholder layer is the last layer of this four-fold approach. They elaborate on the features of the ICT foundations of a smart city with various sensors and devices like traffic lights, smart billboards, GPS systems on buses, lamp posts, air quality monitors, weather stations, security cameras, and even mobile phones carried by citizens. These sensors continuously collect data throughout the city to improve urban life.

In the initial Infrastructure layer, technological resources such as networks, cloud and edge computing, communication systems, and data storage work together to gather, process, and manage all the data collected throughout the city. Due to its complexity and the rapid development of technologies like 5G, infrastructure management becomes essential to increase the efficiency and access for as many citizens as possible. This layer is followed by the Management layer, which includes security measures, monitoring systems, privacy policies, and quality of service (QoS) to ensure

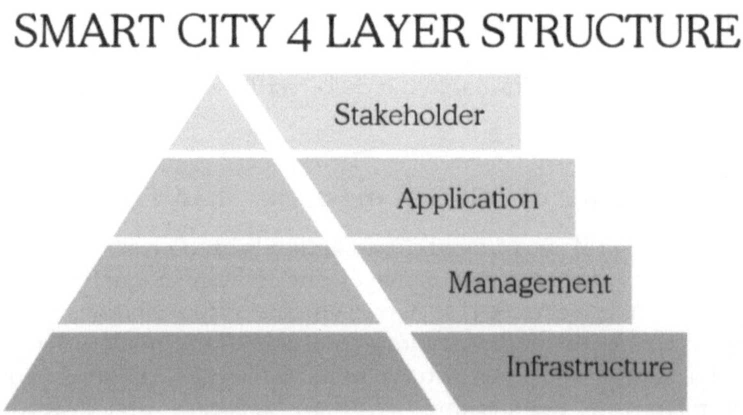

Figure 2.1 Smart city 4 layer structure (Source: Author's Visualisation adapted from Puliafito et al. (2021)).

systems run smoothly by including advanced tools and features, resource management, reward systems for citizens, and processes to enforce rules and regulations. This is followed by the application layer, which focuses on urban areas like the energy sector, building infrastructure, transportation, public safety, waste management, and healthcare. This layer focuses on providing the actual services to the citizens. The final Stakeholder layer includes all the residents, local government bodies, doctors, law enforcement, businesses, telecom companies, and even vehicles. In its essence, it involves all the people and organizations living and interacting with the elements of the city regularly.

2.2 CASE STUDIES

Contemporary applications of SCPS in global metropolitan centers demonstrate how cities use technology to improve efficiency, sustainability, and living standards for their citizens. From real-time data management adaptable infrastructure to energy grids and disaster resilience systems, each is featured in smart cities worldwide, like Barcelona, Singapore, Amsterdam, Tokyo, and Dubai. These cities have implemented SCPS in a way specific to its difficulties and objectives. Table 2.1 provides an overview of the SCPS's main features, the application that has been implemented, and the sustainability objective that it addresses in smart cities.

2.2.1 Real-time data management infrastructure: Barcelona, Spain

Barcelona, Spain, is a city of history and a significant HubSpot for all architecture lovers. Its implementation of SCPS through real-time data management infrastructure for essential services, including traffic monitoring, waste management, and parking management, has transformed it into a sustainable and citizen-friendly urban environment. The city continuously collects data through a network of sensors and Internet of Things (IoT) devices dependent upon Sentilo, an open-source web-based platform for data collection and sharing [15]. These sensors have been effectively implemented in smart meters, position sensors in public transportation, sensors in asphalt, smart bins, and so on to identify real-time changes across the city and adapt systems to the changing conditions (identifying parking spaces and water levels in fountains). For instance, the city's smart lighting adjusts streetlight brightness based on pedestrian activities and time of day, reducing energy costs and helping the city meet European Union (EU) climate goals [16,17]. Additionally, citizen-centric platforms like the "Decidim Barcelona" have enabled its residents to participate in governance

Table 2.1 SCPS application and sustainability objective across smart cities

City	*SCPS main feature*	*SCPS application*	*Sustainability objective addressed*
Barcelona	Real-Time Data Management Infrastructure	Real-time urban monitoring systems [15], adaptive lighting based on environmental factors [16,17], and citizen platforms [18,19]	Enhances energy efficiency, reduces carbon emissions, and increases citizen engagement in urban planning.
Singapore	Integrated Transport Management and Electronic Road Pricing (ERP)	Smart traffic management, predictive analytics for transportation, and urban mobility [20].	Reduces traffic congestion, optimizes public transport, and lowers emissions, supporting the city's sustainable urban mobility vision.
Amsterdam	IoT-enabled Waste and Water management	Sensor-based waste collection [21] and smart water monitoring [22].	Reduces waste overflow, improves water usage efficiency, and supports circular economy principles in waste management.
Tokyo	Smart Energy Grid and Disaster Resilience System	Decentralized energy production [23] and automated disaster response systems [24].	Promotes energy resilience through decentralized grids, enhances disaster preparedness, and minimizes energy wastage for sustainable urban energy management.
Dubai	Advanced Infrastructure & Smart Government	High-speed fiber optic and 5G networks [25], smart traffic and parking management, solar power and district cooling systems, and digital government services [26].	Enhances urban quality of life, optimizes resource use, reduces carbon emissions, and fosters digital governance with citizen engagement.

through project voting and idea-sharing, allowing policymakers to adapt urban strategies based on citizen input and embedding inclusivity within city planning [18,19]. Metrics like energy consumption and air quality are tracked through urban monitoring systems, as well as smart energy grids that increasingly utilize renewable energy sources [9,27], addressing sustainable resource use. Implementation of SCPS, thereby, has helped the city not only reduce resource consumption but also enhance the city's resilience by allowing authorities to respond proactively to challenges such as traffic congestion and pollution hotspots.

2.2.2 Integrated transport management and electronic road pricing: Singapore

Singapore has strategically applied SCPS to improve the efficiency across transportation and healthcare sectors and as such stands a global leader. Transportation, thereby, includes the implementation of the Electronic Road Pricing (ERP) system that manages traffic through dynamic data collected from the sensors, IoT devices, and other cyber systems across the city. As such, the collected data are utilized to adjust tolls in reference to current traffic conditions. This facilitates traffic management that lowers road congestion during peak hours and encourages off peak travel hours [20]. According to Menon and Guttikunda, the ERP system functions through all the vehicles being equipped with an in-vehicle unit (IU). This IU is permanently affixed to the vehicle, is color-coded depending upon the vehicle class, and can communicate with ERP gantry devices through a 2.45 GHz radio frequency. At its core, these internal IUs help in differential charging on the basis of the time and vehicle type, such as higher pricing during peak hours and vice versa. However, emergency vehicles are exempt from such charges. Payments are made through a CashCard inserted into the IU, durable enough to withstand high temperatures in parked vehicles. Periodic 3-month reviews adjust pricing based on prevailing speeds, targeting optimal speeds which are different for city zones and expressways. This adaptive pricing model discourages peak-hour travel and maintains smooth traffic flow, making Singapore's ERP a global benchmark in SCPS-driven urban traffic management [20].

2.2.3 IoT-enabled waste and water management: Amsterdam, Netherlands

By integrating IoT technology into water and waste management, Amsterdam has solidified its circular economy framework and become a preeminent smart city. The IoT sensors can track the bin capacity around the city, providing IoT-enabled trash management and alerting collection providers when bins are full. This technology makes garbage management more sustainable and efficient by enabling optimum collection routes that lower fuel usage, CO_2 emissions, and collection frequency [19,21]. Additionally, the data collected through this help support analytics of waste trends, which facilitates the creation of focused waste reduction plans.

Its smart wastewater management system supports Amsterdam's sustainable objectives [22]. The city can proactively handle water scarcity, contamination, and flooding hazards thanks to IoT sensors that continuously monitor water quality, levels, and flow. During times of high demand or possible flooding, the real-time data allow for responsive adjustments to water flow and usage, protecting drinking water quality, conserving water resources, and lessening environmental effects. In fact, according to a simulation analysis done by Alzahrani et al., IoT-based wastewater management techniques have shown a recycling rate of 96.3% with an efficiency ratio of 88.7% and increased wastewater reuse of 90.8% [28]. By decreasing waste, preserving resources, and lessening environmental stress, Amsterdam's creative use of SCPS in waste and water management demonstrates its position of transitioning toward a circular economy.

2.2.4 Smart energy grid and disaster resilience system: Tokyo, Japan

Tokyo's smart city feature leverages SCPS to improve sustainability and safety, emphasising disaster preparedness and energy resilience. Local energy production and storage are enabled by the city's smart energy grid, a decentralized infrastructure that facilitates renewable energy integration and efficient distribution. By lowering dependency on centralized power, this configuration guarantees a more robust and durable energy supply, which is particularly important in times of emergency [19,23]. The SCPS-enabled grid helps in lowering energy waste across the city with real-time dynamic monitoring as well as automated changes. Similarly, disaster resilience measures are one of the major highlights of SCPS systems in Tokyo [24]. Since natural disasters are very frequent in Japan, disaster response systems are needed to address public safety protocols automatically in reaction to immediate catastrophes such as earthquakes. As such, SCPS integration into the city's infrastructure, in the form of smart buildings and other resilient and adaptive infrastructures, serves as a potential resolution [19]. These infrastructures facilitate stabilization during seismic occurrences so as to minimize resource use as well as damage. Other individual nudges come in the form as automated notifications to residents of the city. As such, SCPS builds resilience while supporting sustainable energy use in cities like Tokyo.

2.2.5 Advanced infrastructure and smart government: Dubai, UAE

Dubai is another global leader that has improved not only various public utility services such as energy, infrastructure, and transport but also majorly its government services. The 5G network as well as the high-speed fiber optic networks provide a strong infrastructure for IoT connectivity, enabling real-time data collection and analysis across urban services [25]. This connectivity helps support SCPS-based functions and enables the increased efficiency of the government to control, maintain, and monitor city operations. For example, the city's smart government function includes a portal of digital services for easier access to government services online, thereby lowering physical visit requirements [26]. Similarly, the city also has provisions for e-voting, which has transformed governance toward higher citizen engagement and transparency.

2.3 GLOBAL COMMONALITIES OF SCPS ENABLED SMART CITIES

2.3.1 Real-time data collection and automation

SCPS enabled cities to function on the basis of real-time data collection and automation. As such, this essential component allows for tracking indicators pertaining to daily urban life ranging from energy consumption and public safety to traffic and waste management through implementation of IoT devices and sensors. For instance, while cities like Barcelona manage traffic congestion and redirect traffic through real-time data, countries such as Singapore use its ERPS to control its traffic flow. The diverse implementation of SCPS leads to a necessity-based solution that adapts to the changing conditions of that particular urban city. Similarly, Tokyo addresses its natural disaster through the SCPS infrastructure that is resilient against such a phenomenon, which highlight the predictive capacity of SCPS for analytics to anticipate problems and address them.

2.3.2 Integration of multiple urban systems

One of the major aspects of a smart city is the multidimensional integrations of different public utility services that have been enabled with SCPS sensors like electricity, transport, and healthcare. The integration essentially helps in increasing the efficiency of resource consumption in these cities as different structures of the city interact with each other through the SCPS-enabled architecture to form a whole coherent ecosystem. For example, Amsterdam has the combined functions of IoT-enabled waste handling systems which work alongside water management for conservation to provide long-term sustainability through multiple sectors. Similarly, Dubai's energy sector is closely linked with its transportation sector, and SCPS is used to control power consumptions based on the real-time dynamic data of the traffic, such as heightened regulation of cooling systems during peak traffic [19,25]. As such, integrating urban systems into a single SCPS design essentially helps improve efficiency by decreasing redundancy, promoting cross-sector collaboration, and increasing resource usage.

2.3.3 Citizen-centric design and engagement

SCPS-enabled cities are distinguished by their emphasis on citizen-centric design, which includes participatory government and real-time responsiveness to citizen requirements. Platforms such as Barcelona's "Decidim Barcelona" allow locals to propose and vote on civic initiatives, promoting participation in urban decision-making [18]. Herein, the SCPS allows for increased citizen feedback as user input for adapting public urban services.

SCPS empowers citizens by making them active participants in urban governance, increasing trust in municipal management, and customizing services to the community's developing requirements.

2.3.4 Focus on sustainability and energy efficiency

One of the primary reasons for implementing SCPS in smart cities is its ability to focus on sustainability. SCPS has helped cities across the globe with their waste management through reduction of garbage, water conservation means through waste water management, and even regulating energy consumption through a transformation to renewable sources of energy. Energy efficiency caused by SCPS implementation can be seen in Tokyo's energy grids which are decentralised and thereby lead to lowered reliance on central energy systems as well as a shift to use of local renewable resources. In fact, it allows certain abundant areas to create and store energy for future purposes. On the other hand, IoT sensors help improve levels of fuel consumption and vehicle emissions and even maintain collection schedules in cities like Amsterdam that rely on the SCPS for sustainable waste management. Thereby, these applications of SCPS assist cities in forming a circular economy that is capable of meeting climate goals that focus on sustainable usage of resources and conservation for the future.

2.3.5 Scalability and flexibility

One common benefit of SCPS implementation in urban infrastructure is their scalability and flexibility, which allow smart cities to use predictive analytics and address future needs. Dubai, for example, has high-speed fiber optic and 5G networks that essentially provide a scalable framework for its infrastructure to accommodate new technologies like AI and advanced data analytics [25]. Additionally, smart cities systems can also avail the benefits of SCPS functions into new geographic areas or further integration of another dimension of public utility function without compromising on the workings of current services [19]. This represents a critical aspect of sustaining life in urban areas as SCPS scalability and flexibility exhibit the potential for addressing issues such as increased urbanization and climate change by ensuring their infrastructure remains capable of supporting future urban needs.

2.4 FORMS OF SCPS IMPLEMENTATION IN URBAN INFRASTRUCTURE

SCPS implementation in urban planning has taken diverse forms across the globe based on the infrastructural needs as well as the characteristics of the place it is being implemented in, which can be classified into three broad categories: retrofitting existing cities, constructing purpose-built smart cities, and nationwide integration. Each of these forms of implementation represents a specific purpose, with nationwide integration being the overarching form where cyber-physical systems are integrated into every essential aspect of the city.

2.4.1 Retrofitting smart cities

Cities such as Barcelona and Amsterdam are well-established as prominent cities across the globe and as such face a dilemma in implementing cyber-physical systems while maintaining their existing urban layouts due to their complex historical infrastructure [29]. To address this, these cities have undergone the process of retrofitting SCPS-enabled devices such as sensors, IoT devices, and smart grids within their preexisting public utility structures. In fact, Barcelona open sources the real-time dynamic data collected from the multiple IoT sensors spread across the city through the platform "Sentilo", which allows for advancement of SCPS integration to changing conditions and independent research into the efficiency of resource management through a multitude of data such as energy use and air quality [15]. Similarly, Amsterdam utilizes SCPS-enabled IoT devices for its waste management and wastewater management utilities to improve upon its sustainability [22].

2.4.2 Purpose-built smart cities

Purpose-built smart cities are designed with SCPS deriving the core of all functionalities, making it an ideal urban environment that follows real changing data as inputs for its administrative procedures. Songdo, South Korea, has been the primary example of a fully SCPS integrated infrastructure-based city wherein data-driven automation includes services ranging from widespread internet access and maintaining occupancy in Smart buildings to municipal functions such as extensive smart traffic control system as well as garbage management systems [30]. This has led to new benchmarks for resource efficiency that would be difficult to implement in cities such as Barcelona, wherein the historical infrastructure represents limited potential for developing a controlled environment [31].

2.4.3 Nationwide integration

SCPS can also be implemented on a national scale by integrating various services and systems across geographical borders much like they are connected in a city. For instance, instead of a particular city using SCPS-enabled systems for the functioning of its public utilities, Singapore utilizes SCPS to handle healthcare, transport, public safety, waste management, and so on throughout the nation. The country's Electronic Road Pricing (ERP) system, for instance, as part of its smart transportation network, uses real-time data to dynamically alter toll rates in response to traffic conditions, efficiently controlling congestion across the city [20]. Furthermore, Singapore's government is constantly working with private-sector technology companies to develop SCPS capabilities, particularly in healthcare, where wearable IoT devices monitor patients' health in real time [11,19]. This countrywide integration illustrates the SCPS's ability to completely revolutionize urban management, making Singapore a global leader in adaptable, smart technology.

2.5 SMART CITY EVALUATION METRICS

Assessing smart city performance is a complex task that requires a multidimensional approach, focusing on the presence of smart technologies and their real-world impact. It is important to highlight that analyzing smart city success is a complex undertaking, and the evaluation criteria may differ depending on each city's individual goals, circumstances, and priorities [32].

2.5.1 Infrastructure

Smart city infrastructure enables digital transformation by combining sensors, communication networks, broadband connectivity, and digital infrastructure [33]. These components provide the foundation for data exchange, enabling real-time monitoring and optimization of urban services [19]. For example, smart grids and transportation system sensors collect data aiding in efficiently managing energy and traffic flows. Robust infrastructure enables scalable and adaptive solutions, critical for long-term sustainability and resistance to disturbances [34].

2.5.2 Governance & policy

Smart cities with effective administration and rules make collaborating easier for government agencies, the business sector, and citizens. Smart cities promote innovation in governance and improvement of policy services by developing cybersecurity, public–private collaborations, and addressing data privacy frameworks [33,34]. Open data platforms of the data collected throughout a smart city's sensors lead to an increase in public participation in decision-making through increased transparency, ultimately forming community ownership [19].

2.5.3 Energy efficiency

Energy efficiency forms the core of sustainability in smart cities, and it is essentially achieved through renewable energy sources, smart grids, and the formation of smart buildings. These allow smart cities to reduce their carbon footprints by reduction in energy use, essentially contributing to environmental resilience [33,34]. The multitiered systems in a smart city work together to achieve carbon neutrality [19]. For instance, the implementation of smart lighting to adjust brightness based on usage patterns helps reduce energy consumption, while smart grids use only renewable energy sources. These together help form the energy efficiency score of a smart city by measuring its energy consumption limits [35].

2.5.4 Transportation

Public transport is one of the most essential forms of transformation that smart cities include. It includes solutions ranging from smart traffic controls for lowering congestion during peak hours and improving accessibility to smart mobility services such as optimizing travel routes and providing real time updates. Optimizing public transport also helps in lowering emissions through a reduction in private transportation dependency alongside promoting sustainable commuting modes to support environment-friendly urban lifestyle [33,34].

2.5.5 Environmental sustainability

The three domains that are primarily addressed through environmental sustainability in smart cities relate to waste management, ecosystem preservation such as wastewater management, and maintaining the air quality index [33]. Effective waste management techniques include lowering of landfill usage, increasing efficiency of recycling, and even optimizing the waste collection routes. Similarly, the air quality index is maintained by tracking pollution, and water management involves conservation as well as waste reduction measures. As such, smart cities also promote green spaces that balance urbanization alongside the ecosystem.

2.5.6 Public service

Smart city public services are improved through technologies of digital services, which also address citizen interaction as well as act as e-governance platforms [33]. E-governance platforms improve individuals' access to key services such as healthcare, licensing, and education. Digital services enhance service delivery, save administrative costs, and increase transparency. Engaging citizens via digital feedback enables responsive adjustments to public services, ensuring they effectively satisfy community demands.

2.5.7 Data management

Data management in smart cities includes data gathering, analysis, and security standards critical for making educated decisions [33,34]. Effective data management enables cities to assess trends, estimate demand, and optimize resources across several sectors. For example, transportation and energy data can help with city planning and resource allocation. Data security guarantees that people's information is secure, sustaining public faith in smart city efforts.

According to Hassebo and Tealab [19],

i. Cities with effective data management strategies have defined data collection, storage, and use policies. They also have strong security measures to prevent unauthorized data access or disclosure.
ii. Cities with medium-level data management practices have some policies and processes, although they may be less comprehensive or well-enforced than high-level cities. They may also have some security measures in place, but they may be less robust than those in major cities.
iii. Cities with poor data management practices have few policies and processes to collect, store, and use data. They may also have few security measures in place.

2.5.8 Quality of life

Quality of life measurements in smart cities analyze how services affect healthcare, education, and public safety [33]. Smart cities boost overall wellbeing by increasing access to healthcare via telemedicine and streamlining education through digital resources. Public safety systems, such as surveillance and emergency response services, improve urban safety. Improved QoL generates a sense of belonging and happiness among people, encouraging long-term involvement and investment in the community [34].

2.5.9 Economy

Smart cities stimulate economic growth by offering new job possibilities, attracting investment, and encouraging innovation [33]. Technology-enabled jobs, startups, and innovation hubs are critical for increasing economic resilience. They also attract global investments by developing tech-based company ecosystems, creating a competitive, innovative urban economy. Technology-driven economic growth allows communities to reinvest in infrastructure, public services, and sustainability programs [19,34].

2.6 LEGAL CONSIDERATIONS OF SECURITY FOR SCPS IMPLEMENTATION

While SCPS implementation in smart cities poses significant opportunities for achieving sustainability goals, it also raises concerns about data privacy and cyber threats to the vast amount of citizen data collected through the interconnected networks, sensors, cameras, and IoT devices. The collected data represent sensitive information vulnerable to data breaches [36], hacking, and unauthorized access, wherein even a single breach could affect an entire city, including emergency services and public safety. For instance, while SCPS implementation in healthcare improves medical efficiency as well as patient care [11], it also raises serious concerns over the cyber and ethical risks relating to patient information, social inequities, accountability, transparency, and so forth [37]. Addressing these challenges requires robust legal understanding, regular cybersecurity audits, and transparent data usage policies.

2.6.1 Liability for cybersecurity breaches

The interconnectivity of CPSs in smart cities poses severe cybersecurity risks, as highlighted by the 2015 Ukraine Power Grid Attack. This attack showcases how weak encryption and security, particularly in *Supervisory Control and Data Acquisition (SCADA)*, can be taken advantage of, creating far-reaching service disruptions and public safety threats. It can cascade into an outage after a primary cyberattack, implying that interdependency in dependent systems can multiply the impact of a chain reaction [38]. For instance, an initial breach of a substation can lead to system failures across interdependent infrastructure in a smart city. The cascading nature of the effect provides an essential reason why CPS needs to be resilient, whereby preventive measures are needed to address both primary and secondary vulnerabilities [39].

2.6.1.1 Legal accountability

Liability for breaches of CPS must be crafted to consider these systems' interconnectivity. The analysis of cascading failures is essential to giving a view to shared liability, given the presence of multiple actors such as hardware manufacturers, software developers, and city authorities that have partial liability in terms of cybersecurity. This cascading outage incident serves as a point of reference concerning the distributed liability aspect that claims that one failing point can easily lead to failure within the whole system [39].

2.6.1.2 Jurisprudential considerations

Effective liability requires a framework of shared accountability similar to the one proposed by the *California Consumer Privacy Act.* Consequently, there is an imperative need for integrated security measures and shared accountability. A framework under the influence of CCPA would force all parties to follow strict security protocols and share responsibility proportionate to each party's control over the CPS components [38]. This strengthens the need for complete security and accountability protocols in CPS cybersecurity, thereby increasing resilience against cascading failures and protecting vital infrastructures in smart cities.

2.6.2 Risk allocation in public–private partnerships

Complex technical dependencies define PPPs in smart city CPS projects. Third-party hardware leads to proprietary software and potentially complex network infrastructures in a project, creating an opportunity for failure. Several risks include data breaches, system malfunctions, and critical interruptions of services, especially if it is healthcare- or energy-related [38]. For instance, *the 2015 Ukraine Power Grid Attack* [38] shows that third-party software and infrastructure vulnerabilities can place the entire network at risk, making well-defined risk responsibilities necessary. Thereby, provisions involving third-party vendors within PPP contracts must be robust, and accountability must be established for their role in the security of CPS to mitigate this risk.

A further case study, *Rio de Janeiro's 2010 IBM Smart City partnership*, reveals additional challenges in PPPs where CPS dependencies expose operational limitations. IBM's predictive algorithms, although advanced, faced setbacks due to undefined technical and contractual boundaries, leading to operational disruptions [37]. This highlights the importance of structuring PPP agreements with precise technical and contractual risk boundaries to reduce liability exposure in smart city projects.

2.6.2.1 Contract law best practices

A PPP contract depends on allocating technical and operational responsibilities of effective risk distribution, including indemnity clauses, hold-harmless agreements, and liability caps, which are necessary for safeguarding stakeholders' interests during a CPS failure. The 2017 project between *the city and Ameresco concerning the Chicago Smart Lighting Project* represents the best practices of a PPP contract as it included clauses that explicitly state obligations on cybersecurity, infrastructure maintenance, and data protection [37]. These are best practices where municipalities and private partners know what to do and how to share responsibilities within a secure operational framework. Equally important are contractual liability triggers with particular relevance to the stipulation of cybersecurity and service uptime standards [37]. Performance-linked triggers would work to maintain accountability as private partners are compelled to provide reliable CPS because they will trigger liability in case security standards are not achieved.

2.6.2.2 Jurisprudential implications

Embedding the "*public trust doctrine*" in CPS agreements, balancing innovation in smart city PPPs, and holding CPS providers accountable to higher standards would protect the public's reliance on such essential services. A case in point is the *Delhi Metro Rail Corporation (DMRC)*, where private operators are bound by contracts that enforce performance and safety guarantees. This would apply to CPS in smart cities, demanding higher private stakeholders' accountability standards, as their considerations significantly concern the public interest at stake [40].

The doctrine can be applied in smart city PPPs to reinforce a dual focus on innovation, public trust, and the idea that private entities should maintain the system's integrity. It would create a legally enforceable expectation that CPS providers should prioritize public safety and infrastructure security, raising their stature as stewards of critical services.

By accepting structured legal frameworks that will address risk allocation, liability triggers, and public trust obligations, smart city PPPs can better handle the risks inherent in CPS projects. It is one of the approaches to ensure that innovative advancements do not compromise public welfare and protect cities.

2.6.3 AI as a legal entity in CPS decision making

Some existing legal frameworks assign a form of limited legal personality to Artificial Intelligence (AI). For instance, in Corporate Law, the doctrine of corporate personality allows a firm to be a "*legal person*" separate from its shareholders and directors who may enter contracts, acquire property, and assume liabilities independent of its equity holders or officers [41]. As provided by the *doctrine of Salomon v. Salomon & Co Ltd (1897)*, separate legal personalities have separate liability and accountability within organizations in a highly complex and complicated manner so that it is not liable to an individual's stakeholders.

This may enable the CPS AI system to become "*functional agents*" that fall within legally defined boundaries when considering AI that autonomously decides and impacts public safety, for example. If AI systems can be granted limited legal personality, then liability for such malfunction or error might be attributed to the AI as an entity based on an insurance or compensation fund rather than merely focusing on the human developers or operators. This may facilitate easier accountability and the streamlining of compensation procedures in such a scenario as the *2018 Uber autonomous vehicle*, which had problems relating to the liability distribution as the AI had decision authority.

2.6.3.1 The European approach: AI as a "Quasi-Legal" entity

The European legislation bodies have begun to realize the novelty of AI. They can present regulatory frameworks that clarify AI's operational responsibility. For instance, *the European Commission's 2021 AI Regulation Proposal* classifies some applications of AI, like autonomous vehicles and healthcare systems under CPS, as *"high-risk"* [42]. This categorization requires strict oversight and places direct accountability on developers and operators, who are mandated to implement safety protocols and monitoring systems. The proposal does not explicitly state full legal personhood for AI but establishes quasi-legal personality through standards that recognize agency and a unique position for CPS.

2.6.3.2 *Liability in AI-driven CPS: the onus of failure and the doctrine of agency*

Determining the onus of failure when AI decisions result in harm raises an important doctrinal question in establishing AI as a juristic person. In legal theory, there's the doctrine of vicarious liability, where the damage caused by a subordinate agent is imputed to the principal, such as the developer or operator. In legal theory, there's the doctrine of *respondeat superior* (meaning "let the master answer") and the related maxim qui facit per alium facit per se ("he who acts through another does the act himself"), where the damage caused by a subordinate agent is imputed to the principal, such as the developer or operator. This traditional approach, grounded in vicarious liability, becomes complicated with AI systems. Here, this framework could develop an additional layer of accountability for AI decisions in circumstances where the life-or-death consequences of those decisions within CPS contexts can be exacerbated by the nature of CPS, mainly if algorithms will make real-time judgments affecting safety [43]. For instance, *Facebook v. Ireland Data Protection Commissioner (2020)* held that "there was data misuse that implicated Facebook as an operator on the grounds of its lack of controlling its platform's AI system" [44].

2.7 CONCLUSION

Smart cities have continued to be popular worldwide over the past few years. These cities offer opportunities to meet sustainability goals by lowering carbon emissions, lowering capital and operational costs, and enhancing quality of life [45]. The infrastructure adopted by smart cities adapts dynamically to changes in its surroundings, including user needs and other infrastructure, to achieve better and improved performance. Since these cities largely depend upon data collected in real-time through various sensors, cameras, and networks throughout their geographical periphery, their strategies benefit residents in many aspects of their lives. However, having data in the nerve center of the functioning of a city means governments must engage and coordinate with private enterprises to collaborate effectively for beneficial outcomes for all stakeholders. However, such collaborations would require legal considerations such as liability for cybersecurity breaches, risk allocation within public–private partnerships, and the potential role of AI as a legal entity in CPS decision-making to protect public interest and continued innovation.

Therefore, smart city networks rely on execution rather than just strategy. It is no exaggeration to say that smart city projects can only be realized when integrated with basic infrastructure. If a smart city is data-driven, robust, dynamic, scalable, and responsive, its stakeholders will reap numerous benefits.

REFERENCES

1. Lee, EdwardA. "Cyber physical systems: design challenges," in *2008 11th IEEE International Symposium on Object and Component-Oriented Real-Time Distributed Computing (ISORC)*, Orlando, FL, USA, pp. 363–369, 2008. https://doi.org/10.1109/ISORC.2008.25.
2. Gubbi, Jayavardhana, Rajkumar Buyya, Slaven Marusic, and Marimuthu Palaniswami. "Internet of Things (IoT): a vision, architectural elements, and future directions," *Future Generation Computer Systems*, 29, no. 7 (2013): pp. 1645–1660.
3. Patil, Trunal, Lara Rebaioli, and Irene Fassi. "Cyber-physical systems for end-of-life management of printed circuit boards and mechatronics products in home automation: a review," *Sustainable Materials and Technologies*, vol. 32 (2022): p. e00422.
4. Puliafito, Antonio, Giuseppe Tricomi, Anastasios Zafeiropoulos, and Symeon Papavassiliou. "Smart cities of the future as cyber-physical systems: challenges and enabling technologies," *Sensors*, vol. 21, no. 10 (2021): p. 3349.
5. Ting-pat So, Albert Ting-pat So and Wai Lok Chan Chan. "Intelligent building systems," in *The International Series on Asian Studies in Computer and Information Science 5*, pp. 1–6; Springer: Berlin/Heidelberg, Germany, 1999.
6. Khaitan, Siddhartha Kumar, and James D. McCalley. "Design techniques and applications of cyberphysical systems: a survey," *IEEE Systems Journal*, vol. 9, no. 2 (2014): pp. 350–365.
7. Zhang, Desheng, Juanjuan Zhao, Fan Zhang, and Tian He. "UrbanCPS: a cyber-physical system based on multi-source big infrastructure data for heterogeneous model integration," in *ICCPS'15: Proceedings of the ACM/IEEE Sixth International Conference on Cyber-Physical Systems*, pp. 237–248, 2015 April. https://doi.org/10.1145/2735960.2735985.
8. Matsunaga, Fernando, Vitor Zytkowski, Pablo Valle, and Fernando Deschamps. "Optimization of energy efficiency in smart manufacturing through the application of cyber–physical systems and industry 4.0 technologies," *ASME: The Journal of Energy Resources Technology*, vol. 144, no. 10 (2022): p. 102104. https://doi.org/10.1115/1.4053868.
9. Upreti, Kamal, Sangeeta Arora, Anupam Kumar Sharma, Adesh Kumar Pandey, Kamal Kant Sharma and Mohit Dayal. "Wave height forecasting over ocean of things based on machine learning techniques: an application for ocean renewable energy generation," *IEEE Journal of Oceanic Engineering*, vol. 49, no. 2 (2024): pp. 430–445. https://doi.org/10.1109/JOE.2023.3314090.

10. Chen, Fulong, Yuqing Tang, Canlin Wang, Jing Huang, Cheng Huang, Dong Xie, Taochun Wang, and Chuanxin Zhao. "Medical cyber–physical systems: a solution to smart health and the state of the art," *IEEE Transactions on Computational Social Systems,* vol. 9, no. 5 (2022): pp. 1359–1386. https://doi.org/10.1109/TCSS.2021.3122807.
11. Kamal, Upreti, Peng Sheng-Lung, Kshirsagar Pravin Ramdas, Chakrabarti Prasun, Halah A. Al-Alshaikh, Arvind Kumar Sharma, Poonia Ramesh Chandra. "A multi-model unified disease diagnosis framework for cyber healthcare using IoMT- cloud computing networks," *Journal of Discrete Mathematical Sciences and Cryptography,* vol. 26, no. 6 (2023): pp. 1819–1834. https://doi.org/10.47974/JDMSC-1831.
12. Lehmann, Axel P. "Sprawling cities, growing risks?" *World Economic Forum,* January 15, 2015. https://www.weforum.org/agenda/2015/01/sprawling-cities-growing-risks/.
13. Albrechts, Louis "Shifts in strategic spatial planning? Some evidence from Europe and Australia," *Environment and Planning A,* vol. 38, no. 6 (2006): pp. 1149–1170.
14. Zenker, Sebastian, Felix Eggers, and Mario Farsky. "Putting a price tag on cities: insights into the competitive environment of places," *Cities,* vol. 30 (2013): pp. 133–139.
15. "Sentilo, the Barcelona Sensors Network, Barcelona Digital City, Barcelona City Council." n.d. Ajuntament.barcelona.cat. https://ajuntament.barcelona.cat/digital/en/technology-service-citizens/technology-sustainable-city/sentilo-barcelona-sensors-network; Bakıcı, Tuba, Esteve Almirall, and Jonathan Wareham. "A smart city initiative: the case of Barcelona," *Journal of the Knowledge Economy,* vol. 4 (2013): pp. 135–148.
16. Barandiaran, Xabier E., Antonio Calleja-López, Arnau Monterde, and Carol Romero. "Decidim: a brief overview," in *Decidim, a Technopolitical Network for Participatory Democracy: Philosophy, Practice and Autonomy of a Collective Platform in the Age of Digital Intelligence*, Springer, 2024: pp. 1–33.
17. Hassebo, Ahmed, and Mohamed Tealab. "Global models of smart cities and potential IoT applications: a review," *IoT,* vol. 4, no. 3 (2023): pp. 366–411.
18. Villa-Arrieta, Manuel, and Andreas Sumper. "Contribution of smart cities to the energy sustainability of the binomial between city and country," *Applied Sciences,* vol. 9, no. 16 (2019): p. 3247.
19. Menon, Gopinath, and Sarath K. Guttikunda. "Electronic road pricing: experience & lessons from Singapore," in *SIM-air Working Paper Series 33-2010,* 2010.
20. Aleyadeh, Sam, and Abd-Elhamid M. Taha. "An IoT-based architecture for waste management," in *2018 IEEE International Conference on Communications Workshops (ICC Workshops)*, Kansas City, MO, USA, 2018, pp. 1–4, doi: 10.1109/ICCW.2018.8403750.
21. Mancebo, François. "Smart city strategies: time to involve people. Comparing Amsterdam, Barcelona and Paris," *Journal of Urbanism: International Research on Placemaking and Urban Sustainability,* vol. 13, no. 2 (2020): pp. 133–152.

22. Alzahrani, Abdullah I.A., Sajjad Hussain Chauhdary, and Abdulrahman A. Alshdadi. "Internet of things (IoT)-based wastewater management in smart cities," *Electronics,* vol. 12, no. 12 (2023): p. 2590.
23. Elvas, Luís B., Bruno Miguel Mataloto, Ana Lúcia Martins, and João C. Ferreira. "Disaster management in smart cities," *Smart Cities,* vol. 4, no. 2 (2021): pp. 819–839.
24. Kasznar, Ana Paula P., Ahmed W.A. Hammad, Mohammad Najjar, Eduardo Linhares Qualharini, Karoline Figueiredo, Carlos Alberto Pereira Soares, and Assed N. Haddad. "Multiple dimensions of smart cities' infrastructure: a review," *Buildings,* vol. 11, no. 2 (2021): p. 73.
25. Khan, M. Sajid, Mina Woo, Kichan Nam, and Prakash K. Chathoth. "Smart city and smart tourism: a case of Dubai," *Sustainability,* vol. 9, no. 12 (2017): p. 2279.
26. Kuyper, Tijn. *Smart City Strategy & Upscaling: Comparing Barcelona and Amsterdam,* Barcelona: Universitat Pompeu Fabra Barcelona School of Management, 2016.
27. Berigüete, Fanny E., Inma Rodriguez Cantalapiedra, Mariana Palumbo, and Torsten Masseck. "Collective intelligence to co-create the cities of the future: proposal of an evaluation tool for citizen initiatives," *Sustainability,* vol. 15, no. 10 (2023): 7956.
28. Cortese, Tatiana Tucunduva Philippi, Jairo Filho Sousa de Almeida, Giseli Quirino Batista, Jose Eduardo Storopoli, Aaron Liu, and Tan Yigitcanlar. "Understanding sustainable energy in the context of smart cities: a PRISMA review," *Energies,* vol. 15, no. 7 (2022): p. 2382.
29. Kuecker, Glen David, and Kris Hartley. "How smart cities became the urban norm: power and knowledge in New Songdo City," *Annals of the American association of Geographers,* vol. 110, no. 2 (2020): pp. 516–524.
30. Kshetri, Nir, Lailani L. Alcantara, and Yonghoon Park. "Development of a smart city and its adoption and acceptance: the case of new songdo," *Communications & Strategies,* vol. 96 (2014): p. 113.
31. Toh, Chai Keong. "Smart city indexes, criteria, indicators and rankings: an in-depth investigation and analysis," *IET Smart Cities,* vol. 4, no. 3 (2022): pp. 211–228.
32. Agarwal, O.P., and Jose A. Gomez-Ibanez. "*Case-Study on Delhi Airport Metro Express: Public Private Partnership,*" Public-Private-Partnership Legal Resource Center, January 1, 2010. https://ppp.worldbank.org/public-private-partnership/library/case-study-delhi-airport-metro-express.
33. Fang, Yufei, and Zhiguang Shan. "How to promote a smart city effectively? An evaluation model and efficiency analysis of smart cities in China," *Sustainability,* vol. 14, no. 11 (2022): p. 6512.
34. Tukasiewicz, Katarzyna, Piotr Pietrzak, Jakub Kraciuk, Elżbieta Kacperska, and Małgorzata Cieciora. "Sustainable energy development—a systematic literature review," *Energies,* vol. 15, no. 21 (2022): p. 8284.
35. Hou, Jinghui, Laura Arpan, Yijie Wu, Richard Feiock, Eren Ozguven, and Reza Arghandeh. "The road toward smart cities: a study of citizens' acceptance of mobile applications for city services," *Energies,* 13, no. 10 (2020): p. 2496.

36. Sharma, Abhishek, Umesh Kumar Singh, Kamal Upreti, and Dharmendra Singh Yadav. "An investigation of security risk & taxonomy of cloud computing environment," in *2021 2nd International Conference on Smart Electronics and Communication (ICOSEC)*, Kerala, India, pp. 1056–1063. IEEE, 2021.
37. Upreti, Kamal, Prashant Vats, and Mohammad Nasir. "Managing the ethical and sociologically aspects of AI incorporation in medical healthcare in India: unveiling the conundrum," in *2023 IEEE World Conference on Applied Intelligence and Computing (AIC 2023)*, Jaipur, India, 29–30 July, 2023.
38. Sullivan, Julia and Kamensky, Dmitriy. Cui, Q. and T. Hertz. "How cyber-attacks in Ukraine show the vulnerability of the U.S. power grid," *The Electricity Journal*, vol. 30, pp. 30–35, 2017. Doi: 10.1016/j.tej.2017.02.006
39. Curtis E. A. Karnow. "The application of traditional Tort theory to modern artificial intelligence technologies," *Berkeley Technology Law Journal*, vol. 34, no. 1 (2018): pp. 33–85.
40. Ahmad, Md Onais, Mohd Abdul Ahad, M. Afshar Alam, Farheen Siddiqui, and Gabriella Casalino. "Cyber-physical systems and smart cities in India: opportunities, issues, and challenges," *Sensors*, 21, no. 22 (2021): p. 7714.
41. European Commission. *Proposal for a Regulation of the European Parliament and of the Council Laying Down Harmonised Rules on Artificial Intelligence (Artificial Intelligence Act) and Amending Certain Union Legislative Acts*, European Commission, 2021.
42. Danaher, John "Robots, law and the retribution gap," *Ethics and Information Technology*, vol. 18, no. 4 (2016): pp. 299–309.
43. Lynskey, Orla "Data protection and AI: what are the new rights and obligations?" in *Life and the Law in the Era of Data-Driven Agency*, M. Hildebrandt and K. O'Hara, Eds. Oxford University Press, pp. 296–314, 2021.
44. Department for Business Innovation and Skills." *Global Innovators: International Case Studies on Smart Cities Smart Cities Study – Case Studies Report*," BIS Research Paper No. 135, October 2013.
45. Huang, B., M. Majidi, and R. Baldick. "*Case Study of Power System Cyber Attack Using Cascading Outage Analysis Model*," 2018 IEEE Power & Energy Society General Meeting (PESGM), Portland, OR, USA, 2018, pp. 1–5, doi: 10.1109/PESGM.2018.8585921.

Chapter 3

Integrating CPS in digital payments 5.0

A study on security risks and evidence-based solutions

Aishwarya Balan, Pratham Nijhawan, Vineet Kumar Ramnani, Kamal Upreti, and Puja Kumari

3.1 INTRODUCTION

The rapid evolution of digital payment systems is now entering the era of Digital Payments 5.0, characterized by deep integration with cyber-physical systems (CPS) and the use of advanced technologies like artificial intelligence (AI), blockchain, and the Internet of Things (IoT). Unlike earlier stages, Digital Payments 5.0 emphasizes creating a seamless, automated payment experience that integrates directly into real-world environments, such as smart cities, industrial automation, and healthcare. Central to this transformation is integrating CPS, which bridge the digital and physical domains through interconnected networks of software, hardware, sensors, and real-time data processing. CPS enhances payment platforms' automation, scalability, and security, enabling seamless financial interactions across various devices and environments. However, this convergence also introduces significant security challenges as the interdependency between cyber and physical components expands the attack surface for malicious actors. Cybercriminals can exploit vulnerabilities in both digital and physical infrastructures, targeting sensitive data, disrupting services, and threatening the overall integrity of payment systems. Recent incidents of cyberattacks on global digital payment networks highlight the growing sophistication of threats, including malware, ransomware, and insider attacks, which are increasingly designed to exploit the complex architecture of CPS-enabled payment systems [1]. As digital payments become integral to modern commerce, safeguarding these systems has become critical. A comprehensive approach to security is required to address the technical vulnerabilities within CPS and the regulatory frameworks needed to mitigate risks. Stakeholders may promote confidence and dependability in the larger financial ecosystem, while ensuring the robustness of digital payment platforms against new threats by comprehending and resolving these security issues. In this context, developing evidence-based security frameworks and

DOI: 10.1201/9781003542513-3

real-time threat detection methods is essential to protecting digital payments and maintaining their integrity in an increasingly interconnected world [2]. Moreover, understanding the regulatory and policy landscapes governing digital payments is crucial as governments and industries must collaborate to create secure, resilient systems capable of withstanding evolving threats. This chapter adopts a systematic review methodology, drawing on a broad spectrum of literature and empirical case studies to assess the security challenges and solutions in Digital Payments 5.0, ensuring a robust foundation for analyzing the evolving threat landscape and the role of advanced technologies in maintaining security.

3.2 INTEGRATION OF DIGITAL PAYMENTS & CPS

Integrating digital payments with CPS marks a transformative development in modern technological landscapes such as smart cities, healthcare, transportation, and industrial automation. CPS involves the seamless interaction between computational elements and physical processes, whereby these processes are monitored and controlled using real-time data produced by sensors and actuators. The convergence of Digital Payments 5.0 within CPS environments enables automated, efficient, and secure financial transactions, providing significant operational efficiency and scalability advantages. For instance, digital payments embedded in smart city infrastructures, such as automated toll collection systems or smart healthcare devices for billing, demonstrate how CPS can optimize payment processes [3].

However, integrating Digital Payments 5.0 with CPS brings unique security challenges. CPS architectures are often decentralized and interconnected, exposing the system to a number of cyber threats, including illegal access, fraud, and data breaches. This risk is heightened by CPS's real-time, data-intensive nature, where continuous communication and data flow are required without disruptions. These complexities make developing and implementing robust cybersecurity frameworks to safeguard digital payment systems within CPS environments imperative. Central to these frameworks is the need to protect the integrity of financial data, ensure secure communication between CPS components, and maintain overall system resilience against cyberattacks [4].

One of the most promising technologies addressing these security concerns is blockchain. With its decentralized ledger system, blockchain enhances the security of digital payment systems in CPS by ensuring data integrity, transparency, and traceability. The decentralized nature of blockchain eliminates the risks associated with centralized control, such as a single point of failure or attack. Transaction notes on the blockchain are immutable, meaning they cannot be altered once confirmed, significantly reducing the likelihood of fraud or data tampering. This makes blockchain ideal for securing digital payment systems where trust and data accountability are paramount [5]. Moreover, blockchain's cryptographic protocols add another layer of protection by encrypting transaction data in transit and at rest, ensuring malicious actors cannot intercept or manipulate sensitive information [6].

Blockchain also enables smart contracts, further strengthening the security and efficiency of payment systems in CPS environments. The conditions of the buyer–seller agreement are encoded into code in smart contracts, which are self-executing agreements. When specified requirements, like the delivery of products or the fulfillment of services, are fulfilled, these contracts automatically initiate payment execution. By increasing operational efficiency and decreasing the need for human intervention, this degree of automation lowers the possibility of fraud or human error. In industrial automation, for example, smart contracts can facilitate machine-to-machine (M2M) payments, ensuring that transactions are completed seamlessly and securely without manual oversight [7].

Blockchain integration into CPS-based digital payment systems can be visualized through a streamlined process flow, where real-time data from CPS devices are verified and secured via blockchain, and smart contracts automatically authorize payments. The process begins with data collection from CPS sensors, which are then transmitted to blockchain nodes for verification. Once verified, smart contracts are triggered, and payments are authorized and recorded on the blockchain ledger. This decentralized transaction process ensures transparency, accountability, and security at each step. Figure 3.1 illustrates this integration.

The decentralized structure of blockchain eliminates any single point of failure, and smart contracts streamline the payment process, enhancing the efficiency and reliability of transactions.

Blockchain's role in securing CPS-driven payment systems extends beyond mere transaction verification. Its distributed framework ensures that transaction data are stored across multiple nodes, which mitigates the

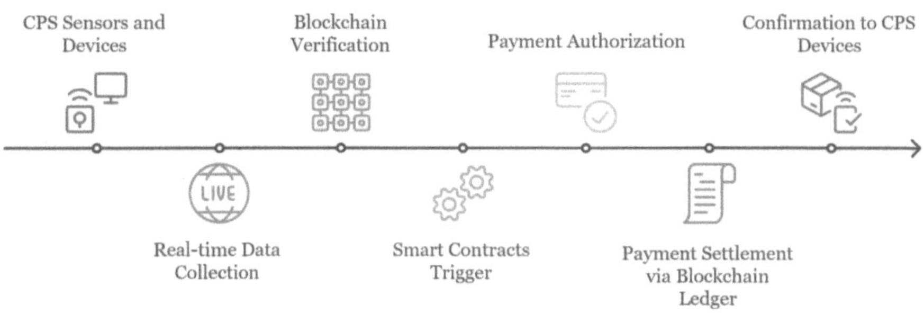

Figure 3.1 Integration of blockchain into CPS-based digital payment systems (Authors' interpretation),

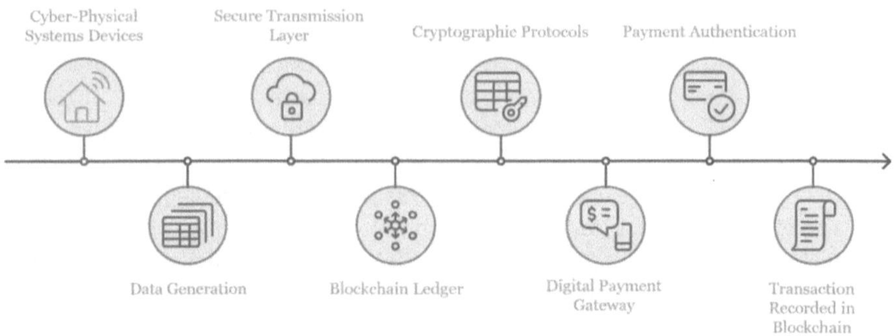

Figure 3.2 Blockchain's cryptographic encryption system (Author's interpretation).

risk of centralized attacks. Moreover, the transparency and immutability of blockchain records build trust among stakeholders, which is particularly valuable in sectors where security and accountability are critical, such as healthcare and industrial automation. In healthcare, for example, digital payments can be securely integrated into medical billing systems, ensuring that payments are processed accurately and in compliance with regulatory requirements [8].

Figure 3.2 illustrates that the blockchain's cryptographic encryption methods further enhance the security framework for integrating CPS with Digital Payments 5.0. These protocols secure the communication channels between CPS devices and digital payment gateways, ensuring that data exchanged between these systems remain confidential and unaltered. Additionally, blockchain's distributed nature enables continuous monitoring of system performance, allowing for detecting and preventing anomalies or cyberattacks in real-time [1].

The combination of decentralized ledgers, cryptographic protocols, and smart contracts ensures that transactions are protected from cyber threats while maintaining the speed and scalability required for modern digital economies. As the CPS continues to evolve, integrating cutting-edge technologies such as blockchain will be essential for securing financial systems in data-intensive, real-time environments [2].

3.3 DIGITAL PAYMENT SYSTEMS ARCHITECTURE

The architecture of digital payment systems forms the backbone of modern financial transactions, especially as they become increasingly integrated into CPS. Digital payment architectures are designed to support the secure and efficient exchange of financial information between users, merchants, and banks through a complex network of technologies. These systems must meet several critical requirements: **scalability** to handle a growing number of transactions,

robustness to ensure system availability and reliability, and, most importantly, **security** to protect sensitive financial data from cyber threats.

At the core of most digital payment architectures is a layered approach that separates functions such as authentication, authorization, transaction processing, and settlement. This layered model ensures that various payment system components can operate independently but in synchronization, allowing for modular upgrades and enhanced security. Typically, these architectures can be divided into four essential layers: the User Interface (UI) Layer, the Application Processing Layer, the Data Management Layer, and the Infrastructure Layer [4]. Each layer is crucial in ensuring payment systems operate efficiently and securely.

1. **User Interface Layer:** This front-end layer interacts directly with users or devices in CPS environments, such as mobile apps or point-of-sale (POS) systems. It facilitates user authentication, transaction initiation, and real-time interaction between the user and the payment system. In the context of CPS, this layer may involve M2M communications, such as in a smart city's automated parking systems, where sensors trigger payments without human intervention.
2. **Application Processing Layer:** This is the core layer responsible for processing transactions. It handles the critical operations involved in digital payments, such as verifying user credentials, executing smart contracts, and managing the logic of the payment gateway. In this layer, blockchain can be incorporated to enhance transaction transparency and traceability. Additionally, real-time analytics can be embedded to detect fraudulent activities and flag suspicious transactions before they are processed [5].
3. **Data Management Layer:** This layer securely stores and manages the data generated during transactions. Given the data-intensive nature of CPS environments, this layer must handle large volumes of real-time data while ensuring its confidentiality and integrity. Encryption techniques and decentralized storage, such as blockchain's distributed ledger, ensure transaction records are tamper-proof and auditable. In CPS systems, where data from physical processes need to be integrated with financial transactions, this layer ensures that the data flows are seamless and secure [6].
4. **Infrastructure Layer:** The underlying hardware and network architecture supports the payment system. In CPS, the infrastructure layer could involve IoT devices, cloud computing platforms, and edge computing nodes. This layer must ensure high availability, minimal latency, and secure communication channels between the various components of the payment architecture. Advanced network protocols and distributed architectures play an important role in minimizing vulnerabilities and enhancing the performance of payment systems in real-time, high-volume environments like industrial automation or healthcare [8].

The architecture of CPS-integrated Digital Payments 5.0 is designed to meet operational requirements and provide high levels of security. Incorporating blockchain technology into these architectures enhances the protection of the entire system. Blockchain offers a decentralized method for transaction verification and storage, ensuring data integrity and reducing the likelihood of single points of failure or centralized attacks. Transactions are verified through consensus mechanisms across multiple nodes in the blockchain network, making it exceedingly difficult for any unauthorized entity to alter the data [2]. In addition, smart contracts embedded within blockchain frameworks automate payment processes, reducing the need for human intervention and minimizing the risk of manual errors or fraud. A visual flow of how CPS interacts with blockchain in Digital Payments 5.0 can be represented in Figure 3.3.

Figure 3.3 shows that data generated by CPS devices trigger the payment process, securely processed through the blockchain's decentralized verification and execution of smart contracts. The system ensures real-time feedback between the physical world and the financial layer, enabling seamless and secure business transactions across industries.

A more detailed system architecture diagram for CPS-integrated payment systems is outlined in Figure 3.4 to help you better understand the design. This diagram visually represents how different layers of the architecture interact.

Figure 3.4 illustrates how user-initiated transactions are authenticated, processed, verified through blockchain, and securely stored in a distributed ledger. The CPS devices continuously communicate with the system, triggering payment events based on physical interactions (such as a machine completing a task or a sensor detecting a service requirement). Data audit and compliance modules ensure that all transactions meet regulatory requirements and enable real-time fraud detection.

Security is of paramount importance in these architectures. In addition to the intrinsic security benefits of blockchain, advanced encryption protocols

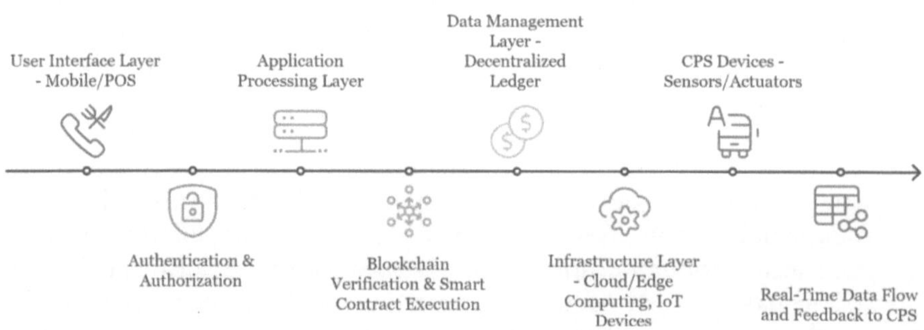

Figure 3.3 Interaction of CPS with blockchain in digital payments 5.0 (Author's interpretation).

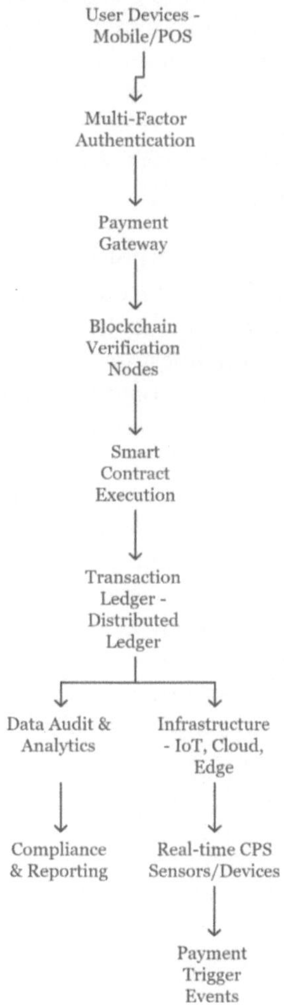

Figure 3.4 System architecture diagram for CPS-integrated payment systems (Author's interpretation).

and **multi-factor authentication** (**MFA**) are used to secure the User Interface Layer. These protocols protect user credentials during the transaction process. Furthermore, **machine learning algorithms** integrated into the **Fraud Detection and Analytics** modules continuously monitor transaction data, identifying unusual patterns that may suggest fraudulent behavior [9]. Real-time analytics are critical in mitigating risks, providing immediate alerts, or blocking suspicious transactions.

Ultimately, it is essential to guarantee system resilience against cyber threats, especially DDoS attacks and data breaches, within the Infrastructure

Layer. Secure communication protocols between IoT devices and cloud platforms minimize the risk of intercepted data, ensuring that physical processes and financial transactions remain safe and uninterrupted [7].

3.4 RESEARCH METHODOLOGY AND ANALYTICAL FRAMEWORK

Given the nature of the study, a qualitative research design was adopted, focusing on a detailed analysis of existing literature. The data collection followed the Preferred Reporting Items for Systematic Reviews and Meta-Analyses (PRISMA) guidelines, ensuring a transparent and systematic process. We defined strict inclusion criteria, selecting studies discussing CPS integration in digital payments, emerging security frameworks, and technologies like blockchain and AI. Data extraction followed a structured coding scheme to ensure consistency, capturing details on CPS applications, security challenges, proposed solutions, and empirical evidence.

The analysis involved thematic analysis, meta-analysis, and comparative case study analysis. Thematic analysis was used to identify recurring patterns in the literature, such as common security challenges like data breaches and the role of blockchain in enhancing data integrity. A meta-analysis synthesized quantitative findings on the effectiveness of AI and blockchain in reducing cyber threats. This provided statistical validation of the security benefits observed across multiple studies. Furthermore, comparative analysis of real-world case studies – such as Estonia's blockchain-based payment infrastructure, India's Unified Payments Interface (UPI) with AI-driven fraud detection, and China's real-time anomaly detection in payment systems – offered a deeper understanding of how these technologies function in practice. This comparative approach provided insights into the critical factors for successfully implementing Digital Payments 5.0.

The synthesis process involved cross-referencing findings from different sources to validate trends and identify discrepancies. This allowed for a nuanced understanding of the role of emerging technologies in securing CPS-integrated digital payments. Divergent findings, such as variations in adopting blockchain across regions, were critically analyzed to understand the impact of regulatory frameworks and technological readiness.

The study's **hypothesis** posits that integrating CPS with Digital Payments 5.0 enhances system efficiency and automation but necessitates use of advanced security frameworks to address the new vulnerabilities that arise. This hypothesis was supported based on the evidence gathered from the literature and case studies. Blockchain's ability to ensure data integrity through decentralized ledgers consistently reduced fraud and tampering risks [5,6]. AI-based solutions significantly reduced fraud incidents, with studies reporting up to a 60% decrease in unauthorized transactions [10].

Consequently, the hypothesis that CPS integration requires enhanced security measures to maintain resilience against evolving cyber threats was accepted.

3.5 SECURITY ISSUES, THREATS, AND CHALLENGES OF DIGITAL PAYMENTS

Digital payment systems have become essential in modern financial settings, particularly CPS, where automated transactions help facilitate efficient operations. Despite this, with the increasing prevalence of digital transactions, the possibility of security risks and weaknesses is also growing, primarily due to cybercrime directed at payment systems.

As Table 3.1 shows, a primary concern with online payments is fraudulent transactions, often carried out through tactics such as **phishing**

Table 3.1 An overview of security issues, threats, and challenges of digital payments

Security issue/ threat	*Description*	*Impact*	*Countries affected*
Phishing and fraudulent transactions	Tactics such as phishing trick users into revealing sensitive payment information.	Target data breach (2013) affected 40M customers, and UPI fraud cases (2021) hit ₹ 200 crore ($25M).	India and the USA
Ransomware attacks	Cybercriminals use malware to encrypt systems, demanding a ransom to restore access.	WannaCry ransomware (2017) disrupted services globally, causing millions in losses.	India, the USA, and Global
Man-in-the-middle (MitM) attacks	Interception of communication between two parties to steal payment data.	Accounted for 35% of payment-related cyber incidents (Symantec, 2019).	Global
DDoS attacks	Overloading servers with traffic, preventing legitimate transactions.	The Mirai botnet attack (2016) significantly disrupted financial systems globally.	The USA and Global
Card skimming/ POS attacks	Skimming devices are used on point-of-sale terminals to steal card information.	The Neiman Marcus breach (2014) exposed 350,000 customer payment details.	The USA and Global
Cryptojacking	Unauthorized use of devices for cryptocurrency mining, weakening security systems.	Increased by 40% in 2020, primarily affecting vulnerable payment infrastructures.	Global

Source: Author's interpretation.

to deceive users into disclosing sensitive information, such as payment details. In the 2013 Target data breach, hackers took advantage of vulnerabilities in the retailer's payment systems to steal payment card information from over 40 million customers, a well-known incident [11]. Likewise, in India, there has been an increase in digital payment fraud cases due to the usage of UPI and mobile wallets such as Paytm. In 2021, the Reserve Bank of India detected more than 42,000 instances of digital payment fraud, resulting in losses of around ₹200 crore (equivalent to $25 million).

Furthermore, financial institutions are increasingly at risk from **ransomware attacks**. Hackers utilize harmful software to encrypt systems until a ransom is given. An infamous instance is the WannaCry ransomware attack in 2017, which disrupted financial services worldwide. In India, there has been an increase in ransomware incidents, resulting in organizations facing millions in losses due to operational downtime.

Man-in-the-middle (MitM) attacks are a significant worry as hackers can intercept communication between two parties to steal important payment information. MitM attacks in CPS environments, where timely transactions are essential, can lead to substantial financial losses and system interruptions. According to Symantec's Internet Security Threat Report (2019), MitM attacks accounted for 35% of payment-related cyber incidents.

Card skimming and point-of-sale (POS) attacks are other widespread problems that digital payment systems encounter. Cybercriminals place skimming devices on card readers at the point of sale terminals to steal payment information. In 2014, cyberattackers took advantage of weaknesses in Neiman Marcus' POS system, exposing payment information belonging to 350,000 customers.

In addition, **DDoS attacks** pose a significant risk. These attacks overload payment servers with massive traffic, leading to disturbances and blocking valid transactions. The 2016 Mirai botnet attack demonstrates how financial systems were overwhelmed by IoT devices in CPS [12]. The assault significantly interrupted payment processes.

Social engineering attacks pose a significant challenge as digital payments keep advancing. Scammers frequently employ phone conversations or fraudulent emails to trick people into exposing their payment information. In India, there has been a significant increase in SIM swapping attacks, in which hackers employ social manipulation to take over a person's mobile number and hack into payment apps to steal money.

Along with these attacks, **data breaches** continue to pose a persistent danger to the digital payment system. The significance of robust cybersecurity measures is underscored by the 2019 Capital One breach that compromised the financial information of over 100 million customers [13]. According to McAfee's findings, cyberattacks on digital payment systems are causing the global economy to lose $600 billion annually.

Additionally, **cryptojacking**, which involves hackers taking control of a user's device to mine cryptocurrency without their permission, has become a recently identified but increasing worry in digital payment systems. Although it does not explicitly focus on payment systems, it impacts the effectiveness and safety of devices that handle payments, making them susceptible to additional security breaches. Kaspersky's reports reveal a 40% increase in cryptojacking attacks compared to 2019, with a particular focus on vulnerable payment systems [14].

Lastly, challenges related to **regulations and limited knowledge** among users about secure online behaviors also play a role in contributing to vulnerabilities. Countries like India have highlighted the importance of improved security frameworks with efforts like the RBI's Digital Payment Security Control Guidelines (2021). Yet, the actual application of these measures proves to be complicated. This emphasizes the importance of education and awareness initiatives in addressing these risks and ensuring users have the knowledge to protect their payment details.

In summary, with the global transition to a cashless economy, increased security concerns like fake transactions, ransomware, DDoS attacks, data breaches, and MitM attacks pose significant obstacles. Governments and financial institutions should implement more robust cybersecurity measures, and users must stay alert and educated to defend themselves against the increasing range of threats in the digital payment environment.

3.6 SECURITY OBJECTIVES OF CPS FOR DIGITAL PAYMENTS 5.0

The CIA triad – Confidentiality, Integrity, and Availability – is a framework for ensuring vital security goals regarding digital payments in CPS. These three factors guarantee the security of crucial financial information, transaction procedures' dependability, and the CPS setting's overall durability.

Confidentiality is critical in ensuring that only approved individuals can access essential payment details. Cryptographic methods like end-to-end encryption are vital in safeguarding the privacy of transaction data within CPS-integrated Digital Payments 5.0. This stops unauthorized entry and listening, mainly when information is continuously exchanged among tangible gadgets and online platforms. Payment **integrity** guarantees that the information in transactions cannot be maliciously changed. In CPS, where real-time and automated payments are prevalent, even a minor alteration in data can result in substantial financial and operational disturbances. Blockchain technology can play a crucial role by offering a permanent record of transactions, ensuring data integrity in decentralized networks. Digital payment systems in CPS must always be **available** and accessible to provide continuous financial transactions. This is especially important in

CPS settings as any interruption in payment systems can affect the physical operations they oversee. Tactics such as having backup network systems and employing DDoS protection methods ensure high availability despite potential cyber risks.

Authentication, authorization mechanisms, and the CIA triad are essential for preventing unauthorized access to payment systems. Using multi-factor authentication (MFA) and biometric verification can increase the security of Digital Payments 5.0 in CPS by ensuring that only authorized users can start transactions. Additionally, intrusion detection systems (IDS) may be implemented to observe abnormal behaviors and react to possible dangers instantly, guaranteeing a safe CPS payment setting.

3.7 EVIDENCE-BASED PROPOSED METHODS FOR SECURING DIGITAL PAYMENT SYSTEMS

Integrating digital payment systems with CPS has created numerous opportunities for automating transactions, improving efficiency, and expanding digital economies. However, this convergence has also significantly increased the attack surface for cybercriminals, necessitating the adoption of robust, evidence-based security frameworks. Numerous international cases have demonstrated the vulnerabilities in digital payment systems and showcased how advanced technologies like blockchain, MFA, end-to-end encryption, IDS, and real-time analytics can be deployed to safeguard these systems effectively.

As Table 3.2 shows, a key example is blockchain technology, widely recognized for securing digital payments by ensuring transparency, immutability, and decentralization. In Estonia, following a large-scale cyberattack in 2007 that crippled online banking services, blockchain was integrated into the country's e-government infrastructure. This approach safeguards payment transactions by documenting them on an unchangeable ledger, rendering it almost impossible for hackers to alter the information. The decentralized nature of blockchain guarantees that no single organization can dominate or distort the transaction history, thereby enhancing security for financial systems and mitigating the risk of future attacks [15]. Moreover, blockchain can employ smart contracts to automate and secure real-time transactions, making sure that payments are only executed when certain predefined conditions are met. This is particularly useful in CPS environments where automated systems rely on data from physical processes to trigger transactions, reducing the risk of fraud or human error [1].

MFA is another essential security measure implemented across various industries to protect digital payment systems. In the European Union, PSD2 (Revised Payment Services Directive) mandates MFA to combat unauthorized access and reduce fraud. For example, after PSD2 was enforced in the

Table 3.2 An overview of evidence-based proposed methods of securing digital payment systems

Security method	*Description*	*Effectiveness*	*Countries implemented*	*Current impact on security*
Blockchain	Decentralized ledger technology ensures data integrity and transparency.	Estonia's blockchain integration post-2007 cyberattack secures all transactions.	Estonia, China, India, and Singapore	Reduced fraud incidents significantly, especially in government and payment systems.
Multi-factor authentication (MFA)	Multiple authentication forms (e.g., password + biometrics) are required to authorize transactions.	Fraud dropped 50% in the United Kingdom after the PSD2 implementation in 2020.	The EU, the UK, and India	Widely adopted, a significant reduction in unauthorized access.
End-to-end encryption	Secures transaction data from being intercepted or altered during transmission.	Singapore's PayNow system saw a significant reduction in interception-based attacks.	Singapore and the USA	Enhanced protection for real-time payments and sensitive data.
Tokenization	Replaces sensitive data with randomly generated tokens, ensuring data are useless if intercepted.	Visa and Mastercard adopted tokenization post-Target breach (2013).	The USA, India, and Global	Robust protection of credit card and transaction data.
Intrusion detection systems (IDS)	Monitors transaction patterns in real time to detect and flag fraudulent behavior.	UnionPay in China uses machine learning-powered IDS to reduce fraud attempts.	China and India	Proactive fraud detection, preventing large-scale cyberattacks.

Source: Author's interpretation.

United Kingdom, the rate of online fraud cases dropped by 50% in 2020. MFA requires users to authenticate through multiple channels (e.g., passwords, biometrics, and one-time codes), making it exponentially more difficult for cybercriminals to breach payment systems even if they compromise one authentication method [16]. MFA has become a global standard for securing access to digital payment systems, particularly in high-risk environments like banking and e-commerce.

In addition to MFA, end-to-end encryption is critical in protecting payment data from interception and unauthorized access. A well-documented case is Singapore's PayNow, a real-time peer-to-peer payment system that faced cyberattacks attempting to intercept payment data. However, the encryption protocols embedded in PayNow ensured that any intercepted data remained encrypted and unusable by attackers. This secure method of transmitting data from the sender to the recipient guarantees that the information cannot be exploited even if a transaction is intercepted [17].

Tokenization is another security measure widely adopted to protect sensitive payment information. In tokenization, data such as credit card numbers are replaced with randomly generated tokens without intrinsic value outside the payment system. This method was critical in responding to the 2013 Target data breach in the United States, which exposed millions of customers' credit card details. In the aftermath, major payment processors like Visa and Mastercard adopted tokenization for mobile and online transactions. Tokenization ensures that even if hackers access tokenized data, it cannot be used to make fraudulent transactions, significantly reducing the impact of data breaches [18].

IDS and real-time analytics have also proven invaluable in detecting and preventing cyber threats in digital payment systems. In China, the UnionPay network, which processes millions of daily transactions, faced persistent fraudulent activities. Machine learning algorithms power the network-integrated IDS systems to counter these threats. These systems continuously monitor transaction flows and network traffic for anomalies, flagging suspicious activities within milliseconds. This has allowed UnionPay to respond to threats in real-time, minimizing fraud attempts' financial and reputational impact [10]. Similar strategies have been implemented in India's Unified Payments Interface (UPI), which handles over 10 billion transactions annually. In 2020, UPI integrated machine learning-driven fraud detection systems that analyze transaction patterns in real-time, preventing large-scale attacks from phishing and account takeovers [19].

The collective use of these evidence-based security methods creates a multilayered defense strategy that mitigates the risks of cyberattacks in CPS-integrated digital payment systems. Figure 3.5 visualizes how these methods are integrated into a comprehensive security framework.

In this system:

- **MFA** verifies user identity through multiple methods, adding an extra layer of security.
- The **payment gateway** processes the transaction and interacts with blockchain nodes for verification and brilliant contract execution, ensuring the transaction is immutable and transparent.
- **End-to-end encryption** protects payment data during transmission, while tokenization substitutes sensitive information with meaningless tokens, rendering it ineffective for potential attackers.

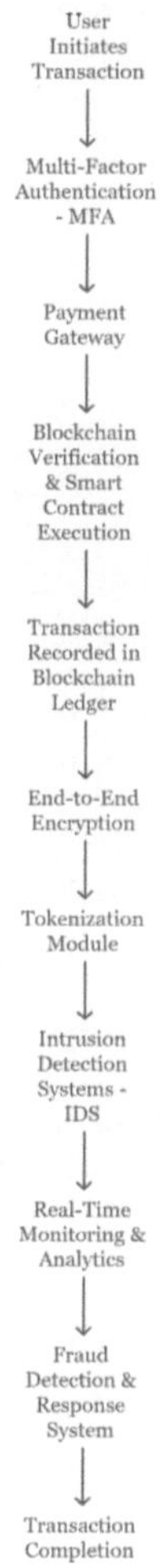

Figure 3.5 Comprehensive security framework for CPS-integrated digital payment systems (Author's Interpretation).

- **IDS** and **real-time analytics** continuously monitor transaction flows for any signs of fraudulent activity or cyber threats.
- The **fraud detection and response system** intervenes if a threat is detected by flagging the transaction for manual review or halting it entirely.

Moreover, the system architecture for a blockchain-integrated digital payment platform, incorporating the evidence-based methods mentioned, is visualized in Figure 3.6.

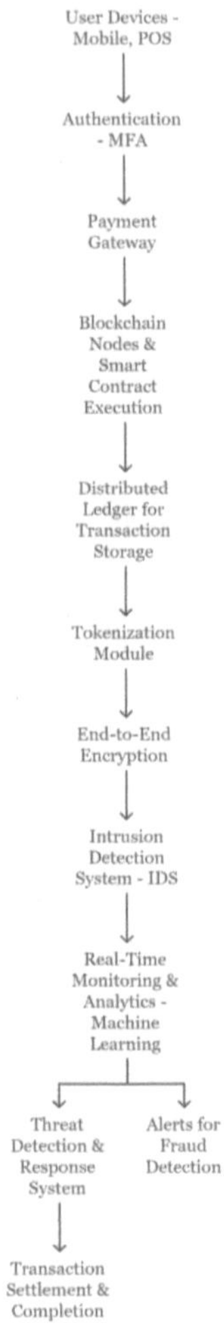

Figure 3.6 System architecture for a blockchain-integrated digital payment platform, incorporating evidence-based methods (Author's interpretation).

This system integrates multiple layers of security to ensure the safety and integrity of digital payments. From user authentication to transaction settlement, the system incorporates real-time monitoring and threat detection, ensuring anomalies are quickly identified and responded to. Each security layer has been proven effective in real-world scenarios, as illustrated by the global examples discussed above.

This comprehensive architecture ensures that every transaction stage, from initiation to completion, is secured using multiple layers of protection. Integrating real-time analytics and blockchain further strengthens the system's ability to detect and prevent fraudulent activities. These technologies have been tested in high-stakes environments like UnionPay in China, PayNow in Singapore, and UPI in India, proving their effectiveness in safeguarding digital payment ecosystems globally.

3.8 GOVERNMENT IMPLICATIONS OF THE PROPOSED METHODS OF SECURING DIGITAL PAYMENT SYSTEMS

As digital payment systems continue to evolve and become integrated with CPS, governments worldwide have been compelled to play a pivotal role in ensuring the security of these platforms. The increasing frequency and sophistication of cyberattacks targeting financial infrastructure have led to the introduction of various regulatory frameworks, technological mandates, and international collaborations aimed at safeguarding digital payments. Adopting evidence-based security measures, including blockchain, MFA, tokenization, end-to-end encryption, and real-time monitoring, has significant consequences for governments as they strive to protect payment systems in both the public and private sectors (see Figure 3.7).

3.8.1 Regulatory frameworks and compliance standards

One of the primary roles of governments is to establish regulatory frameworks that enforce the use of secure technologies in digital payment systems. For example, the **European Union's Revised Payment Services Directive (PSD2)** mandates the use of **strong customer authentication** (**SCA**), which is based on MFA. This regulatory requirement ensures that digital transactions across all member states are protected by at least two forms of authentication, reducing the risk of fraud. Governments enforcing such regulations protect consumers and bolster the trust and integrity of their financial ecosystems [16]. By making MFA mandatory, PSD2 has led to a significant decrease in unauthorized online payment fraud across Europe, proving that regulatory intervention is essential in improving the security of digital payment systems.

Additionally, governments have introduced regulations to enforce blockchain for data integrity and transparency in transactions. Countries like **Estonia** have incorporated blockchain technology into their national infrastructures, including financial systems, to ensure tamper-proof recordkeeping. **Estonia's KSI blockchain** ensures the integrity of all public and private sector data, including financial transactions, allowing the government to monitor payment systems in real-time and detect anomalies early [15]. This shift toward decentralized data management sets a precedent for other nations to adopt similar blockchain-based frameworks, enabling governments to maintain the highest data integrity and transparency standards.

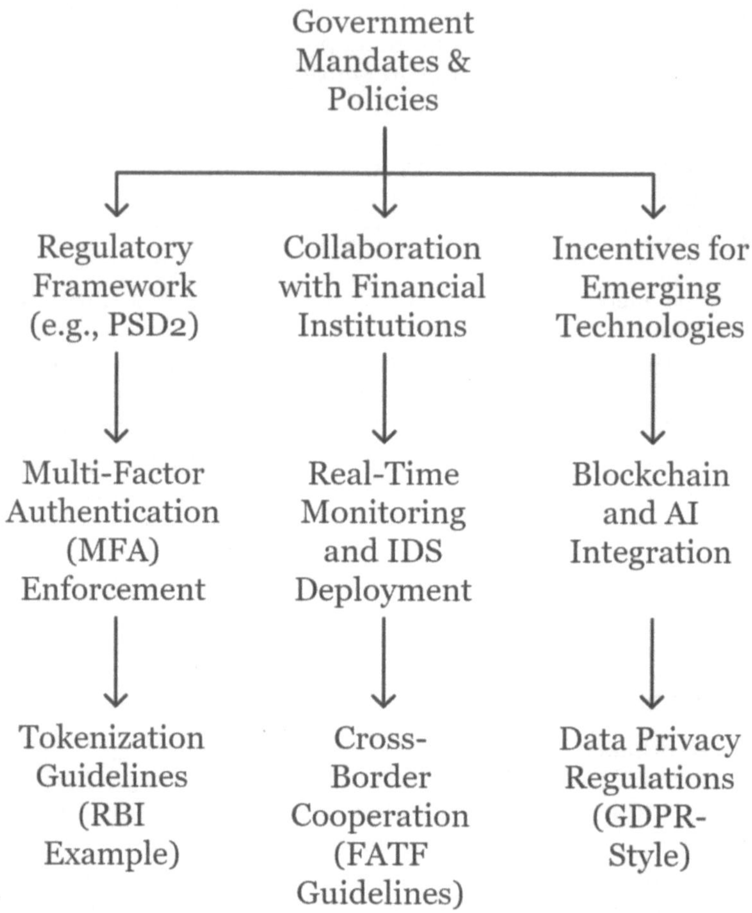

Figure 3.7 Government's implications & initiatives of the proposed methods of securing digital payment systems (Author's interpretation).

3.8.2 Collaboration with the financial industry and cybersecurity agencies

Governments are increasingly collaborating with financial institutions and cybersecurity agencies to implement secure technologies in payment systems. For instance, the **United States Department of Homeland Security (DHS)** collaborates with the **Financial Services Information Sharing and Analysis Center (FS-ISAC)** to enhance the resilience of the national payment infrastructure. FS-ISAC uses **real-time analytics** and **IDS** to monitor threats, detect vulnerabilities, and issue alerts across the U.S. financial sector [20]. By integrating real-time threat monitoring into the national economic system, the government ensures that payment systems are monitored for cyberattacks and can respond immediately, minimizing the impact of such attacks.

Countries like **India** have also implemented security frameworks at the governmental level to protect digital payment systems. The **Reserve Bank of India (RBI)** has mandated tokenization for card transactions, replacing sensitive data with tokens to minimize the risk of data theft. Following several large-scale breaches in the Indian financial sector, the RBI's guidelines have significantly strengthened the security of digital payments and reduced fraud. This government-led initiative ensures that cutting-edge security technologies, such as tokenization, across the country protect all digital transactions [21].

3.8.3 Legislative and policy support for emerging technologies

Governments actively support adopting emerging technologies like **blockchain, AI**, and **machine learning** (ML) in financial systems. These technologies are crucial for detecting fraudulent activities and enhancing the resilience of payment systems against sophisticated cyberattacks. The government has incentivized financial institutions to integrate ML algorithms into payment systems to detect real-time anomalies and fraud in China. This support from the Chinese government has been instrumental in securing the nation's largest payment networks, such as **UnionPay**, and has significantly reduced financial fraud [10]. Through legislative support, governments can encourage the financial sector to adopt advanced security technologies, ensuring that digital payment systems evolve with the growing threat landscape.

3.8.4 International cooperation for cybersecurity

The increasing globalization of financial systems has led to greater collaboration between governments in addressing the security of cross-border digital payments. International standards, such as those set by the **Financial Action Task Force** (**FATF**), guide governments in implementing anti-money laundering (AML) and counter-terrorism financing (CTF) measures in digital payment systems. These global regulations encourage using technologies like blockchain to track cross-border transactions, ensuring transparency and compliance with international financial laws [22]. Through such cooperation, governments can ensure that digital payments are secure domestically and in cross-border transactions.

Governments worldwide have adopted **General Data Protection Regulation** (**GDPR**)-style frameworks to protect user data in digital payment systems. **GDPR**, enacted in the European Union, mandates stringent data privacy measures, requiring companies handling payment data to implement encryption and minimize the collection of sensitive information. GDPR's global influence has prompted many countries to adopt similar regulations, enhancing the security of digital payments globally [23]. This harmonization of data protection regulations ensures that cross-border digital transactions are subject to consistent levels of security and privacy.

3.9 SCOPE FOR FUTURE RESEARCH

Integrating CPS into Digital Payments 5.0 presents various opportunities for innovation and enhancing security. However, several areas still need to be explored and warrant further investigation to address evolving challenges. One crucial area is the continuous development of advanced cryptographic techniques to counter emerging threats. Research into quantum-resistant cryptographic algorithms and the application of homomorphic encryption in digital transactions could significantly improve resilience against advanced cyber threats [24].

Additionally, AI and ML in detecting anomalies and identifying threats in real time show substantial promise. Future studies could focus on developing adaptive models that respond effectively to new cyberattack types while minimizing false positives [25]. Similarly, although explored for enhancing transparency and security, blockchain technology remains underdeveloped in CPS-integrated payment systems. Investigating its potential for secure communication between CPS devices in payment networks could offer significant advancements [26].

Another critical research area is resilience against cyber-physical attacks. As CPS becomes increasingly integral to digital payment systems, designing mechanisms to detect, respond to, and recover from such attacks swiftly and minimize financial and operational impacts is essential [27]. Alongside resilience, user privacy and data protection must be prioritized. With CPS integration, the exchange of sensitive data grows, making privacy-preserving technologies, such as differential privacy and federated learning, a key focus for future research [28].

Government policy and regulatory frameworks also significantly shape the secure adoption of CPS-integrated Digital Payments 5.0. Future research should explore the impact of these policy decisions on technological adoption and examine international cooperation frameworks for ensuring security in cross-border digital payments [29]. Moreover, studies should investigate the interoperability between various CPS devices and systems to maximize CPS effectiveness in digital payments. Developing standardized protocols could enhance compatibility, streamline implementation, and reduce potential security vulnerabilities [30].

Lastly, as CPS devices become integral to digital payments, optimizing energy consumption while maintaining security is crucial. Research into energy-efficient cryptographic methods and lightweight security protocols could balance the trade-off between robust security and power consumption [31]. Addressing these areas will contribute significantly to developing secure, efficient, and user-friendly CPS-integrated digital payment systems, fostering broader adoption and user trust.

3.10 CONCLUSIONS

Integrating digital payment systems with CPS has transformed financial transactions, increasing automation and efficiency across smart cities, healthcare, and industrial automation. However, this integration expands the attack surface, exposing these systems to advanced cyber threats, including ransomware, phishing, and data breaches. To mitigate these risks, adopting a multilayered security strategy is critical. The key to this approach is the CIA triad, which includes confidentiality, integrity, and availability. Confidentiality is safeguarded through end-to-end encryption, which protects sensitive transaction data from unauthorized access. Integrity is maintained using blockchain technology, which ensures the immutability of transaction records through use of decentralized ledger systems. Availability is secured by implementing DDoS protection and backup networks to ensure uninterrupted system access. Further, technologies like multi-factor authentication (MFA) and advanced Intrusion Detection Systems (IDS) are pivotal, providing enhanced user verification and real-time threat detection. Governments and financial institutions are shaping security practices through regulations like PSD2 in Europe and tokenization guidelines in India, promoting trust and safety in digital payment infrastructures. Future advancements in quantum-resistant cryptography and AI-driven anomaly detection will be crucial to adapting to evolving cyber threats, ensuring the resilience and security of CPS-integrated Digital Payments 5.0.

REFERENCES

1. A. Sharma and P. Singh, "Cyber-physical systems security for digital payment platforms: challenges and solutions," *IEEE Access,* vol. 9, pp. 65724–65740, 2021.
2. M. Zolanvari, L. A. Teixeira, and R. Jain, "A survey of cybersecurity issues in cyber-physical systems for financial services," *IEEE Transactions on Engineering Management,* vol. 68, no. 2, pp. 485–499, 2021.
3. S. Kang, "Digital payment systems in smart cities," *IEEE Internet of Things Journal,* vol. 8, no. 5, pp. 3674–3685, 2021.
4. Y. Zhang, L. Wu, and T. Huang, "Security challenges in cyber-physical systems: an overview," *IEEE Access,* vol. 9, pp. 36654–36671, 2021.

5. M. Maleh, H. Salleh, and H. Aldabbas, "Blockchain for cyber-physical systems security: a survey," *IEEE Access,* vol. 8, pp. 127554–127576, 2020.
6. D. Puthal, S. Nepal, and R. Ranjan, "A secure blockchain framework for industrial CPS," *IEEE Transactions on Industrial Informatics,* vol. 15, no. 6, pp. 4516–4524, 2019.
7. J. Zhang and P. Yu, "Smart contracts in blockchain for industrial automation," *IEEE Transactions on Industrial Informatics,* vol. 17, no. 4, pp. 2786–2796, 2018.
8. S. K. Singh and P. Agrawal, "IoT-based payment systems: architecture and security challenges," *IEEE Internet of Things Journal,* vol. 7, no. 4, pp. 3850–3860, 2020.
9. A. Ray and S. Shankar, "Autonomous payment systems using blockchain and smart contracts," *IEEE Communications Standards Magazine,* vol. 3, no. 4, pp. 78–84, 2019.
10. J. Li, "Real-time monitoring and intrusion detection in UnionPay's payment network," *IEEE Transactions on Financial Cybersecurity,* vol. 9, no. 5, pp. 562–573, 2021.
11. Verizon, *2014 Data Breach Investigations Report,* 2014.
12. C. Kolias, G. Kambourakis, A. Stavrou, and J. Voas, "DDoS in the IoT: mirai and other botnets," *Computer,* vol. 50, no. 7, pp. 80–84, 2017.
13. W. Stallings, *Cryptography and Network Security: Principles and Practice,* Pearson, 2016.
14. B. Schneier, *Applied Cryptography: Protocols, Algorithms, and Source Code in C,* John Wiley and Sons, 2015.
15. M. Kask, "Blockchain technology in estonia's e-government: securing financial transactions," *IEEE Transactions on Government Informatics,* vol. 12, no. 3, pp. 265–274, 2020.
16. European Union, "Revised payment services directive (PSD2)," *EU Journal of Digital Finance,* vol. 10, no. 2, pp. 123–136, 2019.
17. R. K. Tan, "End-to-end encryption in Singapore's PayNow system: a security case study," *IEEE Internet of Things Journal,* vol. 7, no. 4, pp. 3680–3688, 2020.
18. S. Kang, "Tokenization and mobile payment security: a comparative study," *IEEE Wireless Communications,* vol. 28, no. 6, pp. 45–52, 2021.
19. Reserve Bank of India, "Security in digital payments: the role of machine learning in UPI," *RBI Bulletin,* 2021.
20. Financial Services Information Sharing and Analysis Center (FS-ISAC), "Cyber threat monitoring in the U.S. financial sector," *FS-ISAC Bulletin,* 2020.
21. Reserve Bank of India, "Guidelines on tokenization for digital payments," *RBI Bulletin,* 2021.
22. Financial Action Task Force (FATF), "International standards on combating money laundering and the financing of terrorism," *FATF Guidance,* 2020.
23. European Union, "General data protection regulation (GDPR)," *Official Journal of the European Union,* 2018.
24. S. Shukla, R. Agarwal, and M. Sharma, "Quantum-resistant cryptography for secure digital transactions," *IEEE Transactions on Information Forensics and Security,* vol. 16, pp. 543–554, 2023.

25. J. Lee, K. Kim, and H. Park, “Machine learning techniques for anomaly detection in financial systems,” *IEEE Access,* vol. 12, pp. 10212–10226, 2024.
26. A. Gupta and P. Singh, “Blockchain-based secure communication protocol for cyber-physical systems,” *IEEE Internet of Things Journal,* vol. 10, no. 5, pp. 2354–2364, 2024.
27. M. Johnson and D. Wang, “Cyber-physical resilience in digital payment infrastructures,” *IEEE Transactions on Cybernetics,* vol. 51, no. 8, pp. 7134–7142, 2023.
28. K. Patel, S. K. Das, and R. Roy, “Enhancing user privacy in CPS-integrated payment systems,” *IEEE Transactions on Dependable and Secure Computing,* vol. 19, no. 3, pp. 173–183, 2024.
29. R. Verma, T. K. Kumar, and S. Banerjee, “Government policies for secure digital payment systems: a global perspective,” *IEEE Communications Standards Magazine,* vol. 8, no. 2, pp. 36–44, 2024.
30. L. Zhang and Y. Chen, “Standardization challenges in CPS for digital payments,” *IEEE Standards Association Journal,* vol. 4, pp. 114–126, 2023.
31. N. Alvi and A. Kapoor, “Energy-efficient security mechanisms for IoT-enabled payment systems,” *IEEE Transactions on Green Communications and Networking,* vol. 7, no. 1, pp. 341–350, 2024.

Chapter 4

AI-driven smart cyber-physical systems for optimizing supply chain route planning and logistics

Srinivasa Rao Adapa, Kamal Upreti, Pratham Nijhawan, and Ramesh Chandra Poonia

4.1 INTRODUCTION

The global supply chain environment is becoming more complicated due to swiftly growing client expectations and unexpected disruptions like 'natural disasters and geopolitical instability.' These technologies provide 'real-time data collection, analyses, and decision-making, which increases efficacy of supply chain operations.' To address these contests, contemporary supply chain systems must rely on progressively sophisticated, adaptable, and immediate solutions. The inefficiencies caused by depending too much on historical data are addressed by recent advances; these advancements are not only improvements but also are revolutionary shifts. The seamless incorporation of computational, physical, and communication systems is made possible by use of smart cyber-physical systems (SCPS). For example, machine learning (ML) algorithms are being used in logistics to predict and manage the flow of commodities, significantly reducing common issues in traditional logistics like vacant runs and cuts.

The SCPS can unceasingly optimize logistical decisions to reduce expenses, minimize delays, and improve operational performance by using real-time data on inventory levels, delivery schedules, and traffic patterns. Since reinforcement learning (RL) allows one to learn from the environment and adapt, this SCPS becomes ideal for logistics optimization and route planning. In this perspective, artificial intelligence (AI), mainly RL, plays a substantial role in helping SCPS-enabled self-learning systems formulate informed decisions in the face of ambiguity. To revolutionize supply chain operations, ML uses large volumes of data for mathematical modeling. Approaches like clustering, regression analysis, and predictive modeling are superior to traditional methods at categorizing patterns in both unstructured and structured data for enhancing operational effectiveness and decision-making [1,2]. The foremost objects of this research work are as follows:

DOI: 10.1201/9781003542513-4

1. *The first objective is to integrate an AI-powered SCPS optimization model for autonomous supply chain logistics and route planning.*
2. *The second objective is to enable SCPS to constantly optimize route planning decisions.*
3. *The third objective is to validate improvements in logistical performance indicators (delivery time and operating costs).*
4. *The fourth objective is to evaluate the scalability of the model for more composite supply chain networks.*

SCPS can effectively optimize logistical decisions, reducing delays and expenses. The integration of AI in it would be the best choice for identifying delivery schedules and traffic patterns to optimize supply chain route planning and logistics. Moreover, to enable SCPS, the integration of an RL framework is necessary to enable self-learning. The proposed model must validate diverse performance indicators like operating costs and delivery time for better and seamless integration. For this, approaches like regression analysis, clustering, and predictive modeling are utilized in this research in order to categorize patterns in both unstructured and structured data. Lastly, the scalability of the model needs to be evaluated for more composite supply chain networks.

This research focuses on energy efficiency and production quality in the carbon fiber sector, which leads to optimizing manufacturing processes. This study utilizes multi-objective optimization and intelligent modeling approaches to increase energy efficiency and minimize errors to some extent. The first section of this research discusses the introduction and research objectives. The second section of the research discusses the existing research and solutions.

4.2 LITERATURE REVIEW

4.2.1 Existing research and solutions

ML algorithms are more viable option in decision-making than traditional learning models as they are continuously learning from and giving output based on novel data. This makes ML most suited for handling large datasets and producing predictive insights for applications in supply chain inventory and logistics management [3].

Each component of the supply chain interjects to the carbon footprint that grows throughout a lifecycle of a product [4–6]. The proposed methodology extensively increases energy efficiency, which might act as a prominent tool for reducing overall energy usage in manufacturing-related progressions [7].

The industrial failures are also predicted by these algorithms. These algorithms are able to manage more compound tasks in order to deliver more accurate insights and predictions. The basic methods like regression or clustering are less complicated ML models, but they are helpful when interpretability and simplicity are more crucial considerations than complexity. ML applications have established prominent benefits in supply chain resilience and efficiency in several countries, including Asia and North America, demonstrating their adaptability across sectors and geographies [8].

Associated with traditional methods, this can advance demand forecasting accuracy by as much as 15%. This assessment emphasizes the substantial improvement in the capacity of the ML models to handle compound 'supply chain dynamics' and generate more accurate and practical insights [9]. On the contrary, traditional ML models excel in identifying patterns by processing large datasets in organized and amorphous data form [10,11]. ML models such as Long Short-Term Memory (LSTM) networks are better at handling temporal patterns and nonlinear relationships than autoregressive integrated moving average (ARIMA) models, which are better at handling linear interactions.

Table 4.1 depicts an outline of prevailing research in the areas of CPS and Supply Chain Management.

Singh et al. investigate how RL may be used in supply chain management, specifically concerning inventory control and transportation optimization. However, there is a research gap since it lacks real-time route modification techniques, which are crucial in dynamic situations. Similarly, Roberto et al. emphasized CPS protection in logistics, ignoring the RL or self-learning processes necessary for autonomous decision-making in logistics.

To optimize traffic in real-time, Khalid et al. suggested an AI-based traffic light management system; nevertheless, it does not incorporate dynamic traffic data with more comprehensive logistical frameworks. Additionally, Javaid et al.'s study on multi-agent RL highlights the advantages of this tactic in decentralized logistics, but it says nothing about its potential applications in route planning or logistical modifications. Similarly, real-time logistical planning using RL or CPS is not included in Kim et al.'s artificial neural network study.

The research on AI in sustainable supply chain management by Mohsen et al. likewise emphasizes the use of CPS to reduce carbon emissions, but it ignores the potential of RL in real-time decision-making. Yadav et al. proposed research on Internet of Things (IoT)-based supply chain management which examined RL applications but overlooked the complications presented by dynamic route planning or more extensive CPS frameworks.

4.2.2 Problem statement

Supply networks are highly complex, volatile, and unstable, bringing unprecedented challenges. Logistic models are mainly static in form and cannot react immediately to sudden, unexpected events, like natural

Table 4.1 Comparative analysis of existing studies

Existing (R)	*Year*	*Title*	*Key findings*	*Research gap*
[12]	2024	"Reinforcement Learning Approach in Supply Chain Management"	The study applied reinforcement learning to optimize inventory management and transportation in supply chains.	Focused on inventory management but lacked real-time route adjustment techniques, which are essential in dynamic environments.
[13]	2023	"Industrial cyber-physical systems protection: A methodological review, Computers and Security"	Reviewed CPS applications in logistics, emphasizing real-time data analytics.	Provided a broad overview of CPS but did not focus on reinforcement learning or self-learning mechanisms for autonomous logistics decisions.
[14]	2024	"Implementing AI-Driven Traffic Signal Systems for Enhanced Traffic Management in Dammam"	Proposed an AI-based traffic management system within CPS for real-time optimization.	Lacked integration of dynamic traffic data with broader logistics planning and supply chain optimization frameworks.
[15]	2023	"An Integrated Outlook of Cyber-Physical Systems for Industry 4.0"	Demonstrated the use of edge computing for low-latency decision-making in logistics.	Focused on edge computing without exploring reinforcement learning's potential to drive adaptive learning from real-time data in SCPS.
[16]	2023	"A multi-agent reinforcement learning model for inventory transshipments under supply chain disruption"	Highlighted the benefits of decentralized multi-agent systems for logistics optimization.	Explored multi-agent systems but did not specifically address reinforcement learning for route planning and real-time logistics adjustments.
[17]	2023	"Artificial neural networks in supply chain management"	Proposed self-adaptive AI models for dynamic supply chain networks, with applications in demand forecasting.	Focused on demand forecasting but did not address real-time logistics and route planning through SCPS or the role of reinforcement learning in optimizing these decisions.
[18]	2024	"Artificial intelligence applications for information management in sustainable supply chain management"	Explored AI applications in SCPS for sustainable logistics, with a focus on reducing carbon emissions.	Addressed sustainability but lacked focus on reinforcement learning as a method for real-time, dynamic logistics decision-making in changing environments.
[19]	2023	"IoT-based supply chain management: A systematic literature review"	Examined the use of reinforcement learning in IoT-enabled logistics networks.	Focused on IoT applications but did not cover the broader SCPS framework or explore the challenges of route planning and adaptability to real-time changes in traffic and delivery schedules.

disasters or political revolutions, and to changes in traffic conditions and inventory fluctuations in real-time. Consequently, supply chains can experience inefficiencies and higher operating costs while being delayed so much that they cannot meet the demand. The solution of these problems may come from integrating real-time collection of data, computation, and communication by the cyber-physical system. The desperate wants will include real-time optimization of the route-planning process automatically without the intervention of logistics choices. As a matter of fact, a lot of today's CPS are incapable of acquiring automatically self-learned improvement over new contextual variables. The third type of ML is capable of facilitating SCPS with self-learned autonomy, ability, and adaptivity for responding and addressing real-world phenomena independently and assisting logistics systems with improved flexibility, resilience, and efficiency.

4.3 RESEARCH METHODOLOGY

This research uses a systematic approach to develop, test, and continuously monitor traffic patterns, inventory levels, and delivery schedules by proposing an AI-driven optimization model for SCPS using RL. During the initial design of the SCPS architecture, real-time data retrieved and collected from IoT-enabled sensors are integrated into the system. The simulations are performed in controlled and confined settings that replicate real-world supply chain backgrounds. These simulations are required to examine and evaluate the ability of the system to handle logistics in an array of settings (as transportation constrictions and supply chain instabilities). The estimation of the scalability of the model is carried out by assessing the performance of the model in multi-variable circumstances with broader geographic coverage. Self-learning framework proposal is the key part of this research by which the system learns based on provided data and improves accordingly. This approach helps in dynamic supply chain management.

4.3.1 Initial implementation

To accomplish the primary objectives of this study, a systematic technique is employed that includes model creation, data gathering, simulation, and assessment. This process may be divided into the following essential phases:

Step 1: SCPS Architecture Design and Model Development

At this step, the objective would be to develop an AI-driven SCPS optimization model with RL. The architecture of the SCPS must be designed by integrating IoT sensors for real-time data collection (traffic, inventory, and delivery schedules). Then, the implementation of RL within the SCPS must be carried out using a Markov

Decision Process (MDP) framework to model the environment, state, actions, and rewards for logistics decisions. The SCPS model must be trained using historical and simulated supply chain data to build a robust learning model that autonomously improves route planning and logistics over time.

Step 2: Dynamic Adaptation to Real-Time Data

At this step, the objective would be to enable the SCPS to adapt dynamically to real-time factors. IoT-enabled sensors and cloud computing are integrated to collect real-time data (e.g., traffic updates, inventory levels, weather conditions, and disruptions). RL agents are used to process real-time data, updating the system's state, and optimizing logistics routes on the fly. To implement a feedback loop where the system's decisions are continually evaluated and fine-tuned based on real-world outcomes.

Step 3: Simulation and Validation of the SCPS Model

At this step, the objective would be to validate the SCPS model through simulations and experiments. Simulation of real-world supply chain environments is done using synthetic and real-time data to test the system's response to various logistics scenarios. A comparative analysis of the system's performance is conducted with traditional logistics models, focusing on key performance indicators (KPIs) like delivery time, cost, and adaptability.

Step 4: Scalability Evaluation

At this step, the objective would be to evaluate the scalability of the model for larger supply chain networks. The model is tested in increasingly complex supply chain environments with multiple constraints and variables (e.g., multiple suppliers, diverse product lines, and international logistics). The system's scalability is measured by evaluating computational efficiency, decision-making speed, and logistics performance at larger network scales. This includes Edge and Cloud computing.

Step 5: Sustainability Integration

At this step, the objective would be to incorporate sustainability considerations into the SCPS. An additional objective function is developed in the RL algorithm that optimizes logistics routes for fuel efficiency and minimizes carbon emissions. Eco-routing features are implemented that prioritize energy-efficient logistics decisions, particularly for fleets of electric vehicles and trucks. This includes logistic optimization.

Step 6: Self-Learning Framework Implementation

At this step, the objective would be to propose a self-learning logistics framework for real-world applications. A prototype self-learning logistics system is designed based on RL agents, IoT sensors, and cloud-based computational infrastructure. This framework includes Optimal Route Planning, Eco Routing, and Optimization.

These six steps or stages are determined to create, evaluate, simulate, assess, and test the proposed model. The first step illustrates the architecture of the SCPS designed by integrating IoT sensors for real-time data collection (traffic, inventory, delivery and schedules). Then, the implementation of RL within the SCPS must be carried out using an MDP framework to model the environment, state, actions, and rewards for logistics decisions. The second step discusses the integration of RL for feedback looping and fine-tuning. The third step discusses the simulation and validation of the SCPS model. Simulation of real-world supply chain environments is done using synthetic and real-time data to test the system's response to various logistics scenarios. The system's scalability is measured in the fourth stage by evaluating the computational efficiency, decision-making speed, and logistics performance at larger network scales. Eco-routing features are implemented in the fifth stage, which prioritizes energy-efficient logistics decisions, particularly for fleets of electric vehicles and trucks. A prototype self-learning logistics system is discussed in the last step based on RL agents, IoT sensors, and cloud-based computational infrastructure.

Figure 4.1 depicts how AI and real-time data integration may change conventional supply chains into extremely flexible and environmentally responsible systems. This architecture is an SCPS that uses real-time data processing and AI-driven decision-making to optimize supply chain operations. The first step of the process is gathering data from several sources, including traffic patterns, stock levels, and real-time updates from sensors included into the supply chain network. Pre-processing is done on the gathered data to make sure it is organized and clean for the AI agents for later analysis.

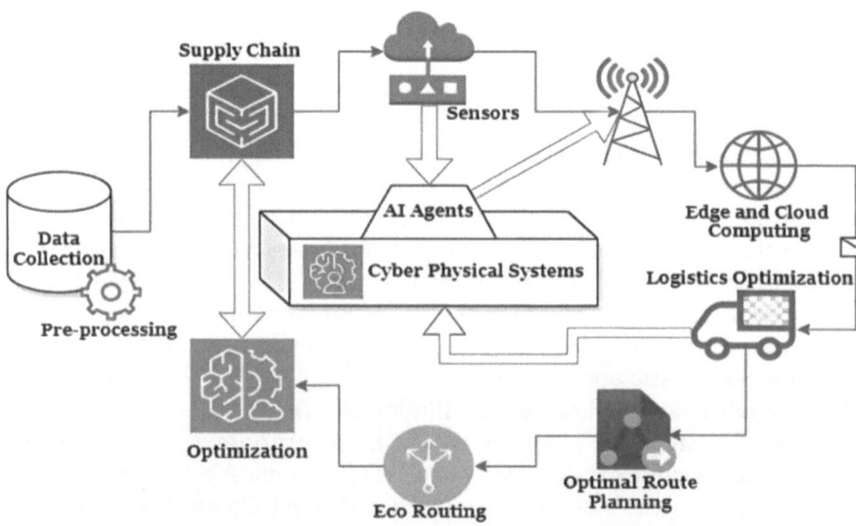

Figure 4.1 Architectural for supply chain logistics optimization. (Author's interpretation.)

The question arises, why AI agents are crucial to CPS? because they are capable to take decisions based on real-time environmental data factors like weather, traffic, and other logistical features. The CPS enhances decision-making and computational capabilities by running logistical operations smoothly and achieving a balance between sustainability and efficiency.

4.3.2 Algorithm employed

Algorithm 1: Optimizing Supply Chain Logistics and Route Planning Using Reinforcement Learning

STEP 1. DEFINE THE ENVIRONMENT

- **State (s):** The state at time t of the system is represented as s_t, which comprises the following:
  - Present Traffic conditions T_t
  - Inventory Levels I_t
  - Delivery schedules D_t
  - Environmental Var. E_t (e.g., disruptions and weather)
- **Actions (A):** The action a_t represents a logistics decision at time t
- **Rewards (R):** The reward function is designed to optimize key logistics performance metrics:
  - Minimize delivery time T_{delivery}
  - Minimize operational costs C_{ops}
  - Maximize fuel efficiency F_{eff}
  - Reduce carbon emissions: E_{carbon}
  - $R\left(s_t, a_t\right) = \alpha(-T_{\text{delivery}}) + \beta\left(-C_{\text{ops}}\right) + \gamma\left(F_{\text{eff}}\right) + \delta\left(-E_{\text{carbon}}\right)$

where, $\alpha, \beta, \gamma, \delta$ are weighting factors.

STEP 2. POLICY (π) AND VALUE FUNCTION (V)

The policy $\pi\left(s_t\right)$ is a mapping from states to actions that amplifies the probable cumulative reward.

$$\pi^* = \text{argmax } E\left[\sum_{t=0}^{\infty} \gamma^t R\left(s_t, a_t\right)\right]$$

The expected cumulative reward estimated by the value function $V\left(s_t\right)$ from the state s_t under policy π :

$$V\left(s_t\right) = E\left[\sum_{k=0}^{\infty} \gamma^k R\left(s_{t+k}, a_{t+k}\right)\right]$$

STEP 3. Q-LEARNING ALGORITHM

The Q-learning algorithm updates the Q-value $Q(s_t, a_t)$ for each state–action pair using the following rule:

$$Q(s_t, a_t) \leftarrow Q(s_t, a_t) + \alpha\left(\left(R(s_t, a_t) + \gamma \ \max Q(s_{t+1}, a) - Q(s_t, a_t)\right)\right)$$

where, α is the learning rate.

STEP 4. ROUTE OPTIMIZATION

For each time step t, the SCPS selects an action a_t grounded on the optimal policy:

$$a_t = \operatorname{argmax} \ Q(s_t, a)$$

STEP 5. REWARD FUNCTION

$$R\ (s_t, a_t) = \alpha(-T_{\text{delivery}}) + \beta(-C_{\text{ops}}) + \gamma(F_{\text{eff}}) + \delta(-E_{\text{carbon}})$$

STEP 6. FEEDBACK LOOP

4.3.3 Algorithm explanation

The algorithm contains six major steps to optimize supply chain operations using AI. The first step includes gathering real-time data from various sensors and sources inside the supply chain, such as traffic patterns, inventory levels, and delivery schedules. The second step is mapping states of Policy ($\pi(s_t)$) and Value Function (V) to actions. The data are equipped for analysis after being pre-processed to eradicate noise and anomalies. The third step is the Q-Learning algorithm, which updates the Q-value for the state–action pair. The algorithm's fundamental uses ML techniques, particularly RL to make decisions or final decisions on its own. The fourth step is route optimization. In this step, the SCPS agent is trained to select the optimal routes considering the inventory constraints and system circumstances, including traffic bottlenecks and delivery dates. The fifth step of this algorithm is the Reward Function. In this step, the model learns and advances by interrelating with the environment and receiving feedback (in the form of rewards/penalties) based on taken decisions (like reducing carbon emissions, delivery times, and operational costs). The final step is the Feedback loop in which the algorithm improves over time, chooses the most direct routes, and adjusts to unlooked-for events like changes in road closures or demands. The key components of this algorithm are State, Action, and Reward.

4.4 RESULTS AND DISCUSSION

4.4.1 Experimental setup

This section contains results describing the simulations and tests performed to validate the usefulness of the proposed system/model. The system was evaluated in certain simulated supply chain scenarios in comparison with traditional models of logistics, with a special emphasis on critical performance indices such as fuel efficiency, delivery time, operational cost, and disruption robustness.

4.4.2 Quantitative data analysis

Table 4.2 shows the performance of the proposed SCPS model as compared to the traditional logistics model. Some of the indicators for assessing logistics operations include average delivery, operational and transportation costs, fuel consumption, and carbon emissions.

The SCPS model is said to be able to lessen the average time of delivery through use of the new model by approximately 9.2 hours from the prior estimate of about 12.5. This is a reduction of 26.4%. As per the results of Table 4.2, the delivery was reduced, which resulted in reducing the expenses from $350 to $275; this is because SCPS allows timely adjustment in routes, time schedules, and also shorter lags, which improves the overall delivery time. The model is said to minimize operational costs and total fuel used via its dynamic routing abilities.

Furthermore, the parameter of fuel efficiency statistics is also commendable. The SCPS model can deliver 9.1 km/L of fuel economy, whereas the conventional model can deliver 7.5 km/L of fuel economy. This results in an increase of 21.3%, which underscores the SCPS's ability to optimize fuel

Table 4.2 Evaluation of SCPS and traditional logistics model with performance metrics

Metric	*Traditional model*	*SCPS (proposed model)*	*Improvement (%)*
Average delivery time (hours)	12.5	9.2	26.4
Operational cost ($)	350	275	21.4
Fuel efficiency (km/L)	7.5	9.1	21.3
Carbon emissions (kg CO_2)	125	98	21.6

consumption by choosing the most efficient delivery routes and minimizing idle time for deliveries. The SCPS also has a positive environmental impact by reducing carbon emissions (see Table 4.3).

Figure 4.2 depicts the comparison between the performance metrics of SCPS in contrast with the traditional logistics model. This adaptability is provided by ongoing traffic data monitoring and dynamic route modifications. At the time of any traffic delays, the technology swiftly and effectively reschedules routes. This method is much more efficient than traditional methods as it optimizes delivery time by rescheduling two routes/day. Traditional methods are unable to do this as they cannot provide this flexibility. This flexibility leads to faster delivery. The proposed approach improves the average delivery time by 26.4% (12.5–9.2 hours), during frequent traffic delays that occur in logistics.

Table 4.3 SCPS adaptability to real-time traffic disruptions

Scenario	*Traffic delay (%)*	*Delivery time (hours)*	*Cost per trip ($)*	*Rescheduled routes (per day)*
Average delivery time (hours)	12.5	9.2	26.4%	2
Operational cost ($)	350	275	21.4%	4
Fuel efficiency (km/L)	7.5	9.1	21.3%	6
Carbon emissions (kg CO_2)	125	98	21.6%	8

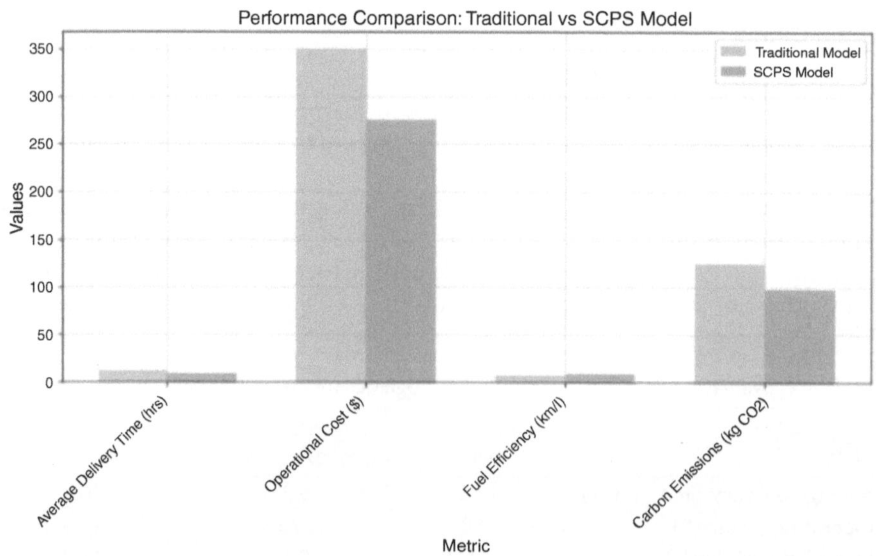

Figure 4.2 Performance metrics comparison SCPS vs traditional logistics model. (Author's Interpretation.)

The algorithm recalculates and selects more fuel-efficient routes during traffic delays, increasing fuel efficiency from 7.5 to 9.1 km/L, a 21.3% increase. The system reschedules four routes daily to reduce expenses even when traffic conditions change. This improves the cost by 21.4%, from $350 to $275. The system is responsible for cost savings as it is proficient in optimizing routes and avoiding excessive fuel due to traffic congestion. The SCPS determines prominent environmental benefits of 21.6% reduction by reducing carbon emissions from 125 to 98 kg CO_2. It reschedules routes eight times a day to reduce needless emissions, which reduces the environmental impact by optimizing routes and minimizing time during traffic abruptions.

Traditional methods are unable to do this as they cannot provide this flexibility. This flexibility leads to faster delivery. The proposed approach improves the average delivery time by 26.4% (12.5–9.2 hours), during frequent traffic delays that occur in logistics. The algorithm recalculates and selects more fuel-efficient routes during traffic delays, increasing fuel efficiency from 7.5 to 9.1 km/L, a 21.3% increase. The system reschedules four routes daily to reduce expenses even when traffic conditions change. This improves the cost by 21.4%, from $350 to $275. The system is responsible for cost savings as it is proficient in optimizing routes and avoiding excessive fuel due to traffic congestion.

This method is much more efficient than traditional methods as it optimizes delivery time by rescheduling two routes/day. It reschedules routes eight times a day to reduce needless emissions, which reduces the environmental impact by optimizing routes and minimizing time during traffic abruptions.

Table 4.4 shows the evaluation of the traditional model and the proposed model with sustainability metrics like energy efficiency, average fuel use, and carbon footprint. These statistics illustrate that the SCPS model performs better than the traditional approach in terms of energy use and environmental effects. Energy efficiency is the vital and most crucial element that is important in optimizing supply chains. The SCPS increases energy efficiency by 20%, from 1.5 kWh/km in the traditional model to 1.2 kWh/km in the proposed model. Figure 4.3 depicts the scenario-wise comparison in the SCPS model.

Table 4.4 Assessment of traditional model and proposed model with sustainability metrics

Metric	*Traditional model*	*SCPS (proposed model)*	*Improvement (%)*
Average fuel consumption (L/km)	0.13	0.11	15.4%
Carbon footprint (kg CO_2/trip)	125	98	21.6%
Energy efficiency (kWh/km)	1.5	1.2	20.0%

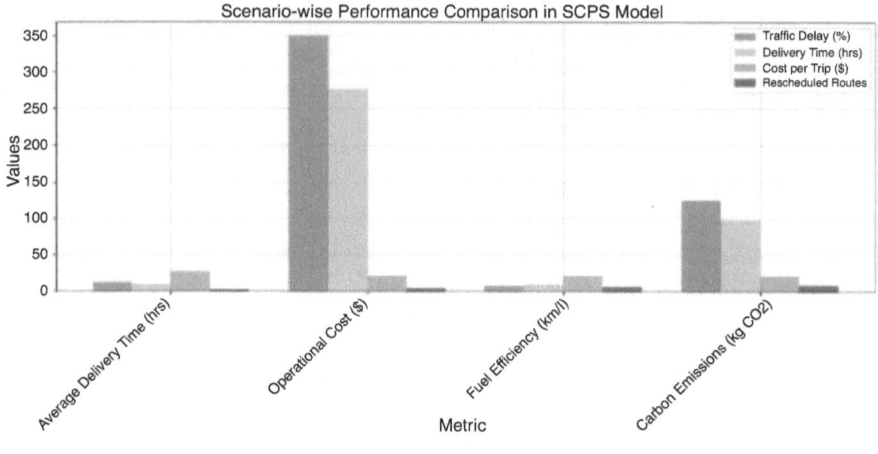

Figure 4.3 Scenario-wise comparison in the SCPS model. (Author's interpretation.)

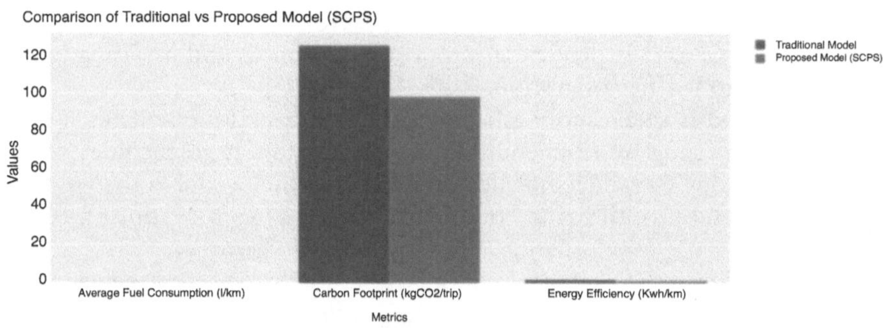

Figure 4.4 Evaluation of traditional vs proposed model (SCPS). (Author's interpretation.)

The average fuel consumption (L/km) is reduced by 15.4% from 0.13 L/km in the traditional model to 0.11 L/km in the proposed model. The carbon footprint of every trip is a significant sustainability metric. The SCPS model reduces carbon emissions from 125 to 98 kg CO_2 per trip as compared to the prior model. SCPS reduces fuel consumption by optimizing routes and preventing redundant delays or diversions, which sequentially reduces carbon emissions and makes the logistics process more environment-friendly. Figure 4.4 illustrates the evaluation of the traditional and proposed models.

These statistics illustrate that the SCPS model performs better than the traditional approach in terms of energy use and environmental effects. Energy efficiency is the vital and most crucial element that is important in optimizing supply chains. The SCPS model reduces carbon emissions from 125 to 98 kg CO_2 per trip as compared to the prior model. SCPS reduces fuel consumption by optimizing routes and preventing redundant delays or diversions, which sequentially reduces carbon emissions and makes the logistics process more environment-friendly. The SCPS increases the energy efficiency by 20%, from 1.5 kWh/km in the traditional model to 1.2 kWh/km in the proposed model. Figure 4.3 depicts the scenario-wise comparison in the SCPS model. The average fuel consumption (L/km) is reduced by 15.4% from 0.13 L/km in the traditional model to 0.11 L/km in the proposed model. The carbon footprint of every trip is a significant sustainability metric. The SCPS model reduces carbon emissions from 125 to 98 kg CO_2 per trip as compared to the prior model. The major objective of this analysis is to assess the traditional model and the proposed model based on diverse metrics like average fuel consumption, energy efficiency, and carbon footprint. The proposed approach can incite more sustainable supply chains by decreasing fuel and carbon emissions that are essential to converge global environmental objectives. Also, the model is flexible in reallocating supply chain variables and dynamic traffic conditions, making it perfect for complex and real-world logistics operations. A reduction in the requirement for human intervention and improvement in the operational efficiency can only be achieved by integrating RL in SCPS.

4.4.3 Simulation results

The SCPS agent (delivery vehicle) will learn to optimize its route based on rewards associated with minimized delivery time, cost, fuel consumption, and carbon emissions.

To create a simulation environment, a grid-like map is created where each grid cell represents a location, and the agent needs to find the optimal route while considering real-time conditions like traffic delays.

Metrics for Evaluation

- Delivery time
- Operational cost
- Fuel efficiency
- Carbon emissions

1. **Environment Definition**
   A grid environment is simulated with traffic delays and random fluctuations in conditions.
2. **Q-Learning Implementation**
   Next, the Q-Learning algorithm is implemented. The agent (vehicle) will learn the optimal path by maximizing its rewards (minimizing delays).

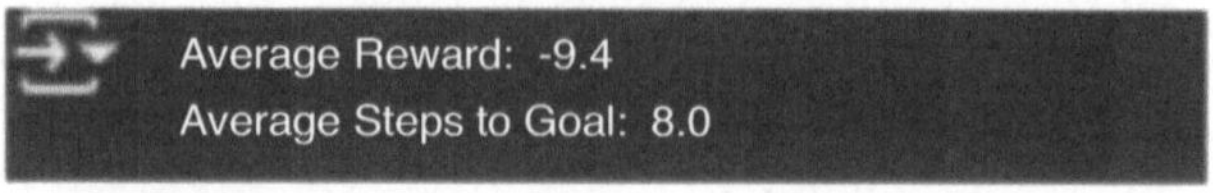

Figure 4.5 Simulation results. (Authors interpretation.)

3. **Evaluation and Simulation Results**

   Once the Q-learning model is trained, the results are evaluated by letting the agent find the optimal route and measure key performance metrics.
4. **Simulated Results**

   Figure 4.5 shows the simulation results as Average Reward and Average Steps to Goal. The Q-learning agent optimizes the route over time, reducing delivery delays, costs, and emissions. You can also track how many steps the agent takes to reach the goal.

   Average Reward: Reflects minimized delays.

   Average Steps to Goal: Indicates how efficiently the agent reaches the destination.

With the help of simulation, the results reflected are Average Reward and Average Steps to Goal, which are –9.4 and 8.0, respectively. Figure 4.5 shows the simulation results as Average Reward and Average Steps to Goal. To create a simulation environment, a grid-like map is created where each grid cell represents a location, and the agent needs to find the optimal route while considering real-time conditions like traffic delays. The SCPS agent (delivery vehicle) will learn to optimize its route based on rewards associated with minimized delivery time, cost, fuel consumption, and carbon emissions. The metrics of evaluation that were used are delivery time, operational cost, fuel efficiency, and carbon emissions. A grid environment is simulated with traffic delays and random fluctuations in conditions. Next, the Q-Learning algorithm is implemented. The agent (vehicle) will learn the optimal path by maximizing its rewards (minimizing delays). Once the Q-learning model is trained, the results are evaluated by letting the agent find the optimal route and measure key performance metrics.

4.5 KEY FINDINGS

This study presents the game-changing mechanism for enhancing SCPS sustainability and efficiency by providing a disruptive paradigm of RL in this area. As per the results of this very research, the operating costs decreased by 21.4%, fuel efficiency increased by 21.3%, carbon emissions decreased by 21.6%, and average delivery time decreased by 26.4%. The results validate the fineness of this approach in lowering expenses, optimizing the sustainability of supply chain operations, and simplifying logistics.

4.6 RESEARCH IMPLICATIONS

The proposed approach can incite more sustainable supply chains by decreasing fuel and carbon emissions that are essential to converge global environmental objectives. Also, the model is flexible in reallocating supply chain variables and dynamic traffic conditions, making it perfect for complex and real-world logistics operations. To reduce the requirement for human intervention and improve operational efficiency, we need to integrate RL in SCPS.

4.7 LIMITATIONS

Supply chain logistics operations may not be able to supply sufficient resources to train the model in a complex and dynamic environment. The model is based on real-time data for inventory levels, traffic updates, and other climatological conditions that could influence the constant accuracy of the model. Issues with data-sharing, communication latency, and decision-synchronization are likely to occur in scaling the global supply chain. Also, the performance of the system may be impacted in scenarios with inadequate data infrastructure. Among the challenges that can occur while scaling or managing the global supply chain are communication latency as well as inadequate data infrastructure.

4.8 CONCLUSION

This study presents an AI-focused SCPS model that optimizes supply chain logistics and route planning utilizing RL. As compared to conventional logistics models, the carried out simulations determine a significant decrease in delivery time, operating costs, fuel efficiency, and carbon emissions. This proposed model has the potential to handle diverse disruptions such as inventory and traffic fluctuations in dynamic and complex environments; this can help supply chain operations to be transformed completely. The significant findings discuss the ability of the system to constantly learn and improve from its environment. The efficacy of the approach in handling complex logistics is determined using extensive modeling. This study presents the game-changing mechanism for enhancing SCPS sustainability and efficiency by providing a disruptive paradigm of RL in this area. As per the results of this very research, the operating costs decreased by 21.4%, fuel efficiency increased by 21.3%, carbon emissions decreased by 21.6%, and average delivery time decreased by 26.4%. These results validate how well the approach lowers expenses, optimizes the sustainability of supply chain operations, and simplifies logistics. With the help of simulation, the results reflected are Average

Reward and Average Steps to Goal, which are –9.4 and 8.0, respectively. Thus, a simulation environment will be developed, where each point on the grid is essentially a location, and an agent needs to navigate an optimum route, while also keeping a check on real-time conditions like how much traffic has slowed its journey. SCPS is the delivery vehicle; it learns based on rewards given with rewards for minimized delivery time and cost in order to optimize the way it takes its route. The metrics of evaluation that were used are delivery time, operational cost, fuel efficiency, and carbon emissions. A grid environment simulates traffic delays and then introduces random fluctuations in conditions. Then, the Q-Learning algorithm is introduced. Here, the agent (vehicle) will learn its way to maximize rewards and minimize delays. Once the Q-learning model is trained, results are evaluated by allowing the agent to find optimal routes and measuring key performance metrics. Perhaps, the SCPS presents a novel model that, potentially, may significantly advance the sustainability, resilience, and effectiveness of the advanced supply chain. Scalability in widely spread networks, issues of complexity in data processing, and the nature of real-time data pose limitations on performance. For this reason, future work needs to focus on making better data gathering techniques as well as their testing with the aim of developing a better algorithm which will ease the training of RL models.

REFERENCES

1. Caton, S., Malisetty, S., and Haas, C. "Impact of imputation strategies on fairness in machine learning," *Journal of Artificial Intelligence Research*, vol. 74 (2022): pp. 1011–1035.
2. Malisetty, S., Ali, H. H., Rastegari, E., and Siu, K. C. "An innovative comparative analysis approach for the assessment of laparoscopic surgical skills," *Surgeries,* vol. 4 (2023): pp. 46–57.
3. Saxena, S., Kumar, V., Chopra, J., Gupta, P., Batra, D., and Tyagi, B. "Enhancing nuclear medicine through innovative ai models: integrating large language models and smoothquant for advanced imaging precision," *Iconic Research and Engineering Journals*, vol. 7 (2023): pp. 1–7.
4. Ma, H., Zhou, Z., Zhang, X., and Chen, X. "Towards carbon-neutral edge computing: greening edge AI by harnessing spot and future carbon markets," *IEEE Internet Things Journal*, vol. 10, no. 18 (2023): pp. 16637–16649. https://doi.org/10.1109/JIOT.2023.3268339.
5. Hua, H., Chen, X., Gan, L., Sun, J., Dong, N., Liu, D., Qin, Z., Li, K., and Hu, S. "Demand-side joint electricity and carbon trading mechanism," *IEEE Transactions on Industrial Cyber-Physical Systems*, vol. 2 (2024): pp. 14–25. https://doi.org/10.1109/TICPS.2023.3335328.
6. Luo, J., Zhuo, W., Liu, S., and Xu, B. "The optimization of carbon emission prediction in low carbon energy economy under big data," *IEEE Access*, vol. 12 (2024): pp. 14690–14702.

7. Xie, L., Zheng, X., Sun, Y., Huang, T., and Bruton, T. "Massively digitized power grid: opportunities and challenges of use-inspired AI," *Proceedings of the IEEE*, vol. 111, no. 7 (2022): pp. 762–787.
8. Yang, M., Lim, M. K., Qu, Y., Ni, D., and Xiao, Z. "Supply chain risk management with machine learning technology: a literature review and future research directions," *Computers and Industrial Engineering*, vol. 175 (2023): p. 108859.
9. Theodorou, E., Spiliotis, E., and Assimakopoulos, V. "Optimizing inventory control through a data-driven and model-independent framework," *EURO Journal on Transportation and Logistics*, vol. 12 (2023): p. 100103.
10. Odimarha, A. C., Ayodeji, S. A., and Abaku, E. A. "Machine learning's influence on supply chain and logistics optimization in the oil and gas sector: a comprehensive analysis," *Computer Science and IT Research Journal*, vol. 5 (2024): pp. 725–740.
11. Elahi, M., Afolaranmi, S. O., Martinez Lastra, J. L., and Perez Garcia, J. A. "A comprehensive literature review of the applications of AI techniques through the lifecycle of industrial equipment," *Discover Artificial Intelligence*, vol. 3 (2023): p. 43.
12. Singh, R., Mandal, P., and Nayak, S. "Reinforcement learning approach in supply chain management: a review," in *How Machine Learning is Innovating Today's World: A Concise Technical Guide*, Arindam Dey, Sukanta Nayak, Ranjan Kumar, and Sachi Nandan Mohanty, Eds., pp. 271–302. Scrivener Publishing, 2024.
13. Canonico, R., and Giancarlo, S. "Industrial cyber-physical systems protection: a methodological review," *Computers and Security*, vol. 135 (2023): p. 103531. https://doi.org/10.1016/j.cose.2023.103531.
14. Almatar, K. "Implementing AI-driven traffic signal systems for enhanced traffic management in dammam," *International Journal of Sustainable Development and Planning*, vol. 19 (2024): pp. 781–790. https://doi.org/10.18280/ijsdp.190236.
15. Javaid, M., Haleem, A., Singh, R. P., and Suman, R. "An integrated outlook of Cyber-Physical Systems for Industry 4.0: topical practices, architecture, and applications," *Green Technologies and Sustainability*, vol. 1, no. 1 (2023): p. 100001. https://doi.org/10.1016/j.grets.2022.100001.
16. Kim, B., Kim, J. G., and Lee, S. "A multi-agent reinforcement learning model for inventory transshipments under supply chain disruption," *IISE Transactions*, vol. 56, no. 7 (2023): pp. 715–728. https://doi.org/10.1080/24725854.2023.2217248.
17. Soori, M., Arezoo, B., and Dastres, R. "Artificial neural networks in supply chain management, a review," *Journal of Economy and Technology*, vol. 1 (2023): pp. 179–196. https://doi.org/10.1016/j.ject.2023.11.002.
18. Yadav, A., Garg, R. K., and Sachdeva, A. "Artificial intelligence applications for information management in sustainable supply chain management: a systematic review and future research agenda," *International Journal of Information Management Data Insights*, vol. 4, no. 2 (2024): p. 100292. https://doi.org/10.1016/j.jjimei.2024.100292.
19. Taj, S., Imran A. S., Kastrati, Z., Daudpota, S. M., Memon, R. A., and Ahmed, J. "IoT-based supply chain management: a systematic literature review," *Internet of Things*, vol. 24 (2023): p. 100982. https://doi.org/10.1016/j.iot.2023.100982.

Chapter 5

Integrating smart cyber-physical systems into urban planning

The future of smart cities

Rituraj Jain, Sitesh Kumar Singh, Damodharan Palaniappan, Kumar Parmar, and Premavathi T

5.1 INTRODUCTION

5.1.1 Overview of smart cities and cyber-physical systems (SCPS)

As urbanization accelerates throughout the world, cities are becoming "smart cities," referring to those who use cutting-edge technologies in support of upgrading quality life for its citizenry. A smart city is a town or a city that uses information and communication technologies to gather, analyze, and take action on information in real-time to improve services, optimize resources, and promote sustainable development. Digital platforms connect infrastructure, governance, and citizen services, so data-driven decision-making, along with driving reduction in inefficiencies, does tend to create more liveable environments—leading examples being smart traffic systems reducing congestion, energy-efficient grids lowering consumption, and sensor-based environmental monitoring [1].

The heart of smart city ecosystems consists of smart cyber-physical systems (SCPS) that couple the known physical processes with computational algorithms made possible by embedded sensors and actuators with networked communication. Such systems connect, in a way, the digital to the physical world by building sensors to collect data from the environment and further use it to autonomously manage urban infrastructure. SCPS will likely include Internet of Things (IoT) devices, artificial intelligence (AI), data analytics, and automation to monitor and control processes in real time. Applications would run the full gamut from transportation management and utility monitoring to enhanced emergency response and public safety [2].

SCPS thus herald a new trend in the management of urban spaces, whereby decisions are automated, and an atmosphere of constant feedback and adaptation is engendered. This makes cities a dynamic system where live data are fed into intelligent systems to self-adjust resources and services provision based on short-term needs. Such changes hold so much importance

 DOI: 10.1201/9781003542513-5

when considering some of the key issues facing urban challenges, such as traffic congestion, energy wastage, pollution, and resource inefficiencies, which make cities more efficient, sustainable, and resilient [3].

5.1.2 Importance of urban planning in the age of SCPS

The role was more of an expert discipline associated with the technical details of spatial designing, zoning, and infrastructural development based on historical data for long-term, static decisions. SCPS has changed the practice of city planning by offering the possibility of data-driven real-time strategies that dynamically account for ongoing shifts in the needs of cities [4]. The traditional approach for urban planning always acts so rigidly in ensuring that no change comes about with regard to the rapid growth of towns or changes in the population demographics and the environmental changes, hence causing inefficiency in resource delivery and public service.

SCPS-enabled urban planning allows more granular and real-time control over urban systems compared to most traditional approaches. Sensors and IoT devices collect significant information concerning traffic flows, energy utilization, and environmental conditions. This information is analyzed for immediate adjustments. It helps cities optimize resource use, decrease wastage, and react to unintended events such as accidents, natural disasters, or surges in public service demand [1]. For instance, SCPS might control traffic signals automatically to lower the congestion or even regulate the power grids in response to real-time consumption patterns much more efficiently than traditional ways.

This integration becomes critical as the populations of cities continue growing, in turn placing a premium on smarter, more sustainable cities. SCPS can help build cities that are not only better connected but also adaptive and responsive to their inhabitants' needs. Such continuous monitoring, analysis, and real-time action improve the utility of urban services and infrastructures in efficient and sustainable ways, leading ultimately to long-term economic, environmental, and social benefits [5] (see Figure 5.1).

5.1.3 Scope and objectives of the chapter

This chapter investigates the integration of SCPS into urban planning for changed aspects of the future smart city. This chapter aims to provide a deep understanding of the scope by which SCPS, through IoT-AI, robotics, and data analytics, can change the whole nature of urban space into smarter, efficient, and sustainable cities.

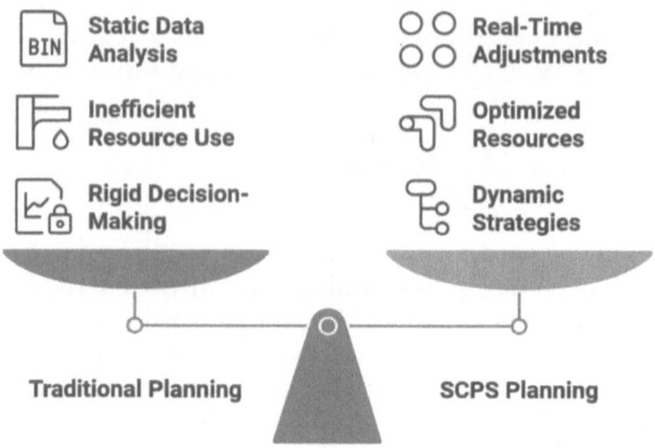

Figure 5.1 Highlights comparisons between traditional and SPCS-enabled urban planning.

This chapter deals with the following critical themes:

- **SCPS Definition and Awareness:** The components of SCPS, such as sensors, communication networks, and data analytics, as well as their role in the advancement of smart cities [6].
- **SCPS in Urban Sectors:** How SCPS integrates all other domains of a city, like transportation, energy management, waste management, and public safety, which in turn makes these systems enhance the efficiency, sustainability, and resilience of cities [7].
- **Urban Design Under SCPS:** The book describes the urban design that can be applied with SCPS, along with the notions of adaptive architecture and self-sustaining systems, as well as AI-based planning tools that optimize services to public [8].
- **City Case Studies:** Contextual examples of some of the cities which were successfully implementing SCPS in its city-planning scheme include Songdo in South Korea, Singapore's Smart Nation initiative, and Amsterdam's Smart City program [9].
- **Challenges and Opportunities:** The technical, regulatory, and financial challenges SCPS will face in opposition to the giant opportunities it will create for making more innovative and sustainable spaces for the city that are even more focused on the citizen [10].

This chapter aims to unfold the transformational potential of SCPS in urban design by highlighting how the integration of cyber-physical technologies can set a route toward smart cities envisioned for the future. The final analysis is about laying groundwork for more intense discussions on how and what exactly cities can unlock the future-proofing potential with SCPS and address numerous challenges and enhance quality of life while securing long-term sustainability.

5.2 UNDERSTANDING SMART CYBER-PHYSICAL SYSTEMS (SCPS)

5.2.1 Defining SCPS: components and framework

SCPS is an integrated approach to managing and optimizing urban spaces by infusing physical processes with digital computation. SCPS works through the integration of seamless sensor, actuators, and other IoT devices with embedded systems and communication networks, forming a critical underpinning for smart cities, as listed in Table 5.1. These device components allow real-time gathering, analysis, and automated decision-making process, thus creating highly responsive environments in urban contexts.

Sensors, Actuators, Internet-of-Things Devices, and Other Embedded Systems in SCPS: The heart and spine of the SCPS consist of sensors that

Table 5.1 Key components of smart cyber-physical systems (SCPS)

Component	*Description*	*Examples*
Sensors	Devices that detect and measure changes in the environment (e.g., temperature, traffic, and air quality).	Traffic flow sensors and air quality sensors
Actuators	Mechanisms that take action based on sensor data, adjusting systems or infrastructure as needed.	Traffic light systems and water management valves
IoT devices	Networked objects with embedded sensors and communication capabilities that connect the physical world to digital systems.	Smart meters, connected streetlights, and smart grids
Embedded systems	Hardware and software combinations within devices, designed to perform specific functions in real-time.	HVAC systems in smart buildings and traffic control systems
Communication networks	Infrastructure that enables data transmission between devices and systems in real time.	5G networks, LoRaWAN, and Zigbee
Data analytics	Techniques and tools for analyzing large datasets, detecting patterns, and predicting future trends.	Predictive energy usage models and traffic flow prediction tools
Artificial intelligence (AI)	Algorithms that autonomously analyze data and make decisions without human intervention.	Autonomous traffic management and smart waste collection

report most of the physical quantities and parameters which include the following: Temperature, air quality and quantity, water and other such water levels, and traffic flow of human beings. A city is typically dotted with sensors for monitoring critical urban infrastructure. The set of instructions, based on the data received from the sensors, is given to the actuators to then implement the actual physical action. For example, in a traffic light system, if the sensor units and the actuator units are added onto it, its signal timings can be adjusted so as to best handle congestion [11–13].

IoT devices—networked objects carrying embedded sensors and communication capabilities—are required to interconnect both the physical and the digital levels in SCPS. The advantage that these devices bring to infrastructure in a city is that it makes infrastructure "speak" with each other and share information at one system with other systems. The example would be making a streetlight intelligent such that it dims up when there is no pedestrian or vehicle in the locality, based on sensor input [14–17].

Within these devices are embedded hardware and software that are specifically designed to execute pre-programmed functions, which constitutes embedded systems in control of the real-time applications. The embedded systems will process the data from the sensor in real time, enabling a quick response in their applications, including system control of opening water management system valves or adjusting HVAC systems in smart buildings for energy efficiency [18–20]. Communication speed and reliability enable systems for real-time coordination [21].

It is the time when such collected information is processed to gain understanding through data analytics because information hoarded with SCPS is given diverse meanings as well as different kinds of analyses. In this activity, machine learning algorithms mine through these large collections for patterns and predicted results about the same, thereby allowing decisions. For instance, predictive analytics will forecast a surge in energy demand using a historical trend and adjust the power distribution [22].

The AI will then constitute the third part of SCPS—an autonomous decision-making factor. AI systems learn constantly from data streams while capable of making sophisticated decisions without any human-centric intervention. Urban functions such as traffic management, waste management, and public safety are served by AI-powered systems to ensure effective and efficient service provision [23].

Figure 5.2 illustrates the key enablers of SCPS, which include sensors, actuators, IoT devices, communication networks, data analytics, and AI. All of these aspects will combine in order to deliver real-time monitoring, automation, and decision-making functionalities for a smart city. All of these aspects give a firm foundation to incorporate digital processes with physical city infrastructure processes.

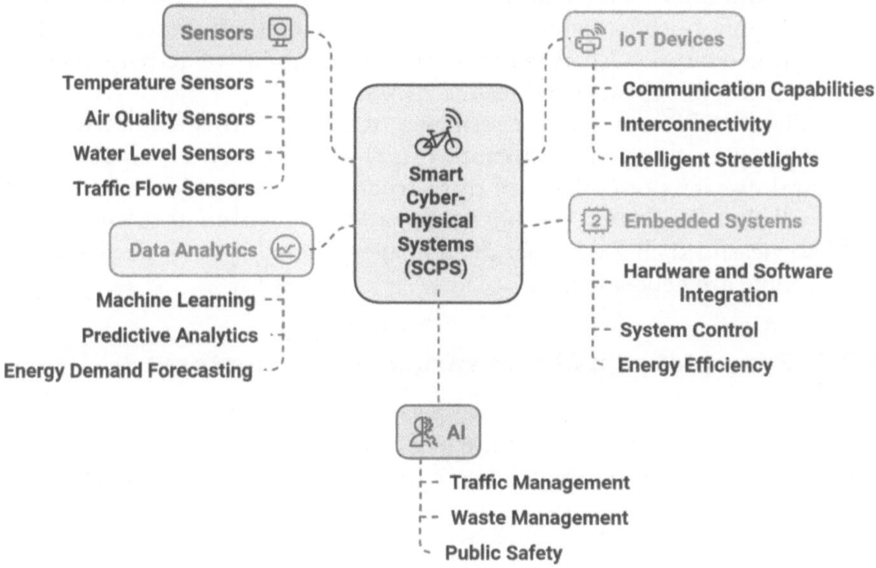

Figure 5.2 Key Enablers of SCPS.

5.2.2 Cyber-physical systems and their role in smart cities

SCPS is at the heart of citywide service monitoring, control, and automation. It enables cities to monitor the continuously changing performance of various systems, be it traffic flow, energy usage, or public health infrastructure, and dynamically adjust those factors to meet the constantly changing needs of the population.

Monitoring: In SCPS, real-time data refers to the fact that city officials and systems are always informed about activities taking place in the urban environment. Environmental conditions, infrastructure status, and resource usage are monitored using smart sensors and IoT devices, which generate streams of data that may be analyzed in real time to take quick decisions. For instance, in a smart water management system, sensors should indicate leaks, thereby reducing excess wastage of water, and remedial action may be taken within the shortest time possible [24].

Control and Automation: SCPS offers an automated method of control over urban systems, thus making them less dependent on human beings. Traffic management can make use of the timing of signals based on real-time movement of a vehicle in order to reduce congestion levels, whereas automatic energy grids can manage supply and demand through the adjustment made in real-time for the distribution of supply. Such methods of control maximize the efficiency of services within a city as well as maintain the sustainability of activity in terms of minimizing waste and resource usages [25].

Resilience: SCPS enhance the resilience of the city through an early alerting system along with automatic response to unknown events like natural disasters or collapse of infrastructures. In flood instances, SCPS will shut floodgates, reroute traffic, and send warnings to emergency services in such a way that the damage inflicted is minimized and the people are kept safe [26].

5.2.3 Examples of SCPS in various urban sectors

SCPS applications vary widely across a variety of sectors that are crucial for urban life such as transportation, energy, waste management, and healthcare. In these, cities see better operational efficiency, sustainability, and resilience.

Transport: SCPS transform urban mobility through devices in the IoT, AI, and automation in transport networks. They enhance real-time information from traffic sensors on control of flow of vehicles, adjusting signal timings, and offering routes to reduce congestion through route advice. SCPS-operated autonomous vehicles travel through cities for a good and safer transport system. For example, smart traffic signals in Copenhagen, Denmark, will stop other channels of traffic if they have precedence over buses and emergency vehicles to reduce delay and smooth out the traffic [27].

Energy: SCPS offers a contribution toward the energy sector in managing smart grids. Sensors as well as IoT devices established in energy distribution systems recognize consumption live, thereby establishing the balance of supply and demand. In relation to predictive analytics, SCPS predicts increases in consumption of energy and readjusts energy supply to prevent overloads from consumption. Smart energy grids and building automation systems such as that found in South Korea's Songdo optimize energy use, reducing total consumption and thus making it sustainable [28].

Waste Management: SCPS are increasingly used to streamline the processes of waste collection and disposal. Intelligent bins can automatically detect when they are filled using sensors and automatically send alerts to waste collection services, even optimizing the routes for collection and thereby avoiding unnecessary trips. Autonomous vehicles for waste collection are now being tested in a few cities, thus reducing the labor intensity associated with the process but at the same time improving the operational efficiency [29].

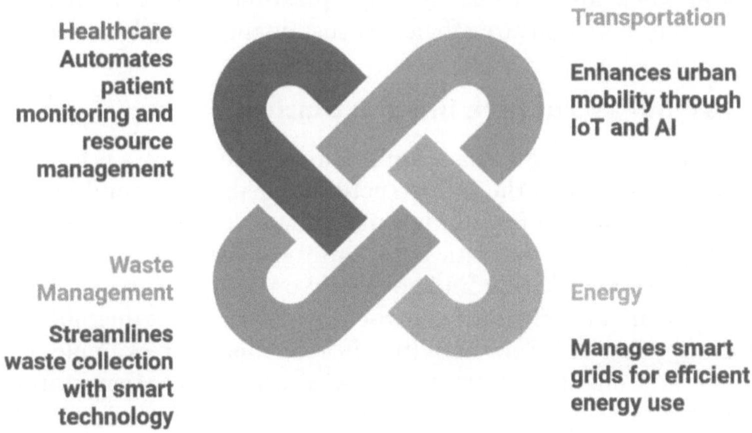

Figure 5.3 Applications of SCPS in different sectors.

Healthcare: In healthcare, SCPS enable smart healthcare systems where patient monitoring will be automated. Data retrieved from the patients' wearable devices can then be analyzed to provide real-time health insights. For instance, smart hospitals can be used to monitor the conditions of medical equipment and manage resources such as medication and staff, as well as follow up with patient's heart rate among other traits, improving general care efficiency [30].

By integrating SCPS into these urban sectors, the city should realize that such integration would provide a more sustainable growth, enhance public services, and provide good-quality life to citizens. SCPS represent one of the layers of critical infrastructure in the making of resilient, adaptive, and smart cities that should meet future requirements of populations [31].

Figure 5.3 illustrates some of the applications of SCPS in different key sectors of an urban space in transportation, energy, waste management, and healthcare. This figure shows how SCPS technologies applied through smart traffic systems, energy grids, automated waste collection, or real-time health monitoring can be utilized for changing service levels from operational efficiency to sustainability, and eventually to resilience in city services.

5.3 KEY COMPONENTS OF SCPS IN URBAN PLANNING

The keyword for urban planning in the SCPS era involves key technologies of IoT, data analytics, machine learning, autonomous systems, robotics, and AI in seamless integration to offer real-time monitoring, predictive decision-making, automation, and optimization of urban infrastructure and services to better ensure efficiency, sustainability, and resilience.

5.3.1 IoT infrastructure in smart cities

The basis is IoT in SCPS when applied to urban space since it provides a backbone from which the infrastructural capabilities required to source and monitor real-time information can emanate. By plugging the IoT across transport, utilities, or other cities, the entire network then turns into a surveillance system for all these and goes fully on autopilot.

IoT infrastructure, therefore, involves sensors and other devices that monitor a myriad of urban activities in real time. For example, environmental sensors might be tracking levels of air quality, temperature, and humidity; traffic sensors may track vehicle flow and congestions, and utility sensors might measure water consumption, energy usage, and the state of waste. All these devices continuously operate and transmit data to central control systems in real time. It helps prevent problems from occurring and enables urban planners and city administrators to respond better in time, detect issues, optimize resource use, and respond in an emergency. For example, smart water meters can detect the existence of leaks in the water distribution system, and city officials are immediately alerted to prevent wastage of water [26].

Public Safety, Transportation and Resources: The applications of the IoT are providing public safety through sensors set up into surveillance systems and streetlights and also installed in emergency services. In transportation, the devices of IoT provide real-time data about traffic conditions, allowing dynamic traffic management systems to adjust traffic signals for lesser congestions. It further included the communication of IoT-enabled vehicles and road infrastructure that communicate with each other in real time. This ensures that traffic flow and safety are optimized while avoiding accidents. IoT provides benefits to resource management, for instance, in the more efficient use of waste containers, energy grids, or water systems in collecting and distributing and using resources, hence saving costs and reducing harmful impacts on the environment. For example, the system for smart waste management in Barcelona has sensors placed in every garbage bin that alert when the bins are full for the waste collectors. This allows them to optimize their route collection while saving fuel [32].

5.3.2 Data analytics and machine learning

IoT devices are generating a huge amount of data that also requires complex data analytics and learning algorithms in order to extract meaningful insights and contribute toward making decisions regarding planning in an urban area.

Big data, in the context of smart urban planning, plays a very important role as big data encompasses all the information ranging from that received by IoT devices, the platforms of social media, GPS systems, and also through public records. And by taking all that in, big data helps understand all the city dynamics. Be it a traffic pattern, energy consumption pattern, or population growth, this has a reflection on the impact of the climate on that city. Data analytics platforms process and visualize such information and enable city planners to make decisions based on evidence. For example, through traffic data analysis, it is possible to establish where traffic congestion points arise and improve roadway design with better outcomes [33].

By making use of machine learning models analyzing historical and real-time data, predictive analytics helps offer future trends to urban planners in optimizing resource allocation. For instance, predictive models can predict the electricity demand associated with certain weather patterns so that an energy company can alter production to avert blackouts. Transportation can also predict peak hours and advise altering routes in decongesting major routes. This would allow for more efficient use of resources and hence lesser wastes and increases in the sustainability of urban systems.

5.3.3 Autonomous systems and robotics

Autonomous systems and robots change the face of urban planning, for instance, whereas the automation of key services gives it a new hue regarding efficiency, although human intervention in labor-intensive tasks has been reduced.

Smart Transport (Autonomous Cars, Drones, etc.): Autonomous vehicles, self-driving cars, drones, etc., are examples of revolutionizing smart transportation systems. Such sophisticated sensory AI and IoT can operate completely without human intervention in any cityscape. In public transport, safety, and reliability, where drivers are not needed, autonomous buses and trains would move people efficiently on fixed routes. These devices from tracking traffic and packages in a smart city to emergency responses provide quick, cost-effective solutions to some of the urban challenges. Autonomous flying taxis in Dubai, for example, are being researched as a way to fight traffic congestion while offering a unique system of public transport [34].

Robotics Applications in Waste Management and Infrastructure Maintenance: Robotics has played an important role by providing automated processes for collecting wastes and maintaining infrastructures. It has been utilized in offering smart waste collection robots, which can move about on their own in the urban environment to pick up trash and recycling without any form of human supervision. Robots are also used in infrastructural maintenance, including road repair, inspection of bridges, and other maintenance procedures on water and energy networks. For example, if a robot that is armed with the sensors of AI and IoT can spot and mend potholes and cracks on roads to avoid further widening of damage, then it is for sure a case in point [35].

5.3.4 Artificial intelligence in decision-making

Artificial Intelligence is at the center of SCPS, which enables cities to automate many decision-making processes and optimize public services as well as infrastructure.

AI-assisted City Planning and Policy Making: With the development of AI, it has increasingly been used to support data-driven urban planners. The AI-supported planning analyzes these planning tools by analyzing complex data which generate models and simulations of different planning scenarios in reference to population growth, changes in transportation needs, environmental changes, and how planners may make decisions that relate to the creation of more resilient and sustainable urban environments. For example, AI algorithms can optimize the layout of public transportation networks by analyzing commuter data and predicting future travel patterns [36].

AI in Optimizing the Efficiency of Public Utilities and Infrastructures: AI is capable of optimizing services such as waste collection, energy distribution, and even an emergency response with AI-based systems. Such AI can predict times for maximum energy usage in advance, thereby adjusting the energy production levels. In transport, AI-based traffic management can, in real time, change the timing of traffic lights to ease traffic congestion and increase road safety levels. AI also strengthens public safety through surveillance footages' analysis and behavior detection, hence making a rapid response to possible threats effective.

This makes the smart integration of IoT, data analytics, machine learning, autonomous systems, robotics, and AI transform cities into more efficient and sustainable environments. These are key components of SCPS that are critical in addressing the challenges of modern urbanization, optimizing resource use, and improving the quality of life for citizens. They will transform cities further as they advance in these technologically integrated forms.

Table 5.2 provides a concise overview of how SCPS is applied in different sectors, reinforcing the detailed explanation provided in the text. It makes the relationships between SCPS components and urban services more tangible.

Table 5.2 SCPS integration in urban sectors

Sector	*Application of SCPS*	*Benefits*
Transportation	Smart traffic management, autonomous vehicles, and traffic flow optimization.	Reduced congestion, improved safety, and lower emissions
Energy	Smart grids for energy distribution, integration of renewable energy sources, and real-time energy monitoring.	Energy efficiency and reduced carbon footprint
Waste management	Smart bins with sensors, optimized collection routes, and autonomous waste collection vehicles.	Reduced waste, optimized collection, and reduced fuel consumption
Water management	Real-time monitoring of water usage, leak detection, and optimized distribution networks.	Minimized wastage and improved water resource management
Public Safety	AI-enabled surveillance systems, real-time emergency response coordination, and smart streetlights.	Faster response times, enhanced safety, and reduced crime

5.4 URBAN PLANNING AND DESIGN IN THE AGE OF SCPS

SCPS brings in the new wave of breaking urban planning and design, thereby heralding an interconnected and data-driven world of cities. SCPS integrated urban infrastructure offers adaptive, efficient, and sustainable city features, which would dynamically adapt to the needs of their citizens.

5.4.1 How SCPS is shaping modern urban design

Modern urban design is changing in order to support the capabilities of SCPS, and smarter adaptive cities have been formed as a result. Contained in this shift are resilient infrastructures that are flexible, sustainable, and based on technology—from urban planners that consider advanced infrastructures like IoT, AI, and automation.

Adaptive Architecture and Smart Infrastructure: The most changed, by far, in the realm of urban design is the use of adaptive architecture. SCPS allows buildings to become "smart" and adopt real-time changes in environmental conditions, usage patterns, and consumption of energy. It has smart buildings with sensor and actuator technology along with AI systems, which adjust the lighting, heating, and ventilation automatically depending on whether anybody is occupying the space or adverse weather. SCPS supports smart infrastructure: Roads, bridges, and utilities are embedded with sensors to monitor their structural health and ensure maintenance in a timely fashion. All urban infrastructure endures longer, and the possibility of catastrophic failures is reduced.

Smart Energy Grids, Sustainable Buildings, and Green Urban Spaces: Smart energy grids are another significant innovation being encouraged by SCPS. Unlike traditional systems, smart energy grids function better due to real-time sensing of demand, integration of renewable sources of energy, and detection of outages. Part of SCPS should be sustainable building practices in terms of energy efficiency, water-saving, and reduced emission that will guide cities to decrease their environmental footprint. For example, buildings with SCPS-enabled energy management systems could take advantages of solar radiation during peak production times and store it during periods of high consumption. In this regard, SCPS plays a crucial role in creating green urban spaces and smart parks and green roofs for which the air quality, soil moisture, and temperature will be monitored to keep the ecosystems of a city healthy and alive.

5.4.2 SCPS integration in transportation networks

Transportation networks are the most obvious beneficiaries of SCPS integration. In fact, such systems are rapidly becoming automated and interconnected, so transport will continue to be safer, more efficient, and less polluting in the future.

Autonomous Public Transport and Smart Traffic Management: Autonomous public transport is the latest trend in a smart city. In this context, self-driving buses, trams, and trains reduce human interferences by working together to achieve success. SCPS allows autonomous vehicles to bring them to complex urban terrain or better interaction with the traffic system. They can react according to the current traffic conditions. For example, smart management of roads can be conducted with the help of IoT devices installed both in road infrastructure and in vehicles—to collect data on optimizing traffic signal timing. The results would be reduced congestion, improved safety on roads, and a better general efficiency of urban mobility systems. Such automation reduces jams, reduces travel times, and improves the overall efficiency of urban mobility systems.

SCPS significantly contributes to minimizing congestion and pollution. Smart cities are very effective in reducing their carbon footprint due to optimizing traffic flows, focusing on autonomous electric vehicles. With real-time data and vehicle density, the monitoring of traffic systems can advise on alternative routes or modes of transport to reduce bottlenecks. The users are also encouraged to opt for shared mobility through SCPS-enabled car-sharing platforms and ride-hailing services; this reduces the number of vehicles on road and subsequently the emissions. SCPS is used in cities such as London and Singapore in congestion pricing, where charges depend on the time spent driving across high-traffic areas, driving users to public transport or routes other than theirs.

5.4.3 Public utilities and SCPS-enhanced urban services

In particular, aspects of public utilities such as water, waste, and energy management were being increasingly automated and made efficient through the application of SCPS.

Smart Water, Waste, and Energy Management: SCPS revolutionizes the way cities manage their resource consumption at the water and waste level and their energy management. Sensor-enabled smart water systems use sensors and IoT devices that monitor water quality, identify leaks, and optimize the distribution of water. The instant usage data on water help avoid wastage of water, besides making urban water networks more efficient in a city. Similarly, intelligent waste management systems deploy IoT-enabled bins that report to waste collection service providers to advise them when these bins are full so that they can reroute their collection, hence lessening fuel use for its waste collection trucks. As again mentioned above, dynamic energy distribution and inclusion of renewable sources such as solar and wind power in smart energy management systems reduce dependence on fossil fuels and pave the way toward sustainability in urban areas.

Cyber-Physical Systems in Enhancing Efficiency of City Services: SCPS promotes betterment of cities based on the efficient operation of their urban services through automation, optimal resource usage, and immediate feedback. For instance, smart streetlights with motion sensors that either turn off or dim their brightness levels depending on the level of activity can save energy and reduce light pollution. SCPS-based emergency service monitoring systems keep track of the locations of emergency vehicles and, consequently, optimize routes in real time, which helps reduce the response time.

5.5 CASE STUDIES OF SCPS INTEGRATION IN GLOBAL SMART CITIES

In different countries, cities have introduced SCPS to support urban planning, improve public services, and create even more sustainable and efficient environments. The cases below have transformed urban life in all these different contexts through incorporation of SCPS.

Case study 1: Songdo, South Korea

Songdo is an example of a city built from the bottom up as a smart city. Thus, it proves that SCPS can be beneficial if followed at an early stage in urban planning.

Smart City Design and Implementation of SCPS in Infrastructure: SCPS pervades Songdo's infrastructure, from its buildings to transportation networks. Sensor and IoT devices have been seamlessly woven into almost every aspect of the city's daily operations—including monitoring everything from traffic flow to energy consumption and more. SCPS-enabled systems in Songdo optimize energy distribution, schedule waste collection through automated collection systems, and monitor real-time environmental factors such as air quality and water supply [37].

Closing Impact on Sustainability and Public Services: One thing that defines Songdo is the sustainability of the city. Providing sustainability in building services, Songdo uses 30% less energy than comparable structures, and an extensive bike-sharing system minimizes automobile use. Waste is handled in a network of pneumatic tubes that transport the waste to processing facilities, with little use of garbage trucks. Coupled with SCPS-enhanced public services, Songdo has become a model of how smart city design can well be improved in efficiency and sustainability.

Case study 2: Singapore's smart nation initiative

One classic example is Singapore's Smart Nation Initiative, which has been used for SCPS in enhancing the quality of urban mobility for citizens, and thereby the quality of life.

SCPS in Urban Mobility and Citizen Engagement: Singapore implements its smart city framework on SCPS by maximizing urban mobility through autonomous vehicles, smart traffic management, and efficient public transport systems. The citizens then get real-time travel information from automated ticketing systems and seamless changing of different modes. Citizen engagement is also ensured through platforms that enable the reporting of issues, access to government services, and participation in the city plan.

Role of Data Analytics in Improving Urban Standards: In Singapore, data analytics play a major role in the smart city strategy. Platforms allowed through SCPS capture data from multiple urban systems. The collected data are then analyzed to identify trends, predict future needs, and optimize services in the city. Through this kind of approach, Singapore has managed to improve almost everything in her healthcare to housing to become one of the most efficient, and in fact, livable cities in the world [38].

Case study 3: Amsterdam smart city

Amsterdam has been one of the cities that lead at the forefront of an innovation smart city, making it keep a very high hope with regard to environmental sustainability and optimization of resources.

IoT and SCPS in Environmental Monitoring and Resource Optimization: Amsterdam utilizes IoT and SCPS very extensively in all its smart city initiatives regarding monitoring factors for air and water quality, energy consumption, and waste levels. In this way, it optimizes resource usage, reduces pollution, and improves the general quality of life among its residence. For instance, the City employs IoT-enabled sensors to monitor their energy grid, thereby ensuring that renewable sources are utilized in the most efficient manner and energy is delivered in such a manner that losses are minimum.

Contribution of SCPS to the Building up of the Urban Resilience and Efficiency: Improving urban resilience in Amsterdam SCPS allows real-time monitoring of critical infrastructure and provision of data to the city officials on emergencies to respond promptly. SCPS has also ensured efficiency in urban services, such as waste collection and water management, thus reducing costs and environmental impact.

5.6 CHALLENGES AND OPPORTUNITIES IN INTEGRATING SCPS INTO URBAN PLANNING

SCPS have tremendous potential toward the revolution of urban planning, rendering cities more efficient, sustainable, and citizen-centric. In return, the inclusion of SCPS in the urban environment brings with it a number of unique challenges and opportunities that need to be addressed in order to fully benefit from the advantages of these systems.

5.6.1 Challenges

1. **Technological, Regulatory, and Infrastructural Barriers**

   Technological Barriers: SCPS will only be implemented in urban development if the underlying technologies are mature and reliable. Examples of such underlying technologies are IoT devices, 5G communications networks, AI, and data analytics. Most of such barriers are due to a lack of standardization and interoperability issues among heterogeneous devices and platforms and legacy interfaces. Second, the deployment of SCPS demands robust data storage and processing infrastructure, which will be lacking in some parts of the urban area because of old infrastructural facilities.

Regulatory Barriers: Rules and regulations about SCPS are still in the formation process. Urban planners and policymakers would have to face a complex web of rules and regulations about data collection, privacy concerns, as well as regulatory issues on AI application in decision-making processes of autonomous systems. Can one claim ownership over the data collected through an IoT network? How should cities regulate the use of autonomous systems in public space? Lack of definite regulatory frameworks can act as a barrier in the deployment of SCPS because it would be bogged down by court cases and bureaucratic delay.

Infrastructural Barriers: Many cities, in fact old ones, are not wired to take in SCPS completely. Upgrading the existing systems so that they could accept SCPS-enabled technologies demands significant investments on the supply of fiber-optic networks, wireless communication systems, and sensor deployment. An integrated infrastructure to let SCPS function throughout the areas at multiple sectors in the city—the transportation system, utility system, and waste management system—can only be possible with high planning and coordination among public and private agencies.

2. **Cybersecurity Concerns and Data Privacy Issues**

   Security Vulnerability due to Cyberattacks: SCPS being inherently data-exchanging between cyber and physical entities are susceptible to cyber-attacks, which also affect cities' critical services such as managing traffic, distributing energies, and emergency response systems. The more interconnected the devices and systems, the more the attack surfaces with a potential for huge financial and reputational losses to a city. Therefore, these require superior protocols for security to be followed in conjunction with continuous monitoring and better encryption techniques to defend against cyberattacks.

   Data Privacy Issues: SCPS depend on data collection of citizens and urban infrastructure. There are many privacy concerns relevant to this problem. The devices employed for IoT in smart cities collect tremendous amounts of data about the movement, behavior, even preferences of citizens. The problem here is how those data are collected, stored, and processed in such a way that respects the rights to privacy of citizens. Cities must have strict frameworks and mechanisms of data governance and transparency to sustain people's confidence in them and to avoid abusing personal data.

3. **Managing the Costs of Large-Scale SCPS Implementation**

   Heavy Initial Investment: Since SCPS is a part of the city planning, it attracts enormous investments upfront. It could be in terms of IoT devices, upgradation of communication networks, or incorporation of AI systems and specialized human resources, massive commitments by the city government end. This might turn out to be a hurdle in the adjustment of budgetary priorities from the choices available for the cities.

   Long-term Maintenance and Upgrades: Once the SCPS is deployed, the ongoing care and periodic upgrading of infrastructures becomes quite expensive. Urban technology is constantly changing, and cities have to ensure that their SCPSs are not only upgraded to avoid obsolescence but also continuous financial commitment, which might strain public resources and require innovative models of funding.

Table 5.3 provides a clear breakdown of the various challenges and summarizes the issues faced in SCPS integration, helping readers see the multiple barriers at a glance.

5.6.2 Opportunities

1. **Innovative Public Services:** SCPS involves the promise of innovated public services. For example, taking a real-time data analytics-driven smart traffic management system can enable better traffic flow

Table 5.3 Challenges in SCPS integration

Challenge type	*Description*	*Examples*
Technological barriers	Lack of interoperability between devices, integration with legacy systems, and reliance on evolving technologies.	Difficulty integrating new IoT devices with older infrastructure
Regulatory barriers	Unclear policies on data ownership, privacy, and AI deployment in public spaces.	Unclear regulations on the use of AI in traffic management
Infrastructural barriers	Outdated city infrastructure not suited for SCPS technologies, requiring costly upgrades.	Fiber-optic network upgrades and smart grid installation
Cybersecurity concerns	Vulnerabilities in interconnected systems, leading to potential cyberattacks on urban services.	Breaches in traffic control systems and energy grid hacking
Data privacy issues	Concerns regarding the collection, storage, and use of personal data collected by SCPS devices.	Personal movement tracking through surveillance systems
Cost of implementation	High initial costs of SCPS deployment and long-term maintenance, upgrades, and cybersecurity measures.	Expensive IoT infrastructure installations and continuous system updates

as well as reduce congestion, in turn, safety on roads. Similarly, with SCPS-augmented telemedicine, there can be remote diagnosis and continuous monitoring of patients, thereby increasing access to medical care in underprivileged areas. With SCPSs, new ways of becoming more efficient citizen-centered services that reflect the moment-by-moment needs of cities will unfold.

2. **Sustainable, Resilient, and Citizen-Friendly Cities:** The biggest opportunity is cities that are sustainable and resilient. SCPS enabled through smart energy grids, optimized water management systems, and effective waste management solutions can optimize resource use and reduce an environmental footprint. For example, smart grids can be designed to distribute energy more efficiently by tallying renewable sources and losses of power. SCPS also ensures resiliency in the sense that they provide cities with real-time data and predictive analytics, which allows for a quick response to catastrophes and long-term planning against calamities brought by climate change.

   SCPS are intrinsically citizen-centered as they were made to enhance the quality of life of urbanites. City authorities can customize their services for maximum benefit to the citizens by using data analytics and AI. For example, SCPS can be used to enhance public safety by rapidly identifying incidents and ensuring emergency services reach the location efficiently. In the transport sector, SCPS guarantees dynamic changes in bus routes and schedules to suit the demand in question, thus reducing wait time and enhancing commuter satisfaction.

3. **Economic and Environmental Benefits of SCPS-Urban Planning:** SCPS-improved urban planning techniques provide some important economic benefits. Cities can reduce operational costs by automating routine jobs, efficient resource allocation, and improving the efficiency of public services. Moreover, it will attract new businesses and industries on account of smart urban environments that are conducive to innovation and economic growth.

Besides the environmental advantages, this is significant since SCPS will help cities achieve their sustainability goals through reduced emission of carbon, water conserving sources, and an increased reliance on renewable energy sources. SCPS, as a case study, can construct intelligent transportation systems through flow optimization through control of traffic flow strategies and promoting electric vehicles that would also reduce fuel consumption and also lead to decreased emissions. In the case of waste management, SCPS would further improve effective recycling and diminish landfill use through enabling municipalities to follow real-time information about levels of waste and collection schedules in real time.

5.7 SCPS AND THE FUTURE OF SMART CITIES

5.7.1 Long-term vision for SCPS in urban development

SCPS in cities is still at a nascent stage, but the long-term vision for SCPS in smart cities is promising. Of course, by 2030 and beyond, SCPS will be actively integrated into future urban landscapes.

1. **SCPS in 2030 and Beyond:** Predictions: In 2030, SCPS-enabled technologies will be ubiquitous in cities. Autonomous transportation networks, self-sustaining energy grids, real-time environmental monitoring, and much more will make up the urban landscape. Infrastructure will be predictively maintained to avoid breaks and, therefore, costly repairs. The cities will increasingly use AI-based decision-making for resource allocations and management of public services and emergency responses.

   The urban landscape will be much more flexible, dynamic, and adaptive in the future. Buildings will be designed with integrated SCPS that let them change about their conditions in their environment, usage patterns, and energy needs in real-time. Public spaces will be outfitted with smart sensors monitoring foot traffic, air quality, and noise levels. That way, more efficient management of urban resources will be done.
2. **Role of Public–Private Partnerships (PPP):** Only through cooperation from governments, tech companies, and academics will SCPS be successfully integrated into urban planning. Key to funding and the implementation of PPP initiatives for SCPS will be the collaboration between government representatives and key providers of technology to support the development of related infrastructure, while research input and expertise in AI, IoT, cybersecurity, etc, will be provided by academic circles.

   Policy frameworks will provide the overall guidelines for developing and implementing SCPS. Governments should establish clear regulations in respect to data privacy, cybersecurity, and AI used in public services, while creating funding strategies that should encourage private investments in SCPS technologies with equal access by all citizens.

5.7.2 Citizen-centric smart cities

1. **Citizen Participation in SCPS-Based Urban Planning Activities:** This is possible only if SCPS engages citizens within the planning and implementation processes. The particular mechanisms for citizen feedback

and problem reporting may be through platforms, such as giving ideas about a possible urban development project or responding to a public services survey through a digital platform. This could then assist cities in becoming more inclusive and responsive to citizens' needs.

The other way cities can embrace SCPS is in social inclusion, where the use of smart technologies attracts and gives more benefits to all its citizens. This might include low-cost internet, a digital literacy program, and SCPS applications, among other benefits to people with disabilities.

2. **Quality of Life and Social Inclusivity:** SCPS may improve the quality of life of the citizenry staying in an urban city with quality public services, less pollution, and high security. For example, SCPS-enabled health systems will be able to provide real-time monitoring of health for senior citizens or those who have chronic diseases for them to receive the necessary care from home. The system will further improve public safety through quick response to emergency cases and decreased crime rates with the implementation of AI surveillance systems.

   SCPS can also be a tool of social integration to ensure that every citizen receives the basic rights with respect to income or location. SCPS-enabled transportation systems can provide non-expensive mobility options that are accessible for the underprivileged, and smart housing initiatives offer energy-saving and low-cost dwelling for those who belong to the low-income group.

5.8 CONCLUSION

Key Points Summary: The integration of SCPS into urban planning offers several advantages: higher public services, greater sustainability, and added citizen participation. Conversely, there are technological, regulatory, and infrastructural impediments while security and data privacy issues arise in the face of SCPS implementation. An overall outlook, however, misses the potential that SCPS may bring to more resilient, citizen-centric, and economically viable cities.

Future Direction:

1. **SCPS Further Research and Development:** SCPS technologies and regulatory challenges require more systematic research. Such aspects will develop better suitability for more reliable and secure IoT devices, frameworks for data governance, and new applications in healthcare, transportation, or energy management.
2. **Policy Recommendations for Cities Implementing SCPS Technologies:** Policymakers should prepare their efforts as regulatory pro-innovation,

yet citizen-protection will be considered. It covers regular formation of privacy regulations governing how data are collected or used, securing standards toward cybersecurity for any SCPS infrastructure, and funding models that encourage public–private funding.

Final Thoughts: SCPS shall redesign the future of the urban ecosystems as smart, sustainable, and citizen-oriented cities. Cities will manage to overcome increasingly pressing challenges in the course of further growth with the help of SCPS. It is possible for cities to design a more resilient, inclusive, and prosperous future by passing over the obstacles created till now and seizing the chances offered by SCPS.

REFERENCES

1. T. Dias, T. Fonseca, J. Vitorino, A. Martins, S. Malpique, and I. Praça, "From data to action: exploring AI and IoT-driven solutions for smarter cities," in *Lecture Notes in Networks and Systems,* Guimaraes, Portugal, 2023, pp. 44–53. https://doi.org/10.1007/978-3-031-38333-5_5.
2. R. Chaganti, D. Gupta, and N. Vemprala, "Intelligent network layer for cyber-physical systems security," *International Journal of Smart Security Technologies,* vol. 8, no. 2, pp. 42–58, 2021, https://doi.org/10.4018/ijsst.2021070103.
3. A. Vanolo, "Smartmentality: the smart city as disciplinary strategy," *Urban Studies,* vol. 51, no. 5, pp. 883–898, 2013, https://doi.org/10.1177/0042098013494427.
4. M. Yousefi, S. M. Hoseini, Y. A. Vatnikov, A. Karamyan, and E. V. Kulikov, "Dietary thymol supplementation promotes antioxidant responses and thermal stress resistance in rainbow trout, *Oncorhynchus mykiss,*" *Animals,* vol. 14, no. 20, p. 2988, 2024, https://doi.org/10.3390/ani14202988.
5. I. Mahmoud, E. Morello, A. Bisello, and D. Kolokotsa, "Embedding technologies for improving nature-based solutions performance and fostering social inclusion in urban greening strategies: augmented NBS for cities," *Urban Forestry & Urban Greening,* vol. 93, p. 128215, 2024, https://doi.org/10.1016/j.ufug.2024.128215.
6. L. G. Anthopoulos, Z. Pourzolfaghar, K. Lemmer, T. Siebenlist, B. Niehaves, and I. Nikolaou, "Smart cities as hubs: connect, collect and control city flows," *Cities,* vol. 125, p. 103660, 2022, https://doi.org/10.1016/j.cities.2022.103660.
7. G. Rzevski, S. Kozhevnikov and M. Svitek, "Smart city as an urban ecosystem," in *2020 Smart City Symposium Prague (SCSP),* Prague, Czech Republic, 2020, pp. 1–7, https://doi.org/10.1109/SCSP49987.2020.9133849.
8. S. Grimmond et al., "Integrated urban hydrometeorological, climate and environmental services: concept, methodology and key messages," *Urban Climate,* vol. 33, p. 100623, 2020, https://doi.org/10.1016/j.uclim.2020.100623.

9. S. G. Hong and D. Lee, "Development of a citizen participation public service innovation model based on smart governance," *Service Business,* vol. 17, no. 3, pp. 669–694, 2023, https://doi.org/10.1007/s11628-023-00536-w.
10. M. S. Csukás and R. Z. Szabó, "The many faces of the smart city: differing value propositions in the activity portfolios of nine cities," *Cities,* vol. 112, p. 103116, 2021, https://doi.org/10.1016/j.cities.2021.103116.
11. D. Tripathi, A. Biswas, A. K. Tripathi, L. K. Singh, and A. Chaturvedi, "An integrated approach of designing functionality with security for distributed cyber-physical systems," *The Journal of Supercomputing,* vol. 78, no. 13, pp. 14813–14845, 2022, https://doi.org/10.1007/s11227-022-04481-9.
12. N. Nasser, N. Khan, L. Karim, M. ElAttar, and K. Saleh, "An efficient time-sensitive data scheduling approach for Wireless Sensor Networks in smart cities," *Computer Communications,* vol. 175, pp. 112–122, 2021, https://doi.org/10.1016/j.comcom.2021.05.006.
13. N. Ali, M. Hussain, Y. Kim, and J. -E. Hong, "A generic framework for capturing reliability in cyber-physical systems," in *ESSE'20: Proceedings of the 2020 European Symposium on Software Engineering,* vol. 2013, pp. 148–153. Association for Computing Machinery, New York, NY, USA, 2020. https://doi.org/10.1145/3393822.3432331.
14. C. M. Chang, G. T. Salinas, T. S. Gamero, S. Schroeder, M. a. V. Canchanya, and S. L. Mahnaz, "An infrastructure management humanistic approach for smart cities development, evolution, and sustainability," *Infrastructures,* vol. 8, no. 9, p. 127, 2023, https://doi.org/10.3390/infrastructures8090127.
15. A. Viswas et al., "SCN1A genetic alterations and oxidative stress in idiopathic generalized epilepsy patients: a causative analysis in refractory cases," *Indian Journal of Clinical Biochemistry,* vol. 40, no. 1, pp. 105–110, 2025, https://doi.org/10.1007/s12291-023-01164-x.
16. M. R. Khan et al., "Analysis of elliptic curve cryptography & RSA," *Journal of ICT Standardization,* vol. 11, no. 4, pp. 355–378, 2023, https://doi.org/10.13052/jicts2245-800x.1142.
17. N. Kannan et al., "Future perspectives on new innovative technologies comparison against hybrid renewable energy systems," *Computers & Electrical Engineering,* vol. 111, p. 108910, 2023, https://doi.org/10.1016/j.compeleceng.2023.108910.
18. K. Upreti, M. H. Syed, M. A. Khan, H. Fatima, M. S. Alam, and A. K. Sharma, "Enhanced algorithmic modelling and architecture in deep reinforcement learning based on wireless communication Fintech technology," *Optik,* vol. 272, p. 170309, 2022, https://doi.org/10.1016/j.ijleo.2022.170309.
19. M. Haque, V. V. Kumar, P. Singh, A. A. Goyal, K. Upreti, and A. Verma, "A systematic meta-analysis of blockchain technology for educational sector and its advancements towards education 4.0," *Education and Information Technologies,* vol. 28, no. 10, pp. 13841–13867, 2023, https://doi.org/10.1007/s10639-023-11744-2.
20. R. Djehaiche, S. Aidel, A. Sawalmeh, N. Saeed, and A. H. Alenezi, "Adaptive control of IoT/M2M devices in smart buildings using heterogeneous wireless networks," *IEEE Sensors Journal,* vol. 23, no. 7, pp. 7836–7849, 2023, https://doi.org/10.1109/jsen.2023.3247007.

21. K. A. Daudov, A. M. Patrusova, and V. V. Nadrshin, “Analysis of narrowband data transfer technologies on the Internet of Things (IoT),” in *IOP Conference Series Materials Science and Engineering,* vol. 1111, no. 1, p. 012016, 2021, https://doi.org/10.1088/1757-899x/1111/1/012016.
22. F. Heymann, H. Quest, T. L. Garcia, C. Ballif, and M. Galus, “Reviewing 40 years of artificial intelligence applied to power systems – a taxonomic perspective,” *Energy and AI,* vol. 15, p. 100322, 2023, https://doi.org/10.1016/j.egyai.2023.100322.
23. M. S. Samsurijan, A. Ebekozien, N. A. N. Azazi, M. M. Shaed, and R. F. R. Badaruddin, “Artificial intelligence in urban services in Malaysia: a review,” *PSU Research Review,* vol. 8, no. 2, pp. 321–340, 2022, https://doi.org/10.1108/prr-07-2021-0034.
24. F. Lorenz et al., “A scalable and dependable data analytics platform for water infrastructure monitoring,” in *2020 IEEE International Conference on Big Data (Big Data),* Atlanta, GA, USA, 2020, pp. 3488–3493, https://doi.org/10.1109/BigData50022.2020.9378138.
25. M. Awais, Q. Abbas, S. Tariq, and S. H. Warraich, “Blockchain based secure energy marketplace scheme to motivate P2P microgrids,” *International Journal of Informatics and Communication Technology (IJ-ICT),* vol. 11, no. 3, p. 177, 2022, https://doi.org/10.11591/ijict.v11i3.
26. D. G. Costa et al., “A survey of emergencies management systems in smart cities,” *IEEE Access,* vol. 10, pp. 61843–61872, 2022, https://doi.org/10.1109/access.2022.3180033.
27. R. Sundar, S. Hebbar, and V. Golla, “Implementing intelligent traffic control system for congestion control, ambulance clearance, and stolen vehicle detection,” *IEEE Sensors Journal,* vol. 15, no. 2, pp. 1109–1113, 2015, https://doi.org/10.1109/jsen.2014.2360288.
28. L. Park, J. Eom, J. Kim, and S. Cho, “Greedy technique for smart grid demand response systems,” *KEPCO Journal on Electric Power and Energy,* vol. 2, no. 3, pp. 391–395, 2016, https://doi.org/10.18770/kepco.2016.02.03.391.
29. K. F. Haque, R. Zabin, K. Yelamarthi, P. Yanambaka and A. Abdelgawad, “An IoT based efficient waste collection system with smart bins,” in *2020 IEEE 6th World Forum on Internet of Things (WF-IoT),* New Orleans, LA, USA, 2020, pp. 1–5, https://doi.org/10.1109/WF-IoT48130.2020.9221251.
30. A. Simonet-Boulogne et al., “Toward blockchain-based fog and edge computing for privacy-preserving smart cities,” *Frontiers in Sustainable Cities,* vol. 4, p. 846987, 2022, https://doi.org/10.3389/frsc.2022.846987.
31. H. Attaran, N. Kheibari, and D. Bahrepour, “Toward integrated smart city: a new model for implementation and design challenges,” *GeoJournal,* vol. 87, no. S4, pp. 511–526, 2022, https://doi.org/10.1007/s10708-021-10560-w.
32. M. Ghahramani, M. Zhou, A. Molter, and F. Pilla, “IoT-based route recommendation for an intelligent waste management system,” *IEEE Internet of Things Journal,* vol. 9, no. 14, pp. 11883–11892, 2021, https://doi.org/10.1109/jiot.2021.3132126.
33. S. E. Bibri, “Data-driven smart sustainable cities of the future: urban computing and intelligence for strategic, short-term, and joined-up planning,” *Computational Urban Science,* vol. 1, no. 1, 2021, https://doi.org/10.1007/s43762-021-00008-9.

34. A. Nikitas, K. Michalakopoulou, E. T. Njoya, and D. Karampatzakis, "Artificial Intelligence, transport and the smart city: definitions and dimensions of a new mobility era," *Sustainability,* vol. 12, no. 7, p. 2789, 2020, https://doi.org/10.3390/su12072789.
35. H.-W. Zhao, "AI assists operation and maintenance of future cities," *Artificial Intelligence Advances,* vol. 5, no. 1, pp. 25–27, 2023, https://doi.org/10.30564/aia.v5i1.5780.
36. G. Dhuuriya, S. Sinha, D. Aroa, T. P. Rai, P. Johri, and D. Arora, "Introduction to artificial intelligence in smart transportation," in *Artificial Intelligence for Future Intelligent Transportation,* pp. 1–32. Apple Academic Press, 2023. https://doi.org/10.1201/9781003408468-1.
37. J. Yang, Y. Kwon, and D. Kim, "Regional smart city development focus: the south korean national strategic smart city program," *IEEE Access,* vol. 9, pp. 7193–7210, 2020, https://doi.org/10.1109/access.2020.3047139.
38. A. Perdana and I. A. Mokhtar, "Seniors' adoption of digital devices and virtual event platforms in Singapore during Covid-19," *Technology in Society,* vol. 68, p. 101817, 2021, https://doi.org/10.1016/j.techsoc.2021.101817.

Chapter 6

Smart cyber-physical systems in healthcare

Revolutionizing modern medicine in RPM and health security

Brijal Pandey, Reeta Tomar, Kamal Upreti, Puja Kumari, Shitiz Upreti, and Ramesh Chandra Poonia

6.1 INTRODUCTION

In healthcare, smart cyber-physical systems (SCPS) enable efficient decision-making by maintaining digital records of patients and utilizing systematic smart feedback mechanisms and biosensors. However, there remains a gap between the demanding application of data technologies for collection and analysis to drive data-informed decisions, which is crucial for understanding the potential benefits and impacts of these systems. CPS has shown adaptability in medical devices, enabling them to respond to the changing health conditions of patients without manual intervention [1].

Accurate real-time monitoring of medical devices is enhanced through CPS, which optimizes operational efficiency. For example, advanced pacemakers, designed as miniaturized implantable devices, help in monitoring and adjusting to heart rates as per the body's requirements. These devices form a network that communicates with both external and internal systems. Wearable sensors check patients' vital signs, sending signals to a cloud-based system that processes the data, detects alerts, and stores information for future analysis. The programming of these materials allows for adaptation based on changes in bodily conditions. The reliance on hardware, such as sensors and monitoring devices, in conjunction with software that processes patient performance data, enhances the reliability of CPS in healthcare settings. The self-adaptive nature of these systems can yield more informed results due to their learning capabilities, allowing insights to be retrieved from historical data. As a result, CPS fosters innovation in the implementation and use of medical devices, providing ultra-high resolutions for the manipulation of materials at the nanoscale, thereby enhancing applications in nanotechnology and environmental monitoring [2].

DOI: 10.1201/9781003542513-6

6.2 REMOTE PATIENT MONITORING THROUGH SCPS

6.2.1 Overview of remote patient monitoring (RPM) systems

Remote Patient Monitoring (RPM) systems have a crucial part in managing the care of patients with chronic or infectious diseases, post-surgical needs, limited mobility, or other conditions that require daily assessment but prevent frequent hospital visits. RPM brings healthcare services to the comfort of people's homes, especially benefiting infants, the elderly, and those with restricted mobility, while also reducing the burden on healthcare facilities. To function effectively, RPM systems rely on robust infrastructure for processing and securely storing complex healthcare data. The primary challenge lies in efficiently delivering this critical information to healthcare providers, who depend on timely and accurate data for patient care. As the demand grew for real-time patient health information, the healthcare industry underwent modernizations to meet the need, developing RPM systems that use advanced technology to monitor patients in remote settings [3].

6.2.1.1 Technological evolution and system design

Traditional patient monitoring involved extensive use of sensor-equipped wires networked to healthcare facilities. However, these setups were limited by large, immobile devices that could only handle few patients. RPM systems addressed this limitation by enabling in-home monitoring, allowing healthcare providers to remotely track a patient's health status without requiring the patient to visit a hospital daily. Over time, RPM systems have evolved to support patients who cannot regularly access in-person care. The main goal of RPM systems is to provide hospital-level monitoring at home or in a patient's workspace, improving convenience and accessibility. This is achieved through a sequence of data collection, filtering, processing, and analysis that enriches the healthcare provider's understanding of the patient's condition. The system offers the potential for early detection of health issues and timely intervention, enabling patients to remain in familiar, comfortable environments.

6.2.1.2 Functionality of RPM systems

RPM systems utilized by wearable devices are equipped with biomedical sensors that track various physiological metrics, including blood pressure, heart rate, oxygen saturation, and hemoglobin levels. These sensors continuously capture real-time data, which are then circulated securely to healthcare providers via remote applications, ensuring information precision and reducing the risk of data loss. Such real-time data allow healthcare professionals to make informed and proactive decisions before the patient's condition worsens. In spite of their advantages, RPM systems face some challenges. Earlier devices relied on low-cost sensors, leading to variability in the accuracy based on the quality of the hardware. There are significant issues for processing and storage of high-end data, which can strain system performance. Additionally, daily data collection generates large volumes of both essential and non-essential information, complicating the findings of actionable insights [4]. The RPM system's reliance on robust infrastructure and efficient data processing capabilities still remains critical to its success. As these systems continue to evolve, innovations aim to improve data accuracy, expand storage capacity, and enhance the speed of information retrieval and transmission.

6.2.2 Integration of RPMS with smart cyber physical system

SCPS create a unified integration of physical and digital worlds, bringing together physical devices with computing and communication technology to facilitate real-time data monitoring and control in healthcare. The combination thus forms a cohesive system, capable of monitoring patient responses with respect to environmental and health changes. SCPS differs from traditional systems, which are generally rule-based. By embedding software into physical monitoring components, the system becomes highly reliable and is capable of adapting to unexpected events efficiently.

A range of physical sensors capture biological signals from patients, including ECG for monitoring physiological behavior, temperature sensors for checking the body temperature, SPO_2 for oxygen levels, and camera sensors for motion. Communication is facilitated through channels such as Bluetooth, ZigBee, or Wi-Fi, creating a network between client devices and remote applications. A communication broker acts as an intermediary for managing data transfer between the client and server, directing the flow of data into processing modules and communicating key information back to the client. The service layer leverages a Hadoop system, distributing data across various nodes to handle large volumes in an efficient manner and maintain reliability by managing node failures. This distribution enables quick processing and scalability [4] (see Figure 6.1).

As data flow in large volumes, the stream processor in the service layer organizes and handles various streams according to preset platform methods. A real-time accelerator module then optimizes the processing by implementing window techniques for the stream processor, ensuring prompt and effective data handling. Queries from users are translated through a query engine into a machine-readable format, simplifying the retrieval of relevant information for analysis and visualization, which is then presented in a clear and reportable form. Incorporating a large data framework requires strong infrastructure that will support real-time mechanisms at multiple

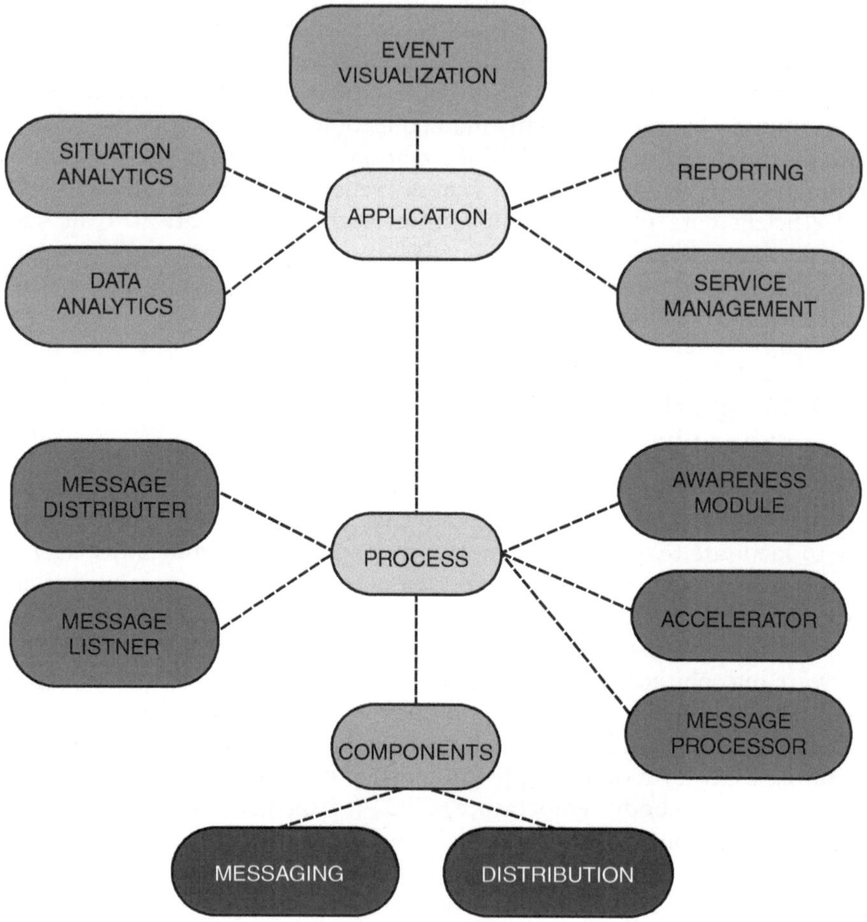

Figure 6.1 Structure for big data in MCPS. (Author's visualization.)

levels. The process begins with accessing data from different sources, filtering out unnecessary information, and organizing relevant data into a hierarchical structure.

A time-window mechanism is employed, managing data based on the arrival time to increase the efficiency in handling real-time data. Data are categorized and processed in parallel by a message processor, which routes it through different streams. Worker threads and analyses then extract relevant insights. An accelerator module reduces delay in processing, enhancing the SCPS's ability by combining data with semantic knowledge for meaningful, actionable insights. In the application layer, these analyses culminate in prioritized reports, allowing RPM to achieve greater accuracy and handle extensive data volumes [4].

By modernizing traditional methods, SCPS frameworks enhance RPM systems, providing valuable insights and improving patient outcomes through accurate and large-scale data handling capabilities.

6.3 CASE STUDY IN RPM

Case study 6.1: RPM in devices

WEARABLE DEVICES: APPLE WATCHES

With the evolving nature of work, increased stress levels are anticipated, requiring the need for monitoring various health factors on a daily basis. Thankfully, medical directives now enable us to keep a track of our stress levels and other health indicators. In today's fast-paced world, digital watches, commonly referred to as smartwatches, such as Apple Watches and Fitbits, have become prominent. These devices draw attention to individuals' health parameters and serve as constant reminders to check on our well-being amid time constraints, reminding us to stay hydrated, displaying heart rates, and monitoring physiological factors. Users often set daily step targets to encourage physical activity. Of course, the services provided vary depending on the device and the software it uses. The primary objective of these devices is fitness monitoring.

Specifically, taking the example of Apple Watches, these devices utilize PPG optical heart rate sensors alongside a three-axis accelerometer, gyroscope, magnetometer, altimeters, ambient light sensors, temperature sensors, ECG, and capacitive touch sensors. The inclusion of these sensors and their capabilities has expanded with each new iteration of the Apple Watch, accompanied by software updates aimed at enhancing the overall accuracy of the collected data [5].

Data are collected passively throughout the time the device is worn by the individual; this collection is then stored in the cloud since the system operates on RPM, utilizing real-time data. Data collection occurs in multiple web-based data warehouses, and the services offer dashboards that facilitate monitoring and visualization of commonly used metrics. If the data stored in the cloud need to be accessed, it can be done through a secure interface, ensuring data security. With RPM systems, the information provided to users is notably more accurate [6].

MEDICAL DEVICES: BLOOD PRESSURE MANAGEMENT

The system of RPM is particularly beneficial for chronic disease management, providing a means to measure disease-specific biometric data, including critical indicators such as blood pressure, heart rate, and oxygen levels. By enabling consistent monitoring, RPM aids in early intervention, helping prevent acute health problems and improving patients' quality of life by reducing recurring hospitalizations. Chronic diseases impose a substantial economic burden since hospital treatments are costly and might place an economic burden on an individual, which he cannot ignore [7].

In a recent study, 6,595 patients with blood pressure and heart-related conditions participated in an RPM program, where each individual received a blood pressure monitor with a specific registered ID that transmitted data securely to a designated portal. Patients were educated on proper blood pressure measurement techniques, and nurses followed-up if readings were outside healthy ranges or data had not been transmitted for 3 days. Monthly meetings allowed for setting health goals and providing guidance on diet, exercise, stress management, and sleep to help control blood pressure. Weekly data averages helped track blood pressure trends, with the study including only those who used the device consistently (at least 30% of days) over a minimum of 90 days. Patients were categorized based on blood pressure levels and divided into groups by RPM duration, spanning from 90 days to over a year.

Of the initial 6,595 participants, 4,370 patients (66.3%) had uncontrolled blood pressure based on their first seven readings. After approximately 289 days into the program, this number decreased by 18.3%, with 2,648 patients showing improved control. Notably, 2,476 patients (38%) began the study with stage 2 hypertension, but after around 290 days, this was reduced to 1,261 patients, reflecting a 19% improvement. Additionally, the average systolic blood pressure across all patients decreased from 135 mmHg at the program's start to 127.8 mmHg, a 7.3-point improvement [8].

The study also assessed how RPM use duration impacted patients' mean arterial pressure (MAP), a critical measure of blood pressure health. For the 4,370 patients with initially uncontrolled hypertension, the average MAP improved by 8.5 mmHg over the study period, indicating a significant reduction. These findings suggest that RPM, combined with regular counseling and integrated electronic health records, can positively impact high blood pressure management. Consistent RPM usage appears to yield cumulative benefits over time, making it a potentially valuable tool for long-term healthcare improvements. Furthermore, as RPM is covered by many insurance plans and eliminates the need for in-person visits, it offers a practical approach, making healthcare more accessible [8].

Case study 6.2: Chronic disease management

The RPM system has proven effective in enhancing clinical outcomes and medical compliance, especially in the self-management of patients with chronic diseases. It is widely used within home healthcare services to assist patients in managing long-term conditions, although isolating and accurately allocating costs can be challenging when incorporated into other healthcare services. This system is economically friendly for users, aiding in diagnosis and enabling the monitoring of one or multiple vital signs [7]. It provides significant benefits for elderly patients diagnosed with chronic diseases, who require continuous monitoring. The case study focused on tracking essential vital signs needed for managing chronic diseases. For example, in monitoring a patient with diabetes, it is important not only to track glucose levels but also to check for other parameters such as blood pressure and respiratory rate as these may affect organ health. This comprehensive system monitors various indicators that could be vital in assessing a particular health issue [9].

Lightweight biosensors, which attach directly to patients, enable continuous monitoring and are particularly beneficial for elderly patients who may not respond quickly to device alerts, like low battery. In such cases, caregivers are immediately notified. These sensors operate on Zigbee technology, specifically designed for short-range communication. This technology suits compact devices, allowing them to transmit information to the Central Control Unit (CCU) with minimal power loss. Under Zigbee, data are continuously gathered through sensors and transmitted to the CCU, which serves as the "brain" of the RPM system, facilitating real-time data processing and analysis. Patient data are then transmitted to a cloud server within the healthcare network, enabling detailed monitoring based on advanced processing and long-term data storage.

In cases of emergencies, this network system can alert relevant parties through triggers or warning messages. Another efficient advantage of this system is its ability to function independently in a situation of no internet availability. Monitoring can continue locally through the Clinical Decision Support System (CDSS), which may use simple knowledge bases, such as patient medical histories or specific health requirements based on recent vital signs, to provide basic advice or alerts [10] (see Figure 6.2).

6.4 HEALTHCARE SECURITY IN SCPS

6.4.1 Healthcare security under SCPS

In today's fast-growing world, data for medical automation rely heavily on the internet, enhancing information communication. Technological advancements in methods such as the Internet of Things (IoT) have spurred smart innovations across many sectors, including healthcare. Healthcare is one of the world's fastest-evolving sectors, leveraging these technological innovations to improve management, quick intervention techniques, remote patient monitoring, and broad data accessibility [11]. However, given the intricate network of interconnected systems, a significant amount of medical information is exposed to the internet, which poses high security risks.

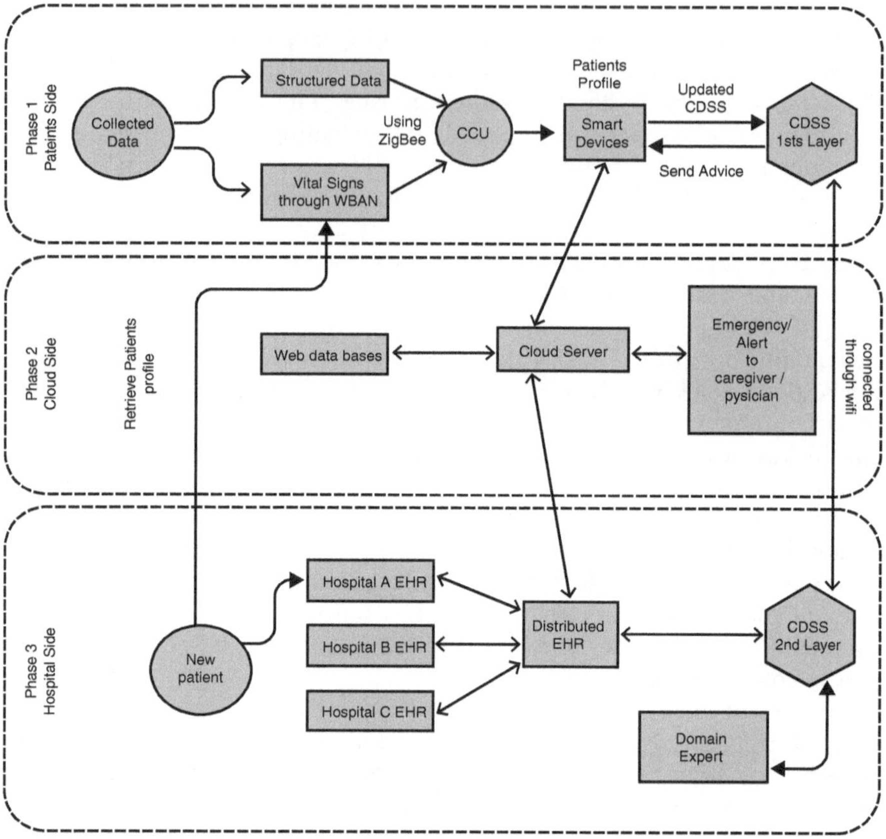

Figure 6.2 Monitoring of chronic disease in patients. (Author's visualization.)

6.4.2 Sources for possible threats in healthcare security

Medical data are often confidential, serving numerous critical purposes, and ensuring its security is essential. These data can include controlled, authorized databases or electronic patient records. When sharing information across parties, healthcare providers often rely on the internet, which can introduce various security threats or bugs [12]. External security breaches are a notable risk, with hackers targeting information systems in medical organizations to access sensitive and confidential data.

Internal security threats can be harmful and often harder to detect. These can arise from individuals within the organization who hold authorized access to sensitive data but may misuse this access with ill intent. Since these malicious insiders have access to internal systems, detecting their actions can be particularly challenging. Statistics reveal that about 70% of threats stem from such internal sources, as found by the CERT (Computer Emergency Response Team) [12]. The potential harm from these insiders increases with their position and level of access within the organization.

Further security risks involve the intrusion of healthcare systems via spam messages or viruses. This could include emails containing malicious attachments or links that damage internal systems. When spam accumulates in large volumes, it can slow down or disrupt regular functions. Identifying spam can be especially challenging as it can come from various unknown sources globally. With so many emails reaching the system, it may become increasingly difficult for the system to function smoothly as these spam messages can cause glitches and slow down the performance. These hazardous spam texts or emails pose severe threats to healthcare operations, making robust security protocols crucial [12].

6.5 CASE STUDIES IN HEALTHCARE SECURITY

Case study 6.3: Blockchain for secure data sharing

Blockchain can be described as an absolute ledger that enhances the process of recording transactions and tracking assets in a business network. It consists of a collection of data that can be expanded, where these collections, referred to as blocks, include transaction details. Each block contains the hash of the following block, linking it with the previous one in the chain. These blocks are connected through encryption, meaning that any alteration to even a single block would require changing the cryptographic hashes of all subsequent blocks, a process that demands substantial computational power. This key feature makes blockchain an extremely reliable method for data transactions as each part of the network holds a complete copy of the blockchain, allowing for quick detection of inconsistency [13].

Decentralization is crucial to blockchain's functionality; no single authority has control over the content within the blockchain. In healthcare services, blockchain is applied to ensure system safety and security by facilitating secure data transfers, data protection, and logistics management.

Blockchain employs a multilayered approach to enhance security in healthcare by integrating medical systems with cryptographic techniques, such as attribute-based and identity-based encryption, which control data access by specifying who can view the information. The decentralization principle maintains confidentiality by securing data from unauthorized access. Blockchain's consensus algorithm governs data access, storage, and sharing, requiring network approval for changes and thereby reducing risks. Research suggests that integrating principal component analysis (PCA) with blockchain addresses storage concerns by minimizing the blockchain's storage demands, which reduces the network resource load. For improved diagnostic accuracy, blockchain solutions have eased Health Information Exchange (HIE). Additionally, smart contracts within the blockchain system eliminate intermediaries and harness immutability to manage data from IoT devices in healthcare.

In insurance management, blockchain combined with homomorphic encryption enables data calculations without decryption, enhancing claim processing and patient record management security.

New innovative techniques are being employed to enhance healthcare privacy using electronic health record (EHR)-based blockchain methods. Elliptic curve cryptography (ECC) encrypts data to prevent unauthorized access to reduce the risk of Distributed Denial of Service (DDoS) attacks. Inscription with attribute-based access (ABA) secures data sharing, reducing costs while improving care quality. Ciphertext policy attribute-based encryption (CP-ABE) limits data access to authorized users only, while pseudonym-based encryption (PBE-DA) anonymizes patient identities. Additional privacy measures include bilinear keyword mapping for search privacy protection [13]. Ethereum's blockchain ensures patient records are tamper-proof, and the Data Preservation System (DPS) maintains indefinite data accessibility. Attribute-based signatures (ABS) allow confidential data sharing between healthcare providers, and proxy re-encryption keeps private data encrypted across all nodes. Together, these cryptographic solutions enhance data privacy and provide controlled access, ensuring data integrity.

Case study 6.4: Cybersecurity strategies in smart hospitals

Authentication is a login credential used by any user for any authentication process to gain access, ensuring security by mandating the verification of these credentials. The process involves verification to prove that the user is authentic. Technologies are constantly innovating and adapting to changing environments, with the IoT being a crucial system implemented to improve authentication across healthcare units.

MULTI-FACTOR AUTHENTICATION

An extra layer of privacy is provided under Multi-Factor Authentication (MFA), which requires multiple forms of validation to access an account. For instance, after entering a password, the user must answer a security question or provide with a physical characteristic such as fingerprints, voice recognition, or facial scans [14]. Often, after entering a one-time password (OTP) sent to a mobile device, another layer of security is added through an OTP sent to the user's email. Since healthcare information is sensitive and more prone to misuse, we essentially need multiple factors to validate a user's identity. Therefore, information gathered by IoT devices must be safeguarded during transmission. When MFA techniques are integrated with IoT devices in healthcare, they enhance authenticity and security for both professionals and patients by keeping the information intact [14].

STRATEGIES TO PROTECT CRITICAL PATIENT DATA FROM CYBER THREATS

MFA is vital for enhancing security, especially in environments transitioning toward password-free methods. While social media may enhance users' sense of security, it can inadvertently provide cybercriminals with clues to bypass MFA. Therefore, organizations must prioritize adopting MFA solutions to strengthen their defences against cyber threats.

Funding and resource allocation play a critical role in administering public health. There is a pressing need for funding mechanisms that support the upliftment of authenticated systems, such as compensating hospitals for costs associated with enhancing their security measures in response to increasing cyber threats, particularly during crises [11].

Healthcare organizations need to remain vigilant and adaptive. Regular evaluations of cybersecurity policies with enhanced training programs for staff can mitigate the risk associated with handling sensitive medical data, ensuring that healthcare services remain without any disruption.

The rise of phishing attacks, which reportedly increased by 60% during the pandemic, highlights vulnerabilities in the healthcare sector's cybersecurity posture. Cybercriminals exploited pandemic-related fears to deceive healthcare workers into clicking malicious links. The situation underscores the urgent necessity for healthcare organizations to educate their staff on recognizing and effectively responding to phishing attempts [14].

6.6 COMPARATIVE ANALYSIS OF RMP AND HEALTHCARE SECURITY

RPM and healthcare security are important components of an SCPS for building a strong healthcare network. By incorporating blockchain, IoT-based, RPM systems address the challenge of managing vast volumes of data streams while ensuring data security and patient privacy. Blockchain enables patient anonymity and confidentiality, safeguarding sensitive information. The decentralized nature of blockchain also enhances the resilience and reliability of IoT-based RPM systems, ensuring that data are transmitted securely and remain protected [13].

Criteria	*Remote patient monitoring*	*Healthcare security*
About	It is system that monitors patients' health remotely, and the data measured are real-time-based.	A set of measures and technologies that helps in maintaining the privacy of regarding a patient's health.
Primary focus	Physiological data are collected through biomedical sensors in order to evaluate the wellbeing of patients outside of the hospital setting.	Ensuring confidentiality, security, and maintaining integrity of the patient's data, managing for potential threats.
Key technologies	• Wearable devices (Fitbit and apple watches) • Sensors • (IoT) Internet of things • Cloud computation	• Blockchain technology • Hashing schemes • Encryptions • Access of control
Commonality features	• Collection of real-time data • Censoring of data • Data analysis • Remote communication	• Protection of data • Authorized access management • Determining threats
Challenges	Implementation cost is high since the system building requires infrastructure and software. Many existing RPM devices may not work with current healthcare system, causing device interoperability	Bringing a balance between accessibility and security Since the data are sensitive, managing the system can be complex.
Opportunity	Expansion in telehealth and through innovation in technology better feature-based devices.	Integration with AI for threat detection since systems are complex and require better assistance.
USP	Reduced commutation; hence, dependency of patients increased. Improved the management of chronic disease, especially in patients who require constant monitoring, hence empowering the patient.	Under the healthcare security system, patients are assured of the security of their privacy, safeguarding sensitive healthcare information and protecting it against unauthorized access.
Future Innovations	RPM can further innovate through integrating with AI for predictive analysis through the forecasting method.	The security system needs to innovate in a way that patients' rights are incorporated as well, and they may be achieved through advanced encryption methods.

Key Similarities: Both remote patient monitoring (RPM) and healthcare security are critical in modern healthcare, focused on real-time data management and patient-focused solutions.

Key Differences: RPM is primarily concerned with patient monitoring and improving care, whereas healthcare security is focused on safeguarding data and meeting regulatory standards.

Shared Challenges and Opportunities: Both areas encounter implementation challenges and stand to gain from advancements in technologies like AI and advanced analytics.

6.7 CHALLENGES IN IMPLEMENTING SCPS IN HEALTHCARE

The integration of SCPS in healthcare through software-intensive techniques in wearable, medical, and monitoring devices undoubtedly enhances convenience in individuals' lives. However, this innovation also has its own set of challenges. As software carries vast amounts of patient information, a robust mechanism is essential to ensure safety and prevent malfunctions that threatens an individual's privacy. Therefore, a strong procedure must be in place to ensure protection without flaws, with risk minimization and strict verification and validation of data. A highly dependable software system is necessary, and early-stage verification is required for reliability in devices to meet requirements more efficiently [15].

Maintaining data accuracy and security becomes critical when different devices interchange real-time data. Interoperability, thereby, becomes extremely important when dealing with the substantial information generated from data transfer in a secure and efficient manner. Although some standards exist for overall adoption and compatibility of devices, vital signs are interconnected. Various parameters contribute comprehensively to a patient's health condition. The system must understand these signs parallel to the patient's medical history and other environmental factors. Only with such efficiency can a system provide timely responses to emergencies [15]. This can be achieved through closed-loop system where patient feedback is directly transmitted to device functions. However, balancing such complexity is extremely challenging without compromising the accuracy.

Additionally, since these medical devices are used across all age groups, a significant portion of this population includes the elderly. Thus, an interface that is user-friendly, with easy troubleshooting options, interactive modules, and error prevention features for assistance is needed. For faster adoption of these systems, device design should consider the needs of both caretakers and patients as they will be the primary users of these services.

While a closed-loop system is important in administering real-time data, it should be implemented with caution as different individuals may react differently to treatments depending on their complex medical situations involving multiple conditions. Treatments need to be customized, varying with different control systems to achieve effective results and high accuracy. This customization can be accomplished by studying various situations associated with diverse health conditions [15].

Despite blockchain's efficiency and ongoing innovations, challenges still persist. The first is cost. Additionally, for training and testing procedures,

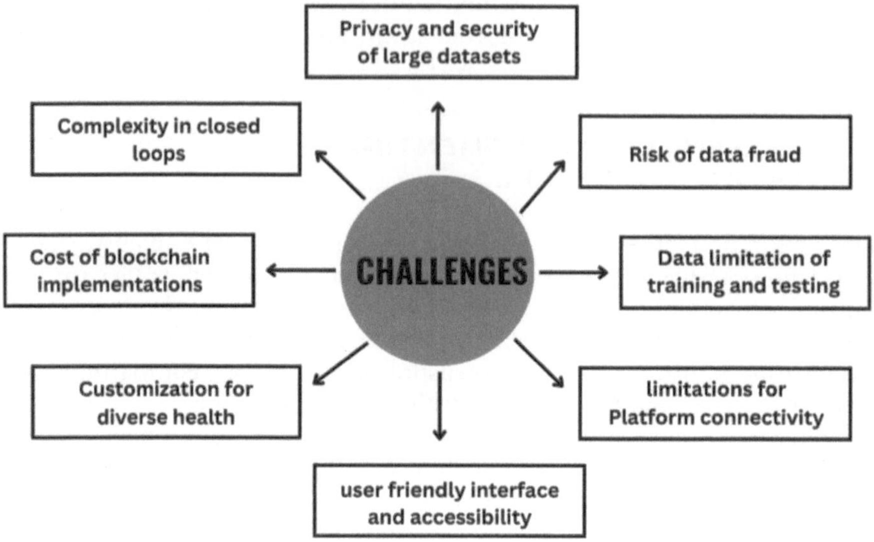

Figure 6.3 Challenges while implementing SCPS in healthcare. (Author's visualization.)

data are often limited, lacking the variety needed for comprehensive simulations. Limited data pose issues for multimedia applications, particularly those requiring image recognition, potentially affecting blockchain's efficacy. Platform limitations represent a further challenge; advanced blockchain components may require a "gateway" to connect with sensors, complicating the setup. Storage constraints also exist, although fog computing can alleviate some of this by storing data off-chain [13].

Another significant concern is the societal aspect as blockchain's restrictive access may lead to the exclusion of underprivileged groups, potentially widening healthcare inequality. The societal impact of blockchain remains under-researched, and limited accessibility might lead to data fraud risks by excluding those without adequate resources from the healthcare system (see Figure 6.3).

6.8 FUTURE TRENDS AND INNOVATIONS IN SCPS FOR HEALTHCARE

For future innovations, there is a need to engage in experiments utilizing machine learning algorithms to streamline the analysis process. The objective is to facilitate the automatic classification of gait phases, which can advance research in detection, prediction, and prevention. Nevertheless, future versions will consider implementing a balanced, two-way communication system where physical devices can give feedback to the user.

The volume of data transferred from the cyber system to the configured physical devices may differ based on the type of feedback provided, such as auditory, visual, or haptic feedback with taps, sensations from touching, or even with vibrations.

Numerous initiatives have been undertaken to enhance the performance of RPMS. Thus, the primary design challenge in RPMS is to prolong the lifespan of sensor devices while ensuring reliable and timely message delivery. Effective RPMS will enhance clinical decision-making, promote self-management, and improve compliance with care plans. This could allow payers (patients) and providers (healthcare providers) to lower healthcare costs [16].

6.8.1 Collaboration of SCPS with artificial intelligence (AI)

AI-driven RPM combines front-line hardware and software to guarantee secure, dependable, and effective patient monitoring. These systems track vital signs like heart rate, pulse, respiratory rate, blood pressure, and oxygen levels, all of which are vital for good health and use machine learning and deep learning to identify and predict them. Wearables and sensors driven by AI continuously gather health data, allowing for real-time monitoring and warning medical professionals of anomalies. IoT connectivity sends data to a cloud infrastructure for processing, where AI algorithms find patterns to support early intervention and individualized care. The system would guarantee general data protection regulation (GDPR) compliance and data protection, ensuring that the collected data are protected, individuals are respected for their privacy, and system meets legal requirements for data handling and privacy [17]. While AI provides a layer of predictive insights to anticipate future health events, such as identifying early signs of disease or deterioration, SCPS primarily concentrates on real-time monitoring and control.

By examining each patient's unique data, medical history, and behavior, AI makes personalized recommendations possible. They are better at detecting minute departures from typical patterns that could point to some medical problems. Real-time monitoring is possible with SCPS, but AI will improve this by spotting complex correlations and variances in big datasets. AI can make data more actionable by facilitating clearer interpretation [18].

6.9 CONCLUSION

The medical system has a crucial role globally, demanding constant innovation and enhanced security to ensure that services remain accessible to patients, developing systems for early disease detection and maintaining individual privacy through secure data management. In this context, integrating SCPS with healthcare applications has proven instrumental in delivering accurate solutions, particularly by leveraging real-time data. During challenges such as disease outbreaks or pandemic situations, RPM has been proven vital, connecting patients with healthcare providers and enabling continuous care.

With SCPS, we observed inclusion of the physical and cyber domains, seamlessly interconnected through a closed-loop system powered by cloud computing and IoT in handling of extensive datasets. RPM with SCPS in wearable and medical devices has simplified healthcare access for patients in remote areas, who require constant monitoring. Another essential aspect of healthcare is data management as vast amounts of sensitive information are generated, requiring robust security solutions. Tools like blockchain and closed-loop systems have been used to strengthen data security, although some challenges with physical cybersecurity in SCPS still remain to be addressed. Combining SCPS with AI and machine learning tools can further elevate the performance, particularly by identifying abnormal patterns caused by errors or any potential cyber threats. This AI-augmented approach will enhance healthcare across various sectors, particularly in data security, improving patient experiences and adding to predictive, adaptive as well as personalized dimensions of healthcare monitoring. These actionable insights ultimately will elevate patient care and contribute to a better resilient healthcare system.

REFERENCES

1 S. A. Haque, S. M. Aziz, and M. Rahman, "Review of cyber-physical system in healthcare," *Int. J. Distrib. Sens. Netw.* vol. 10, no. 4, p. 217415, 2014.

2 R. Verma, "Smart city healthcare cyber physical system: characteristics, technologies and challenges," *Wireless Pers. Commun.* vol. 122, no. 2, pp. 1413–1433, 2022.

3 K. Boikanyo, A. M. Zungeru, B. Sigweni, A. Yahya, and C. Lebekwe, "Remote patient monitoring systems: applications, architecture, and challenges," *Sci. Afr.* vol. 20, p. e01638, 2023.

4 D. Min, "Medical cyber physical systems and bigdata platforms," 2013, p. 106.

5 G. Y. Lui, D. Loughnane, C. Polley, T. Jayarathna, and P. P. Breen, "The apple watch for monitoring mental health–related physiological symptoms: literature review," *JMIR Ment. Health* vol. 9, no. 9, p. e37354, 2022.

6 Y. Liao, C. Thompson, S. Peterson, J. Mandrola, and M. S. Beg, "The future of wearable technologies and remote monitoring in health care," *Am. Soc. Clin. Oncol. Educ. Book*, vol. 39, p. 115, NIH Public Access, 2019.

7 K. R. De Guzman, C. L. Snoswell, M. L. Taylor, L. C. Gray, and L. J. Caffery, "Economic evaluations of remote patient monitoring for chronic disease: a systematic review," *Value Health* vol. 25, no. 6, pp. 897–913, 2022.

8 W. Smith, B. M. Colbert, T. Namouz, D. Caven, J. A. Ewing, and A. W. Albano, "Remote patient monitoring is associated with improved outcomes in hypertension: a large, retrospective, cohort analysis," *Healthcare* vol. 12, no. 16, p. 1583, 2024.

9 D. Peretz, A. Arnaert, and N. N. Ponzoni, "Determining the cost of implementing and operating a remote patient monitoring programme for the elderly with chronic conditions: a systematic review of economic evaluations," *J. Telemed. Telecare* vol. 24, no. 1, pp. 13–21, 2018.

10 N. El-Rashidy, S. El-Sappagh, S. M. R. Islam, H. M. El-Bakry, and S. Abdelrazek, "Mobile health in remote patient monitoring for chronic diseases: principles, trends, and challenges," *Diagnostics* vol. 11, no. 4, p. 607, 2021.

11 A. Almalawi, A. I. Khan, F. Alsolami, Y. B. Abushark, and A. S. Alfakeeh, "Managing security of healthcare data for a modern healthcare system," *Sensors* vol. 23, no. 7, p. 3612, 2023.

12 C.-H. Liu, Y.-F. Chung, T.-S. Chen, and S.-D. Wang, "The enhancement of security in healthcare information systems," *J. Med. Syst.* vol. 36, pp. 1673–1688, 2012.

13 P. K. Ghosh, A. Chakraborty, M. Hasan, K. Rashid, and A. H. Siddique, "Blockchain application in healthcare systems: a review," *Systems* vol. 11, no. 1, p. 38, 2023.

14 T. Suleski, M. Ahmed, W. Yang, and E. Wang, "A review of multi-factor authentication in the Internet of Healthcare Things," *Digit. Health* vol. 9, p. 20552076231177144, 2023.

15 A. K. Tyagi and N. Sreenath, "Cyber physical systems: analyses, challenges and possible solutions," *Internet Things Cyber-Phys. Syst.* vol. 1, pp. 22–33, 2021.

16 F. Arafsha, F. Laamarti, and A. El Saddik, "Cyber-physical system framework for measurement and analysis of physical activities," *Electronics* vol. 8, no. 2, p. 248, 2019.

17 F. Tsvetanov, "Integrating AI technologies into remote monitoring patient systems," *Eng. Proc.* vol. 70, no. 1, p. 54, 2024.

18 K. Upreti, P. Vats, M. S. Nasir, M. S. Alam, F. I. Shahi, and M. S. Kamal, "Managing the ethical and sociological aspects of AI incorporation in medical healthcare in India: unveiling the conundrum," in *2023 IEEE World Conference on Applied Intelligence and Computing (AIC),* Sonbhadra, India, 2023, pp. 293–299. doi: 10.1109/AIC57670.2023.10263931.

Chapter 7

Smart transportation and autonomous system integration

Sitesh Kumar Singh, Bader Eddin Alasali, and Khuloud Nasser Al Balushi

7.1 INTRODUCTION TO SMART TRANSPORTATION

Smart transportation is a departure from the way that we traditionally conceive and function our systems of transportation. It means applying technology and other novel approaches to improving the efficiency, safety, and sustainability of our transportation networks. In achieving these goals, smart transportation uses emerging technologies and data-based methods to tackle the most pertinent of challenges with traditional transport systems including congestion, emissions, and availability, among others.

7.1.1 The need for smart transportation

The field of traditional transportation systems is lagging behind the rising demand to rapid urbanization and population growth worldwide. Urban congestion wastes time, uses too much fuel, and causes pollution. In addition, the increasing demand for transportation through gas-infested engines has been increasing CO_2 emissions of greenhouse gases that result in more climate changes [1].

This is when the importance of smart transportation solutions makes complete sense in every way. Smart transportation systems certainly have been developed around the use of technology, data analytics, and connectivity for better management and operation of infrastructure in order to reduce costs, decrease its environmental footprint, or simply increase level services [2].

7.1.2 Core components of smart transportation

Smart transportation refers to a combination of various components and technologies which, when combined together, are capable of transforming the way people or goods move. One critical element of smart transportation is encouraging more sustainable transport modes like electric and hybrid cars [3]. Creating a cleaner transportation system is part of the goal as these vehicles produce less emissions and can reduce fossil fuel dependence.

DOI: 10.1201/9781003542513-7

The intelligent traffic management system (ITMS) is the key component in smart transportation to handle and control real-time data by using algorithms that help optimize parking and signals modes of operation on signalized roadway networks. These systems can come up with traffic-based detections, signal timing modifications, and much more valuable information to help drivers take realistic strategic decisions in their travel plans [4].

One more revolutionary advancement in intelligent transportation systems is autonomous vehicles. These autonomous vehicles have the power to change our perspective of transportation as it provides more safety, efficiency, and access. With the help of artificial intelligence and sensor technologies, vehicles drive by themselves, talk to one another while on the road, and respond in real time according to traffic conditions.

Also, public transit apps and ride-sharing services are changing the way people can reach transportation. Users plan their trips, order services, and pay with these digital platforms (mobility orchestrators), making it easier for multimodal transport alternatives while decreasing the adoption of private vehicles.

7.1.3 Benefits of smart transportation

Smart transportation solutions have many advantages for the individual as well as the society. One of the many things they have over traditional bus capers is that these reduce travel time and congestion. Smart transportation serves to improve traffic flow; reinforce public transit and carpool services, which in turn reduces the number of cars on the road; and allows people to spend less time in traveling from one point to another [5].

In addition, smart transportation leads to cleaner air and carbon reduction. Electric vehicles must also play a major role in reducing the carbon footprint of transportation systems, ensuring cleaner air and healthier environments for those communities [6].

Improved mobility is another important advantage of smart transportation, especially with regards to the elderly and disabled. Smart transportation systems, on the other hand, make transport options easily available and accessible to all, thus ensuring their full participation in social and economic activities.

7.2 THE ROLE OF AUTONOMOUS SYSTEMS IN SMART TRANSPORTATION

A relatively controversial subject of discussion is the role that autonomous systems play in smart transportation and how they can change traffic management, travel behavior, or the purpose/modes we select. Self-driving vehicles, or autonomous systems, can change the landscape of the transportation system by sensing its environment with little to no human input by using GPS and AI capabilities [7]. Many splendored roles of autonomous systems in smart transportation range from improved road safety and efficiency to traffic decongestion, instant tracking, and real-time monitoring service—through reforming the commercial mode. One of the biggest advantages is that autonomous systems can help reduce the occurrence of road accidents (which are mostly caused by human error) and thus contribute to enhanced safety on roads [8]. Based on statistics, human factors distracted by driving-impaired web, such as errors in judgment, are a well-documented factor contributing to road accidents. With the ability to integrate high-tech sensors and AI algorithms, autonomous systems could drastically lower accident frequencies by eliminating human errors as well as improve transportation safely on a global scale. Along with safety, automated systems contribute significantly to better efficiency in transportation as well. Such systems can optimize route planning, reduce idling times, and help in smoothing traffic flow through real-time data analysis.

Utilizing AI algorithms and connected infrastructure, autonomous vehicles can potentially neither waste travel time nor fuel economy [9]. Driving to a city center can improve the efficiency of transportation networks. By communicating with one another and infrastructure components, autonomous vehicles can help ease traffic flow to prevent or alleviate bottlenecks. This could allow for smoother acceleration and deceleration, reducing congestion through smooth flow of traffic as a result (leading to better driving experience/enhanced emission production). In addition, autonomous systems allow traffic to be managed and monitored in real-time, producing valuable data for transportation authorities and city planners [7]. Utilizing the data captured by autonomous vehicles and analyzing it through intricate analytics, transportation agencies are able to identify traffic trends, infrastructure-used patterns, as well as areas in which they may not perform markedly. Real-time data plays an important role for taking decisions relating to infrastructure development, traffic management strategies, and urban planning initiatives.

From a larger perspective, however, autonomous systems fit into smart transportation, and it is that they have the potential to revolutionize the transport industry. The march of autonomous technologies, meanwhile, is poised to reshape business models and customer experiences like nothing else since the mass production assembly line reshaped urban landscapes about a century ago. These are all areas where we believe that the intersection of autonomy with enterprises like ridesharing, mass transit, and even last-mile delivery may lead to novel mobility solutions that better serve new societal demands. The arrival of autonomous systems will, in addition, pave the way for new industries, like vehicle-to-infrastructure communication, mobility-as-a-service platforms, and urban mobility analytics [10].

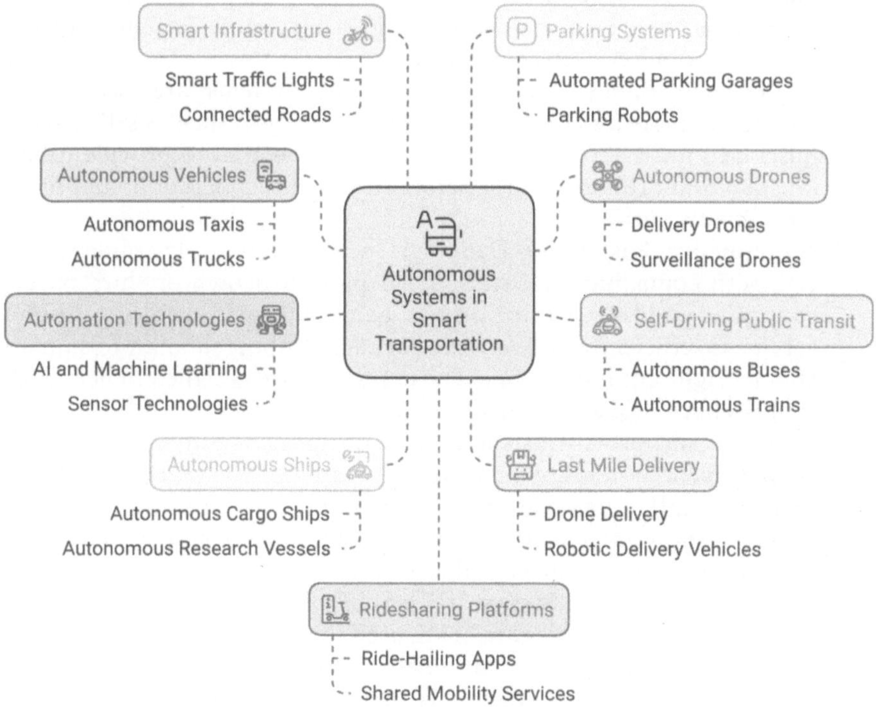

Figure 7.1 Autonomous systems in smart transportation.

7.2.1 Autonomous systems in smart transportation

Smart transportation refers to a range of technologies and applications designed to make the transport experience more seamless, safer, and cost-effective, and autonomous systems have an essential part in that equation, as shown in Figure 7.1.

They are instrumental for revolutionizing mobility as well as improving transportation networks on a large scale. Some of the common applications for smart transportation are as follows:

a. **Autonomous Vehicles:** Self-driving cars, trucks, and other autonomous vehicles use advanced sensors, GPS, and artificial intelligence to move from place-to-place without driver input. The potential of these vehicles to revolutionize how people and goods move is enticing; their supposed advantages include improved safety, reduced congestion, and greater access for all.
b. **Autonomous Drones:** This includes unmanned aerial vehicles (UAVs) or drones with autonomy for different transportation purposes like surveys, cargo delivery, infrastructure inspections, and more airborne

and emergency responses. By using autonomous drones, logistical operations receive further optimization and the performance of the transportation service is enhanced.

c. **Self-Driving Public Transit:** Self-driving systems are also being adopted in more and more forms of public transit, such as self-driving buses or shuttles. These systems have the potential to provide urban and suburban residents with better access, less congestion, and more mobility options.
d. **Autonomous Last-Miles Delivery:** In autonomous last-mile delivery, both companies and start-ups deploy their own or third party-controlled goods—carrying drones, gadgets, or self-operating ground robots to efficiently deliver the package without human-to-human contact right until our doors. Those systems are designed to help with supply-chain efficiency and speed up delivery times.
e. **Smart Transport Infrastructure:** It includes Autonomous Traffic Management Systems, which uses AI, big data, and sensors to obtain real-time traffic information [11] for optimizing multi-traffic operations at various intersections using dynamic control strategies in order to minimize congestion as well as enhance security.
f. **Autonomous Parking Systems** [12]**:** State-of-the-art sensors, artificial intelligence, and robotics are being used to enable vehicle autonomous self-parking solutions and automate valet service within car parks. The objective of these systems is to make the most efficient use of parking space and provide drivers with an enhanced experience during the process.
g. **Autonomous Ridesharing and Mobility-on-Demand Platforms** [13]**:** Autonomous ride-sharing and mobility-on-demand platforms are services that use self-driving vehicles to deliver on-demand transportation to users. The goal of these platforms is to provide convenient, low-cost, and efficient mobility service for city-limits or suburban commuters.
h. **Autonomous Traffic Signal Control Systems:** Like fast-paced autonomous traffic, useful new technology systems are being incorporated for optimizing the signals to innovate its timing and prioritize public transit of multimodal transport. This leads to better traffic flow, reduces waiting time, and increases the overall safety of intersections.
i. **Autonomous Taxis and Shuttle Services:** Companies have been developing autonomous taxi/shuttle services, allowing for point-to-point transportation within urban areas. This work provides a layer of service that could allow convenient, flexible mobility options but with less need for individual vehicle ownership.
j. **Autonomous Traffic Monitoring and Analytics:** Autonomous technology is leveraged for live traffic monitoring, information capture, and data analysis to enable providing valuable intelligence on the mobility trends of parkway offices' congestion focus zones about every other law

enforcement agency. The output from these insights informs decision-making for infrastructure planning, traffic management, and urban development initiatives.

k. **Autonomous Roadside Assistance and Maintenance Vehicles:** Utilize roadside assistance, including autonomous roadside service vehicles carrying tools for minor repairs, as well as autonomous maintenance vehicles tending to duties such as road surface inspection; signboard cleaning after gusts of wind raise dust from desert lands; and vegetation control, an ongoing job that never faces shortage of tasks.

These use cases illustrate the different types of autonomous systems used in smart transportation and how this type of technology can help advance combined mobility solutions in effective, efficient, and sustainable ways. With the availability of technology enhancements, autonomous systems certainly play an increasingly prominent role in our urban and interurban transport landscape going forward.

7.2.2 GPS utilization for autonomous systems

The Global Positioning System (GPS) is used in transportation by autonomous systems to improve navigation and ensure decision-making and operational efficiency of the vehicles [14].

A few important ways that autonomous systems use the GPS technology are shown in Figure 7.2 and described as follows:

a. **High Accuracy Positioning:** GPS helps driverless cars to know their exact geo-location on the surface of this planet. That information is critical for autonomous systems to determine their location in the transportation network, navigate accurately throughout it, and adjust driving behaviors according to different traffic situations. The data collection enhances the performance of autonomous vehicles, thanks to internal mapping systems that use GPS and other detection technologies (camera/lidar) to navigate.
b. **Route Optimization and Planning:** Along with navigation in real-time traffic, road closure information allows autonomous systems to plan and optimize routes. GPS data allows autonomous vehicles to change their route on-the-fly so that they can steer clear of traffic congestion, find the quickest routes, and provide more accurate arrival times. This increases the efficiency of transportation and reduces passenger travel times and cargo delivery.
c. **Geofencing and Boundary Recognition** [15]: With the help of GPS technology, autonomous systems can geolocate, creating virtual barriers or geofences that delimit operational zones. Geofencing further supports safer, more secure, and regulation-appropriate operation of autonomous vehicles within predetermined geographic areas or

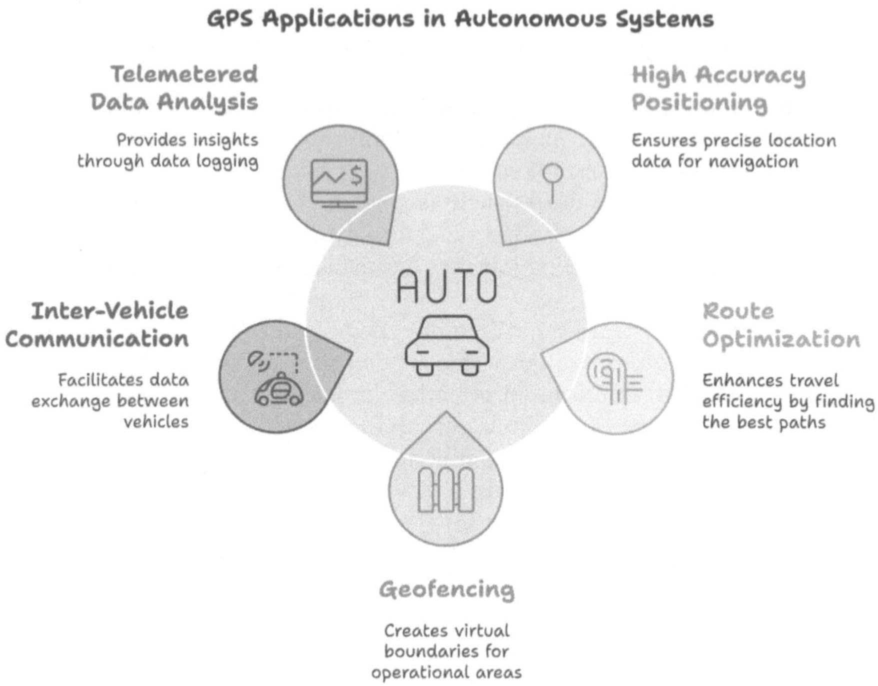

Figure 7.2 GPS applications in autonomous systems.

routing instructions. Also, the GPS data could be used as a recognition for boundaries to allow seamless and safe navigation within specific zones either in confined areas like urban centers or industrial estates among other restricted zones.

d. **Inter-Vehicle Communication:** The presence of GPS allows for inter-vehicle communication between autonomous systems, where they can exchange positional data to agree on next moves and maximize traffic flow. Autonomous functionality can provide enhanced situation awareness, reduce risks of collisions, and enable cooperative driving strategies through the use of GPS-based communication protocols to all work together, leading to safer transportation networks.
e. Logging and analyzing Telemetered Data [16]: Vehicle trajectories and speed data (and geo-positioning as provided by the GPS) are needed for Digital-CAM-type recording. For example, they store high-fidelity travel history using GPS data, provide insights into driving behavior (for better performance), enable fleet management, and do predictive maintenance of trucks. By applying this data-driven approach to the ongoing learning of autonomous transportation operations, we can further advance safety and reliability.

Thus, GPS integration is imperative for the proper functioning of autonomous systems within transportation, which allows these vehicles to navigate and define routes, communicate with other units in motion on streets, or maintain situational awareness. With each evolution of GPS technology, its ability to work seamlessly with self-driving systems will drive significant changes in the future of smart transportation and urban mobility.

7.3 ELECTRIC AND HYBRID VEHICLES IN SMART TRANSPORTATION

Electric vehicles (EVs) and hybrids have become a smart choice as electric cars and EV's in energy-efficient, sustainable, and environmental transportation. These vehicles have a substantial role in reducing emissions, encouraging energy diversity and driving sustainability of transportation systems [17], as shown in Figure 7.3 and as described below:

a. **Emissions Reduction:** EVs and hybrid vehicles reduce greenhouse gas emissions as well as those of local air pollutants to a beneficial effect on both outdoor urban public health and the atmosphere. EVs mitigate the environmental impacts of transportation and encourage a cleaner, more sustainable mode through smarter use, enabling smart propulsion technology (electric or hybrid) to replace traditional internal combustion engine vehicles.
b. **Diversity of Energy and Mineral Resources Efficiency:** EV or plug-in hybrid electric vehicles (PHEV) reduce liquid fuel requirements, enhancing the integration of renewable energy in the transport sector, while also alleviating the dependency on oil. They have been adopted in order to promote national energy independence, as well as for the interest of improving local air quality and mitigating climate changes. 3. Smart Charging Infrastructure: The rise in electric and hybrid vehicles is fueling the market for smart charging infrastructures that consist of public and home charging stations as well as networked solutions. This will provide convenient solutions for EV charging and widespread availability of sustainable transportation.
c. **Electric Vehicle Grid Integration and Demand Management:** EVs can be integrated into smart transport networks through vehicle-to-grid (V2G) technology. The vehicle-to-grid integration supports demand response, energy storage, and grid stabilization to manage a large amount of distributed but controllable stationary smart transportation stored electrical potential in an efficient manner.
d. **Smart Transportation Initiatives:** It integrates with electric and hybrid propulsion in autonomous fleets, such as minor delivery robots, shuttles, or autonomous taxis. Combining autonomous technologies

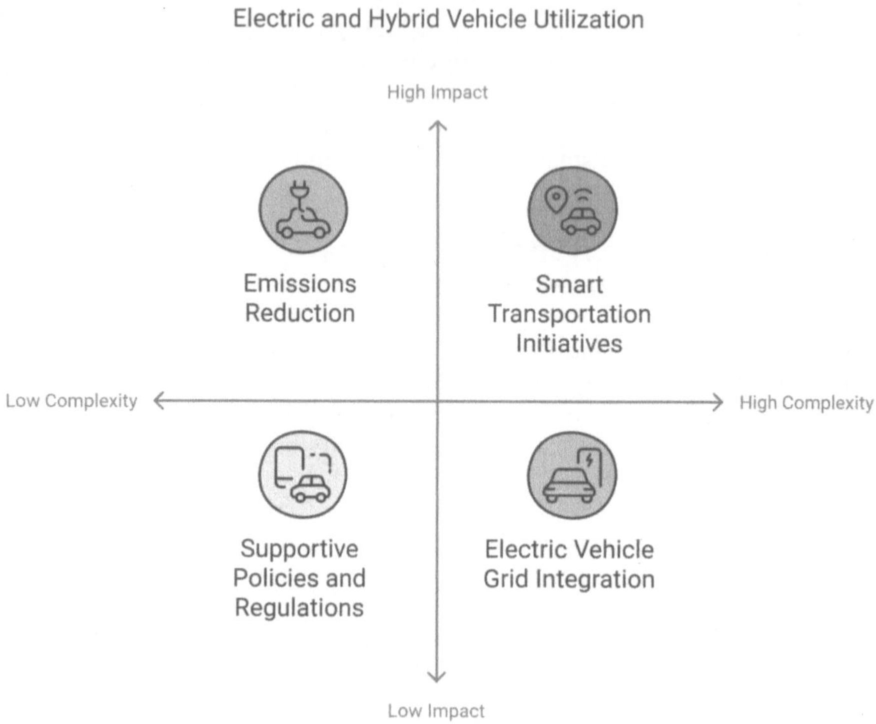

Figure 7.3 Utilization of electric and hybrid vehicles in smart transportation.

and electric/hybrid powertrains enables sustainable, efficient mobility services at lower operational costs with a smaller environmental footprint.

e. **Electric and Hybrid Vehicles Data-driven Optimization:** EV and plug-in electric vehicle (PHEV) drivers leave data trails in terms of energy consumption, driving habits, and charging behaviors. These data can help decision-makers shaping local transportation planning strategies, infrastructure investments, and energy demand services to optimize the deployment of compact EVs and hybrid vehicles within smart transport ecosystems.
f. **Supportive Policies and Regulations:** These include public policies that incentivize the adoption of EVs and plug-in hybrid-electric vehicles and emissions standards for new vehicle sales by supporting markets primed to deploy emerging ZNEV (zero net emission vehicles) technology with investment in ZNEV technology optimized solutions. Such measures of support influence the proliferation and mainstream acceptance of electric and hybrid vehicles under smart city transportation strategies.

EVs and hybrid vehicles are a key part of smart transportation, providing environmentally friendly options in mobile alternatives that employ renewable energy sources. It means better uses of cleaner energy, reduced emissions, and smarter, more bullet-proof transportation systems that take into account environmental concerns. With this rapid evolution in the automotive industry toward sustainable mobility, EVs and hybrid vehicles will define our landscape of personal transportation [18].

7.3.1 Hybrid vehicles vs sustainability

Among the benefits of hybrid vehicles with respect to sustainability, these cars aid in environmental preservation as well promote energy conservation and less ecological footprints. Few of the main green benefits of hybrid cars are given below [19]:

a. **Less Greenhouse Emissions:** This is because hybrid cars, especially plug-in hybrids, can travel on electric power alone for at least part of the journey. It reduces fuel consumption and tailpipe emissions drastically, thus producing lower levels of greenhouse gases in comparison to a traditional internal combustion engine vehicle. Thus, hybrids contribute to the battle against climate change by reducing how much CO_2 is generated from transport.
b. **Energy Efficiency:** These vehicles are designed in such a way that they minimize energy usage as they run using an internal combustion engine and electric motor. Regenerative braking and energy management systems in hybrid vehicles capture, store any otherwise wasted energy from during breaking, and promote deceleration to help increase its efficiency overall while reducing the need for non-renewable fuel resources.
c. **Better Gas Mileage:** Generally, hybrid vehicles have better gas mileage compared to conventional cars, especially under city driving and stop-and-go conditions. Since hybrids rely on an electric drive at low speeds and acceleration of fossil fuels consume much less, they promote efficiency and can reduce our very real reliance upon energy sources, which are limited in nature.
d. **Noise Emission:** The utilization of electric power in hybrid vehicles, especially during low-speed operation and idling, results in less noise pollution by a positive effect on urban areas. That results in a better quality of life for residents and may lead to improved public health outcomes as well.
e. **Public Health:** A decrease in emissions of hybrids resulting to drivers, as well as other people in the cities, where there is restricted circulation due to traffic jams and reduced air quality benefits. It also involves reducing the impact of pollutants like nitrogen oxides (NOx) and particulate matter, which have been linked to respiratory conditions and other health problems that are exacerbated by poor air quality.

f. **Technological Progress and Enlightenment (Awareness):** The emergence of hybrid vehicles also promotes the improvement in electric propulsion, energy storage, and vehicle electrification. Additionally, the proliferation of hybrids has brought additional attention to sustainable mobility solutions and propelled considerable advancements in alternative fuel tech development.
g. **Shift to Electric Mobility:** Hybrid vehicles are the middle step in between traditional internal combustion engine vehicles and fully electric ones, by giving consumers a glimpse into the world of electrification. With this transition, EVs are slowly integrated into the transportation infrastructure, which in return contributes to long-term sustainability and reduced dependency on fossil fuels.

Benefits of hybrid vehicles include reduced greenhouse gas emissions; improved energy efficiency/fuel consumption over conventional technologies (especially for PHEVs); less local air and noise pollution which poses public health risks (in comparison to ICEs) as well some reduced dependence on oil compared particularly with HEVs; technological integration measures required for the ubiquitous deployment of electric mobility in combination with major changes to the electricity generation mix. These benefits have helped hybrids become an integral part of sustainable transportation solutions and are aiding in tackling pollution problems caused by traditional transport.

7.3.2 Electric vehicle integration vs environmental impact

Whether guiding the path for EVs to slip into transportation systems and more, an outlook of environmental consequences from various angles over the transition to EVs is provided [20]. Some of these key environmental impacts are represented in Figure 7.4 and described as follows:

a. **Greenhouse Gas Reduction:** One of the best things about EVs is that they help reduce greenhouse gas emissions, including a drastic reduction in carbon dioxide (CO_2), which accounts for around 95% of fleet CO_2 levels. As a result, these types have lower emissions than conventional internal combustion engine vehicles since they produce zero tailpipe emissions. This reduces the rate of climate change and makes urban areas with excessive vehicle concentration less polluted.
b. **Change in Emissions:** EVs do not produce tailpipe emissions; however, the electricity used to charge EVs does produce. If an appreciable percentage of electricity generation remains dependent on oil and natural gas or, worse still coal, environmental benefits could thus be compromised. Still, as the grid moves toward cleaner power sources like renewables, this will help lower emissions coming from EV charging over and top of that.

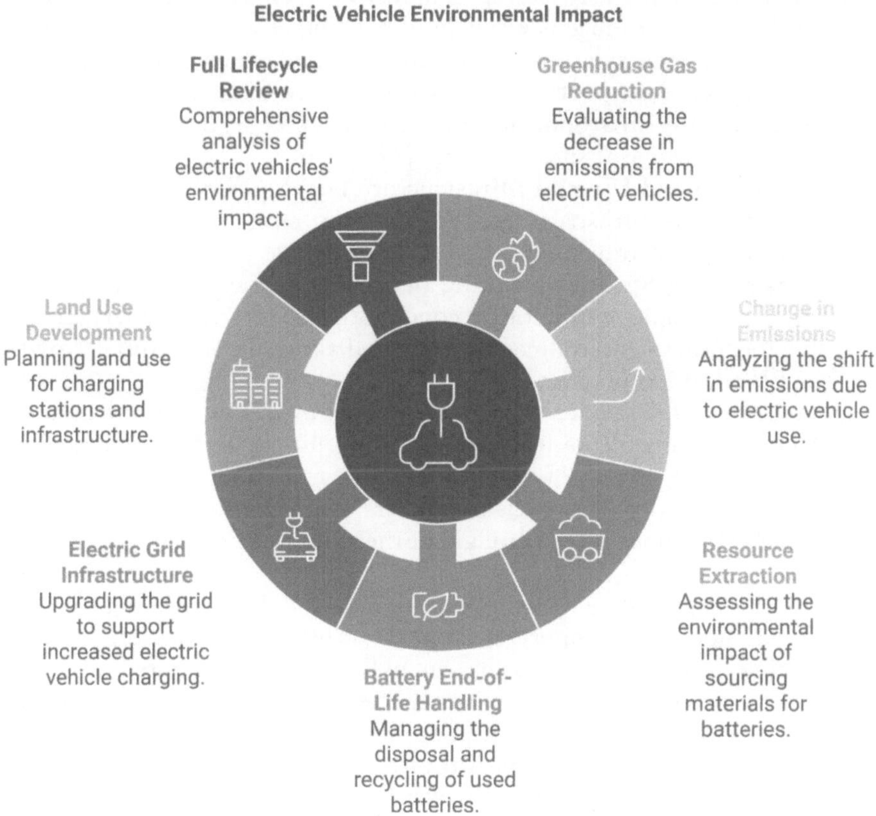

Figure 7.4 Environmental impact by electric vehicle integration.

c. **Resource Extraction:** Mining processes are employed to extract raw materials like lithium, cobalt, and nickel used in the production of EV batteries. Mining activities have adverse environmental impacts including habitat disruption, soil and water pollution, as well as loss of biodiversity. Battery production requires sustainable sourcing practices and recycling initiatives to diminish the overall environmental footprint.
d. **Battery End-of-Life Handling:** Recycling and disposing of EV batteries, especially recycling, creates environmental problems. Improperly disposed batteries can leach toxic chemicals into the environment and contribute to e-waste pollution. In order to address the environmental concerns when EV batteries are eventually disposed of, building an eco-recycling system and various strategies for a secondary service or reuse continue with more intensity [21].

e. **Electric Grid Infrastructure:** As numerous EVs are integrated into the grid, it can introduce new strain to existing electricity infrastructure and increase energy consumption and demand peaks. Smart grid technologies can optimize energy usage, balance loads on the grid, and minimize the environmental impacts of EV charging by controlling when EVs arrive at sites to charge [22].
f. **Need for Land Use and Infrastructure Development:** The growth of EV charging infrastructure, ranging from new public charging stations to fast chargers, must add to some degree of land occupation or construction. Strategic planning and proper application of sustainable design principles can mitigate concerns over land use impacts, ensuring the natural environment is preserved alongside substantial growth in EV adoption practices.
g. **Full Lifecycle Review:** The climate impact of EVs must also be understood in terms of the full lifecycle from mining and manufacturing, to use phase, to end-of-life disposal. By better understanding the life cycle impacts of EVs, decision-makers can learn more about how to make them most green throughout their lifecycle.

Although EVs streamline many aspects of everyday life, including a reduction in emissions and support to a better transportation mode, environmental concerns such as focusing on pollution extent related with EV deployment (production), battery management, grid integration, and infrastructure development have not been dealt with comprehensively. The environmental benefits of EVs in smart transportation systems can be realized by adopting holistic and eco-friendly ways to deploy EVs inside their charging infrastructure.

7.4 INTELLIGENT TRAFFIC MANAGEMENT SYSTEMS

ITMS systems provide a smarter way of managing transportation networks and combine information, communications, control, and electronics technology for optimized traffic flow to reduce congestion and accidents [23]. By integrating traffic signal control systems, surveillance cameras and data analytics platforms in a seamless manner with each other, ITMS gathers real-time information on automatic analysis of behavior pattern recognition, such as the collection is responsive to actual affixed conditions. Using state-of-the-art technologies like Internet of Things (IoT), Artificial Intelligence (AI) and Machine Learning, ITMS provides Predictive Traffic Modeling, Adaptive Signal Control Incident Detection, and Personalized Route guidance. ITMS optimizes the flow of traffic, allowing for fewer bottlenecks and safer roads by utilizing technology to regulate speed (including automated enforcement) and improving incident management, which allows maintenance crews to uncover issues before they cause serious problems [24]. Relying more on smart systems is another way that helps urban areas be greener. With ITMS comes lower emissions from cars idling in congestion or spinning uncontrollably out of control at freeway speeds, paving a path toward better air quality and contributing environmental benefits, such as encouraging green driving, while catering to environment-friendly roles like reducing pollution levels drastically over time.

Further optimized parking availability in the precinct are smart parking solutions that improve traffic congestion, reduce its impact on urban roads, and therefore rationalize better use of their respective parking infrastructure as well; combined approaches like ITMS are used to streamline intelligent traffic management included with data-driven aspects of ITCS across developed megacities thus far [25].

7.4.1 Components of intelligent traffic management system

ITMS are composed of several connected devices that integrate to streamline traffic movement, enhance safety, and make the entire transportation network more efficient. Components of an ITMS are as follows [26] and are represented in Figure 7.5:

a. **Traffic Signal Control Systems:** These systems use advanced algorithms along with real time to change their timers in order to control and manage traffic based on the prevailing conditions. These signal systems are known as adaptive because the way they control traffic signals is designed to improve center-turned intersection performance by dynamically adjusting the phasing and timing.
b. **CCTV Cameras:** They are used for monitoring and managing traffic operations such as real-time surveillance, detection of incidents, or enforcement for violating traffic violators. Video surveillance systems like CCTV cameras provide real-time images that can be processed for better traffic control and road safety.
c. **Vehicle Detection Sensors:** These are designed to detect the presence of vehicles so that they can be counted, and traffic flow may furthermore happen. Most common types of sensors used to extract vehicle counts and occupancy are inductive loops, infrared, radar, and video detection systems.
d. **Variable Message Sign (VMS):** VMS shows the status of traffic, lane closures, and road advisories. These signs make it easier to communicate with drivers and help safely steer them through obstacles that disrupt traffic.

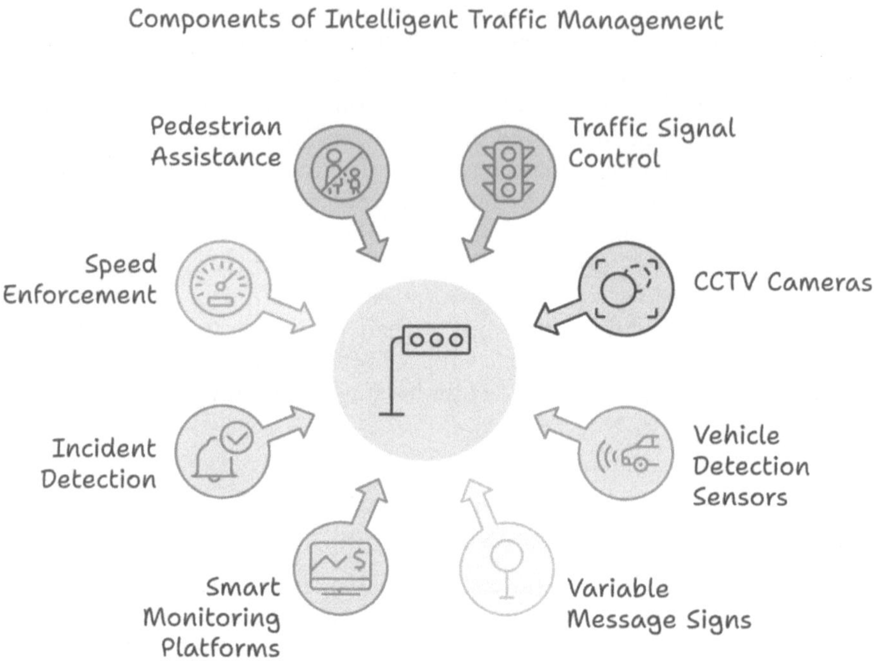

Figure 7.5 Components of an intelligent traffic management system.

e. **Consolidated Smart Traffic Monitoring & Data Analytics Platforms:** These are platforms that bring all the traffic data from heterogeneous sources such as sensors, cameras, and connected vehicles. Decode them for meaningful information present through different visualizations. Traffic engineers use data analytics tools to track key traffic patterns, trends, and KPIs that help them make valid decisions.
f. **Incident Detection Management System:** Used in detection of incidents, in real-time quite similar to the traffic incident we were talking about (accident, road close down, and breakage). Such systems enable quick response and coordination with first responders to reduce traffic delays, which is necessary for keeping road users safe.
g. **Automated Speed Enforcement Systems:** These systems use technology like radar and Lidar to track vehicle speed from cameras and measure enforcement of speeding regulations. Automated speed enforcement helps increase adherence to the speed limit, improve road safety, and reduce collisions.
h. **Pedestrian Crossing Assistance Systems:** This equipment helps pedestrians by providing visual and audio help in crosswalks or pedestrian signals, as well as some peculiarities of this system. Pedestrian cross assistance systems increase the visibility and communication between

pedestrians and drivers and lead to rare less non-related pedestrian accidents.

i. **Smart Parking Solutions:** Providing an interface of sensors, mobile applications, and data analytics within the ITMS to find available parking spaces for drivers in their area so as not only to facilitate better usable guidance but also reducing congestion due to inefficient search.
j. **Emergency Vehicle Preemption Systems:** These systems allow emergency vehicles to be prioritized at traffic signals by preempting the standard signal operations. Emergency Vehicle Preemption Systems are used to leverage the road intersections for providing an emergency vehicle with a high precedence level on manual control, thus allowing faster response time that minimizes risks created by critical situations.

Incorporation of these elements in ITMS helps cities build the best combination for optimized traffic operations, leading to a safer environment along with reduced congestion and overall greater efficiency toward a smarter future-proof urban landscape.

7.4.2 Intelligent traffic management system integration with smart transportation

Experience says that smart transportation has an incomplete definition without integrating ITS into a unified system by connecting ITMS to provide easy integration for achieving safer and sustainable urban mobility [27]. When incorporated with the other components of smart transport infrastructure, ITMS provides a way to ensure cities have good connectivity and throughput. The below specified ways are how ITMS can be integrated with smart transit through which traffic has been managed across various modes such as roads, buses, subways, rail, ferries, and tollbooths within many service owners on operators their timetable organizations [28] as shown in Figure 7.6.

Being able to merge ITMS with other smart transportation efforts is crucial for the meaningful development of our future urban mobility systems so that we finally reach this idea of seamless running of green cities

7.5 AUTONOMOUS VEHICLES AND THEIR IMPACT

Self-driving cars, also known as autonomous vehicles, represent a revolutionary type of technology in the field of transportation [29]. These groundbreaking devices represent incredible progress when it comes to robotics and artificial intelligence that is combined with electronic sensing systems and cutting-edge automotive engineering [30]. The outcome would be game-changing commuter safety and economic accessibility to

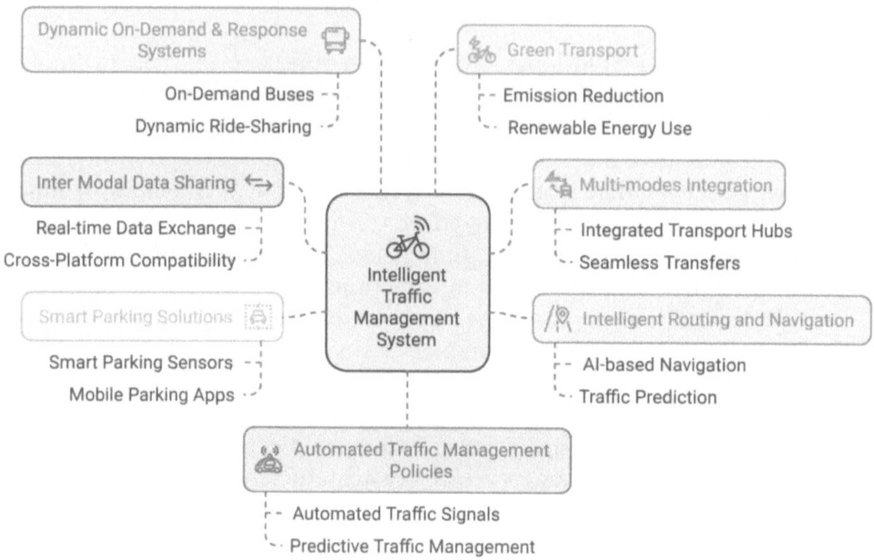

Figure 7.6 Intelligent traffic management system integration with smart transportation.

any destination improved even more with our country's increasing macro-congestion and micro-transportation issues. The red-hot space of killer robotics has gathered continued and focused attention from technology companies, automakers, regulators, and the public through their deployment issues that are indeed multifaceted. The rise of autonomous vehicles has sparked a fascinating journey into the future, as complex artificial intelligence algorithms to the ethical and regulatory framework that underpin their operation are being developed to define what mobility means. This introduction will delve into the technological basis, societal consequences, regulatory hindrances, and future perspectives of autonomous vehicles by examining the numerous facets that make up this revolutionary invention, which has the potential to revolutionize our current transportation landscape.

7.5.1 Autonomous vehicles impact on traffic congestion

While autonomous vehicles could help or exacerbate traffic congestion, they depend largely on adoption rates as well as the deployment strategies implemented and infrastructure development along with regulatory frameworks [31]. Read on to find out the main ways autonomous vehicles can affect traffic congestion as follows and as represented in Figure 7.7:

Such consequences could arise as the result of various technological automation, societal change (such as increased driving rates), policy-related

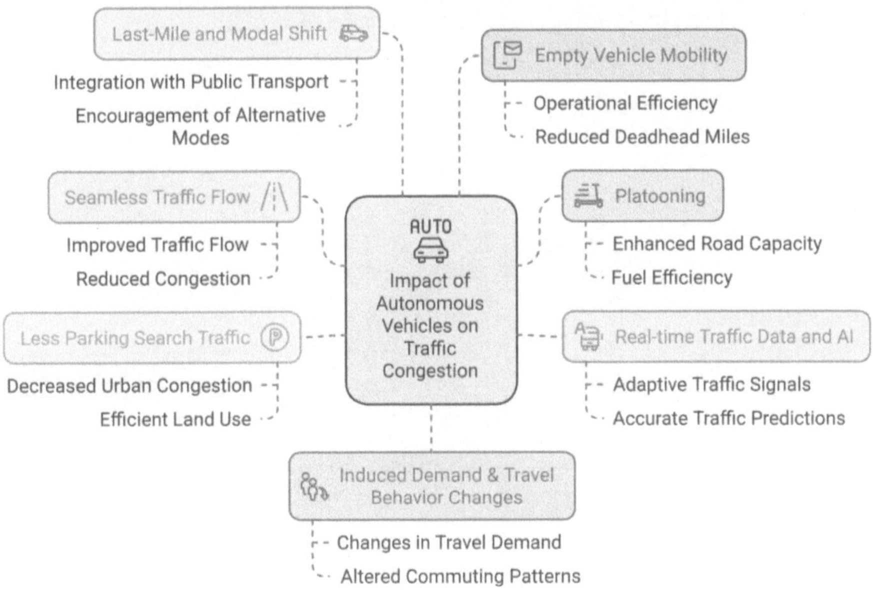

Figure 7.7 Autonomous vehicle impact on traffic congestion.

changes, or urban design strategies. Despite the ease with which autonomous vehicles may alleviate congestion through improved traffic flow, dynamic routing and better understanding of system connectedness consideration of these positive externalities is important to maximize the overall effects on congestion reduction as well as a broader vision for an efficient, sustainable and resilient transportation future. Cities must work within a holistic framework of smart transportation solutions that integrates these technologies and addresses the challenges they will face, namely, how to use autonomous vehicles as part of an integrated system to solve congestion, rather than simply adding them on top.

7.6 PUBLIC TRANSIT APPS AND RIDE-SHARING SERVICES

The digital transformation of urban mobility is supported by the assortment of public transit apps and ride-sharing services available to passengers, giving them access to a diverse range and more flexible spectrum depending on their personal preferences [32]. A summary of their strengths and advantages is given as follows:

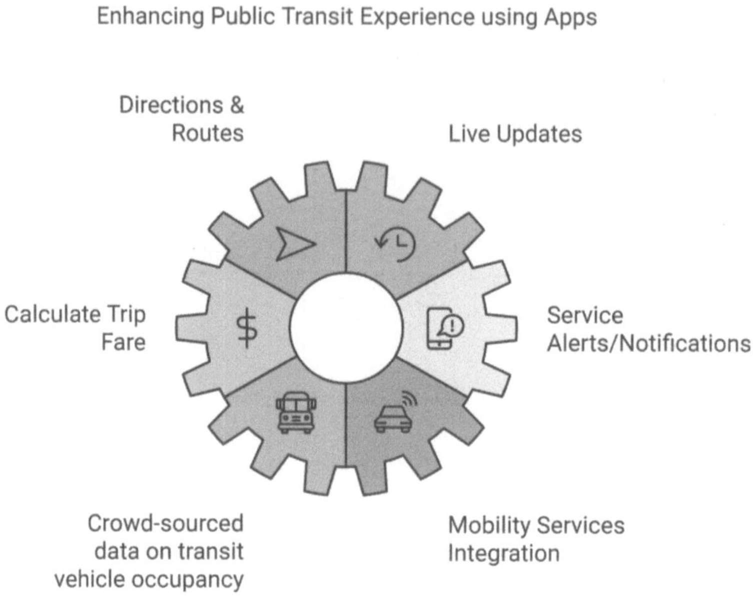

Figure 7.8 Public transit app for better public experience.

Public Transit Apps:

Transit apps are the backbone of tools and services that public transit riders need to be able to most effectively navigate a city's transportation system [32], as shown in Figure 7.8 as key features.

Ride-Sharing Services:

Using dedicated apps, ride-sharing services offer on-demand transportation that link drivers providing private vehicles to passengers [32], as represented in Figure 7.9 with some components.

Integration and Synergy:

When public transit apps and ride-sharing services are combined, it can result in a complete mobility solution for its passengers, as represented in Figure 7.10 in the form of potential integration components.

Public transit apps address the needs of every urban mobility use-case, a well-oiled ride-sharing service fills in the gaps left by their limitations, while both must build short feedback loops to fix and enhance themselves. Public transportation & community-based solutions make app integration easier for the public transit industry as municipalities look to increase efficiency and access, all while minimizing emissions. Ride-pooling is unlikely to change the economy of scale in transit design [33].

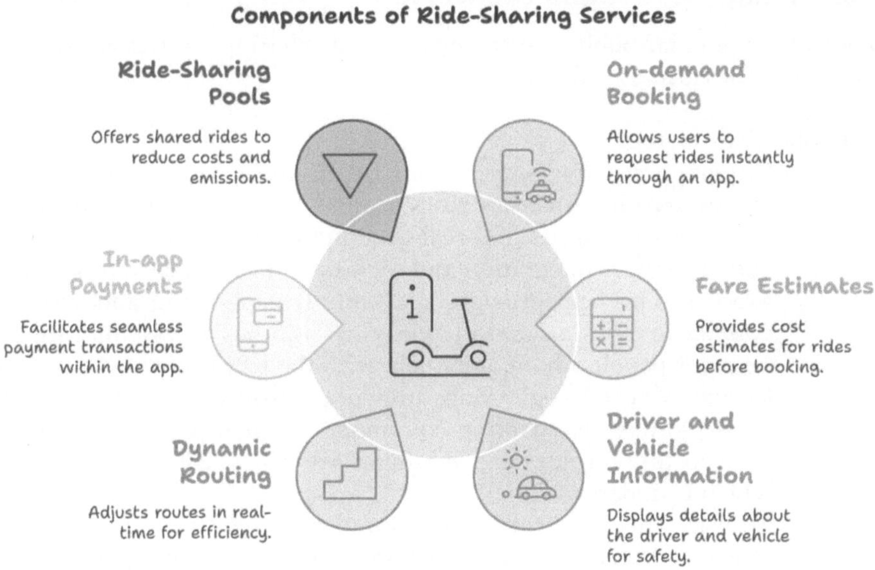

Figure 7.9 Ride-sharing services for better public experience.

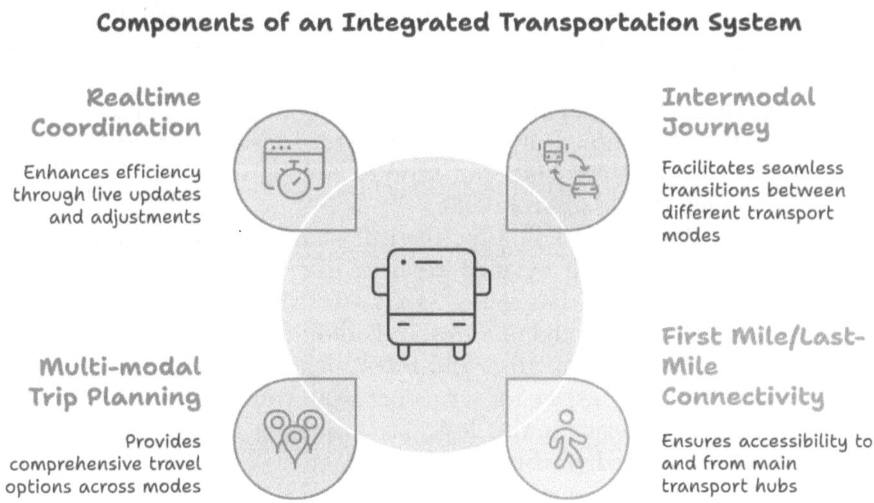

Figure 7.10 Integration and synergy for a better public experience.

7.6.1 Examples around the world

Some of the popular public transit apps and ride-sharing services across the world are described as follows:

Public Transit Apps:

a. **Citymapper:** Citymapper is one of the most widely used public transit apps in numerous cities around, featuring advanced route planning capabilities and real-time information for your journey and bike-share integration and multimodal trip suggestions [34].
b. **Moovit:** This app provides real-time trip planning and service updates in cities around the globe. It even has accessibility functions for people who require help with the UI [35].
c. **Google Maps:** Google Maps provides transit directions in many cities worldwide, enabling commuters to plan trips with public transportation routes and schedules as well as real-time bus/train schedule updates.
d. **Transit:** This is an easy-to-use app displaying public transit real-time information, route planner and multimodal trips options, and bike-sharing integration in dozens of cities around North America, especially New York city/Europe/Australia [36].

Ride-Sharing Services:

a. **Uber:** Uber is one of the most known ride-sharing options worldwide; it has been rolled out to the majority of cities and countries across the globe. The app features the ability to book rides, share trip information, and a variety of transportation options that include UberX, UberPOOL, and Uber Black [37].
b. **Lyft:** Lyft is a popular ride-sharing app in the United States and some international cities, which provides on-demand rides, shared rides, as well as bike and scooter rentals in the United States through their App store [38].
c. **Didi Chuxing:** Dominating the ride-sharing scene in China, Didi Chuxing offers a bevy of services like 56-seat buses alongside shared "pop-up" bus routes and is accessible using some combinations across their moniker roster: Didi Express, Didi Premier (ride-sharing with Audi A6s), and rides such as taxis under messages reading all to inspire for what sort a pill can keep customers loyal when car prices are this low. The company serves provides popular city-based transportation alternatives in many cities across China and continues to grow globally [39].
d. **Grab:** Popular in Southeast Asia, Grab is a more customary addition to a list that includes Uber and Lyft. Grab serves as a ride-hailing, food delivery and transportation provider app [40].

e. **OLA:** Ola is one of the leading ride-sharing services in India, which provides various modes of transport for point-to-point travel and also allows outstation rides through its platform. The app also includes auto-rickshaws and bike-sharing services.
f. **BlaBlacar (International):** BlaBlacar is an online carpooling service in many countries for people who travel long distances, sharing rides together and seats on the price of gasoline [41].

The above samples are just some of the myriad public transit apps and ride-sharing services available in many parts of the globe to give users easy ways for moving from one place or another using efficient, fast, affordable transport solutions.

7.7 ENHANCED MOBILITY FOR THE ELDERLY AND DISABLED

In the transportation space, connecting smart technology and autonomous systems has the power to contribute significantly toward improved mobility for senior citizens or people with disabilities by delivering convenient (user-friendly), reliable, and inclusive individualized public transit [42]. These technologies can help in improving the mobility of these people in the following ways:

a. **Accessibility and Inclusive Design:** Accessible design elements already exist in public transportation, such as low-floor boarding, wheelchair ramps, and visual/audio announcements, that can be included into smart transpiration systems to make both transit vehicles and AVs more hospitable for those with mobility needs.
b. **Personalized and on Demand Service:** Integration with autonomous vehicles and smart transportation platforms can provide scheduled or on-demand transit services that are tailored to the unique requirements of elderly and disabled passengers, delivering convenient mobility options for all.
c. **Real Time Information and Assistance:** Integration with smart city technology includes real-time information and assistance, trip planning, and service alerts; user-friendly interfaces help individuals facing disabilities to navigate the transportation networks.
d. **Safe and Comfortable Travel:** Autonomous vehicles could enable a safe and comfortable transportation experience, with valuable facilities like door automation, flexible seating transitions among available layouts, and appropriately designed ride dynamics to cater for passenger's needs.

e. **Autonomous Shuttle Services:** Involves implementation of autonomous shuttles in cities and neighborhoods to improve transit options for seniors and disabled last-mile connections, reducing dependence on traditional fixed-route services.
f. **User Interfaces and Assistive Technologies/Accessibility:** These can also include assistive technologies and user interfaces like voice commands and braille displays, used in trains imitating the smart features of buses today where commuters with disabilities or elderly citizens are enabled to use almost any bus line feature known.
g. **Supporting Community Integration and Social Inclusion:** In the case of elderly and disabled persons, enhanced mobility options can help achieve better community integration with social inclusion as they require good transportation infrastructure for key services like home-based healthcare visits and recreational activities, including more accessibility to public parks etc.

Smart transportation and autonomous systems that address these design concerns can help people move more out of their homes, make them able to much better contribute to public life, and enjoy a greater overall sense of independence in daily routines.

7.8 FUTURE PROSPECTS AND TECHNOLOGICAL ADVANCEMENTS IN SMART TRANSPORTATION

The smart transportation of the future promises to deliver a host of transformational technological breakthroughs that will redefine how people and goods alike are moved around cities and between city centers. The future of smart transportation, some transformative initiatives, and technological advancement in smart transportation are represented in Figure 7.11 and described as follows:

a. **5G and V2X Communication:** Extensive implementation of 5G networks will allow a communication platform which would ensure optimal networking between the vehicles, infrastructure, and other road users to help in sharing real time data efficiently [43], thus improving safety levels and making transportation smarter.
b. **Advanced Driver Assistance Systems (ADAS):** Increasing adoption of ADAS, consisting of adaptive cruise control and automatic parking systems, among others improving road safety, reducing congestion, and enhancing driver comfort to drive higher levels of vehicle automation [44].
c. **Electric and Autonomous Vehicles:** Wider adoption of electric vehicles (EVs) in addition to the refinement of autonomous driving technologies will help limit emissions, reduce energy consumption in transportation, and shape new mobility ecosystems toward sustainability.

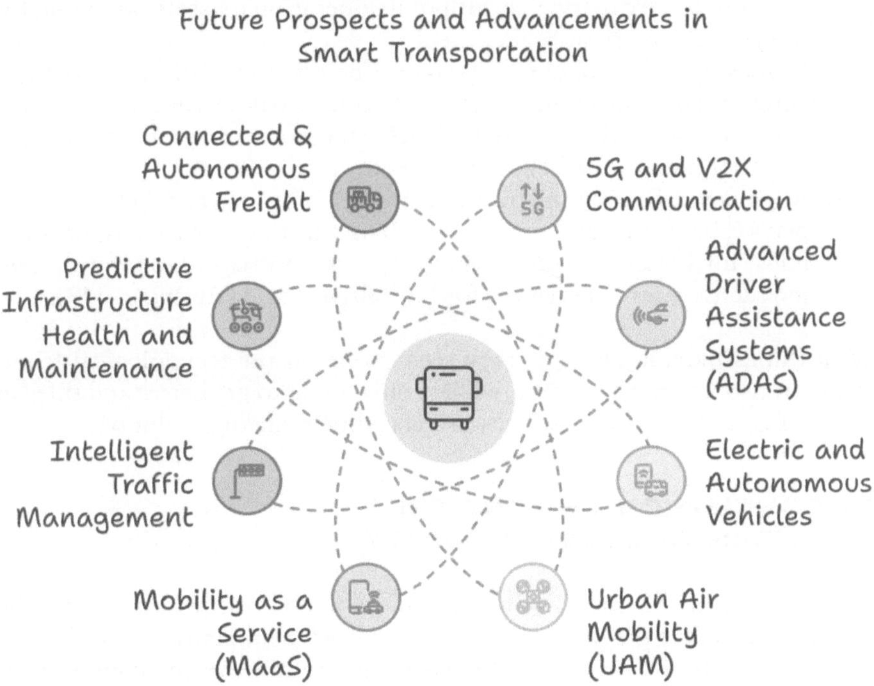

Figure 7.11 Future prospects and technological advancements in smart transportation.

d. **Urban Air Mobility (UAM):** An array of new commercial offerings which range from eVTOL (electric vertical takeoff and landing) aircraft to automated air drones come alive in the trailblazing realm created by UAM. This presents a novel approach for lighter than air travel, enabling rapid transportation to spread over extensive cityscapes, as well as cargo transportation that could be catered using renewable energy systems with unmatched efficiency [45].
e. **Mobility as a Service (MaaS):** The future of MaaS platforms and mobility ecosystems, but most crucially integrated mobility services to power convenient multimodal transportation experiences with an intuitively designed end-to-end solution that connects public transit, ridesharing, micro-mobility and autonomy in context [46].
f. **Intelligent Traffic Management:** These are intelligent and supposedly easy-to-use traffic management systems, run by ML/AI algorithms which smooth the flow of vehicular movements and predict and mitigate jams or congestion (if any) in advance, thereby efficiently increasing the productivity of transport networks.
g. **Predictive Infrastructure Health and Maintenance:** By deploying sensors, IoT devices as well as edge computing capability to public transportation infrastructure customers will be able to constantly

monitor the "real-time" health of its operational assets, allowing for predictive repairs or replacement.

h. **Connected & Autonomous Freight:** The emergence of connected and autonomous vehicles in commercial vehicles will disrupt freight transportation providing smart logistics, superior supply chain management solutions, as well as last-mile delivery.
i. **Augmented Reality in Transportation:** Using AR and heads-up displays (HUDs) inside vehicles, as well as within various transportation hubs, for better navigation support; access to real-time information; interactive user experiences for both drivers and passengers [47].

With things moving along as they seem to be on the technological front, smart transportation systems are likely only going to get better and safer in terms of getting more cars off city streets while reducing pollution.

7.9 CONCLUSION: SHAPING THE FUTURE OF GLOBAL MOBILITY

In summary, intelligent mobility and autonomous traffic systems will be key in how people get around tomorrow. Leveraging innovative technologies, data analysis to enhance efficiency, and an unwavering commitment to sustainability/environmental consciousness as well as equity in access these breakthroughs are poised to address pressing challenges for urban mobility within cities whilst also revolutionizing intercity travel. The intersection of smart transportation and autonomous technology opens up many new opportunities, such as better traffic control, lower emissions levels, less dependence on elderly/disabled drivers, and the potential for integrated multimodal systems that make commuting easier. Moreover, the move toward electrification and automation of vehicles, as well as the emergence of urban air mobility services and maturation of MaaS platforms, is expected to reshape transportation ecosystems globally toward greener more integrated people-centric sustainable mobility environments. Looking forward, this may have a particular focus among public and private stakeholders on the necessary collaborations through 5G to extend inclusive V2X communication interoperability standards that are globally relevant while maintaining safety, cybersecurity, and environmental stewardship in our transportation future. We need to make sure that smart transportation benefits are available for everyone by focusing on equity and social inclusion, especially among marginalized communities. This generational transition presents us with an opportunity to transform the next century of global mobility through innovative advancements coupled with industry-friendly regulation and policy grounded in evidence-based research that preserves citizen sovereignty and promises safer, smarter sustainable transportation networks for a new era. The community can embrace this future together as we harness cutting edge technology to empower people-build economic resilience-protect social well-being globally, While the work is far from done, it showcases that China is taking vital steps forward in smart transportation technology and unbeatable system integration one step at a time to move everything toward intelligent connection of mobility sensibly stitched together seamlessly with needs for quality life and ecosystem preservation.

REFERENCES

1. Shuhua Xu, Chuanwang Sun, Nian Liu, Road congestion and air pollution – analysis of spatial and temporal congestion effects, *Science of The Total Environment*, vol. 945, 2024, p. 173896, https://doi.org/10.1016/j.scitotenv.2024.173896.
2. Mohamed Elassy, Mohammed Al-Hattab, Maen Takruri, Sufian Badawi, Intelligent transportation systems for sustainable smart cities, *Transportation Engineering*, vol. 16, 2024, p. 100252, https://doi.org/10.1016/j.treng.2024.100252.
3. B. C. Ataman, U. E. Igbeka, A. I. Emenime, C. C. Kwasi-Effah, F. Max-Eguakun, Role of Hybrid electric vehicles for sustainable transportation, *NIPES: Journal of Science and Technology Research*, vol. 3, no. 4, 2021, pp. 376–382. doi: https://journals.nipes.org/index.php/njstr/article/view/666.
4. James C. Spall, Daniel C. Chin, Traffic-responsive signal timing for system-wide traffic control, *Transportation Research Part C: Emerging Technologies*, vol. 5, no. 3–4, 1997, pp. 153–163, https://doi.org/10.1016/S0968-090X(97)00012-0.
5. A.H.M. Mehbub. Anwar, Abu Toasin Oakil, Smart transportation systems in smart cities: practices, challenges, and opportunities for Saudi cities, in Belaïd, F., Arora, A. (eds.), *Smart Cities. Studies in Energy, Resource and Environmental Economics*, Springer, Cham, 2024, https://doi.org/10.1007/978-3-031-35664-3_17.
6. Jianqiang Fan, Xiaosha Meng, Jiaxin Tian, Conghui Xing, Chao Wang, Jacob Wood, A review of transportation carbon emissions research using bibliometric analyses, *Journal of Traffic and Transportation Engineering (English Edition)*, vol. 10, no. 5, 2023, pp. 878–899, https://doi.org/10.1016/j.jtte.2023.09.002.
7. Memoona Sadaf, Zafar Iqbal, Abdul Rehman Javed, Irum Saba, Moez Krichen, Sajid Majeed, Arooj Raza, Connected and automated vehicles: infrastructure, applications, security, critical challenges, and future aspects. *Technologies*, vol. 11, 2023, p. 117. https://doi.org/10.3390/technologies11050117.
8. Deema Almaskati, Sharareh Kermanshachi, Apurva Pamidimukkula, Autonomous vehicles and traffic accidents, *Transportation Research Procedia*, vol. 73, 2023, pp. 321–328, https://doi.org/10.1016/j.trpro.2023.11.924.

9. Divya Garikapati, Sneha Sudhir Shetiya, Autonomous vehicles: evolution of artificial intelligence and the current industry landscape. *Big Data and Cognitive Computing*, 8, 2024, p. 42. https://doi.org/10.3390/bdcc8040042.
10. Mohamed Mezghani, Jinhua Zhao, How autonomous vehicles can be integrated with public transport systems for urban mobility, *Urban Transformation*, October 16, 2024. https://www.weforum.org/agenda/2024/10/how-will-autonomous-vehicles-shape-urban-mobility/.
11. Mohammed Qader Kheder, Aree Ali Mohammed, Real-time traffic monitoring system using IoT-aided robotics and deep learning techniques, *Kuwait Journal of Science*, vol. 51, no. 1, 2024, p. 100153, https://doi.org/10.1016/j.kjs.2023.10.017.
12. Younggon Jo, Jeongmok Ha, Sungsoo Hwang, Survey of technology in autonomous valet parking system, *The International Journal of Automotive Technology*, vol. 24, 2023, pp. 1577–1587, https://doi.org/10.1007/s12239-023-0127-1.
13. Breno A. Beirigo, Rudy R. Negenborn, Javier Alonso-Mora, Frederik Schulte, A business class for autonomous mobility-on-demand: Modeling service quality contracts in dynamic ridesharing systems, *Transportation Research Part C: Emerging Technologies*, vol. 136, 2022, p. 103520, https://doi.org/10.1016/j.trc.2021.103520.
14. Haizhu Xu, Duanyang Geng, Zhixian Fan, Dexi Wu, Meizhou Chen, An automatic vehicle navigation system based on filters integrating inertial navigation and global positioning systems, *Machines*, vol. 12, 2024, p. 663, https://doi.org/10.3390/machines12090663.
15. Yury Shevchenko, Ulf-Dietrich Reips, Geofencing in location-based behavioral research: methodology, challenges, and implementation, *Behavior Research*, vol. 56, 2024, pp. 6411–6439, https://doi.org/10.3758/s13428-023-02213-2.
16. Rodgers, Arthur R., Robert S. Rempel, and Kenneth F. Abraham, A GPS-based telemetry system, *Wildlife Society Bulletin (1973–2006)*, vol. 24, no, 3, 1996, pp. 559–566, https://www.jstor.org/stable/3783345.
17. Jay Squalli, Greening the roads: assessing the role of electric and hybrid vehicles in curbing CO2 emissions, *Journal of Cleaner Production*, vol. 434, 2024, p. 139908, https://doi.org/10.1016/j.jclepro.2023.139908.
18. Petr Pavlínek, Transition of the automotive industry towards electric vehicle production in the east European integrated periphery, *Empirica*, vol. 50, 2023, pp. 35–73, https://doi.org/10.1007/s10663-022-09554-9.
19. Richard P. Langford, Kenneth Gillingham, Quantifying the benefits of the introduction of the hybrid electric vehicle, *International Journal of Industrial Organization*, vol. 87, 2023, p. 102904, https://doi.org/10.1016/j.ijindorg.2022.102904.
20. Anahita Jannesar Niri, Gregory A. Poelzer, Steven E. Zhang, Jan Rosenkranz, Maria Pettersson, Yousef Ghorbani, Sustainability challenges throughout the electric vehicle battery value chain, *Renewable and Sustainable Energy Reviews*, vol. 191, 2024, p. 114176, https://doi.org/10.1016/j.rser.2023.114176.
21. Mengyuan Chen, Xiaotu Ma, Bin Chen, Renata Arsenault, Peter Karlson, Nakia Simon, Yan Wang, Recycling end-of-life electric vehicle lithium-ion batteries, *Joule*, vol. 3, no. 11, 2019, pp. 2622–2646, https://doi.org/10.1016/j.joule.2019.09.014.
22. Arvind R. Singh, Pradeep Vishnuram, Sureshkumar Alagarsamy, Mohit Bajaj, Vojtech Blazek, Issam Damaj, Rajkumar Singh Rathore, Fahd N. Al-Wesabi, Kamal M. Othman, Electric vehicle charging technologies, infrastructure

expansion, grid integration strategies, and their role in promoting sustainable e-mobility, *Alexandria Engineering Journal*, vol. 105, 2024, pp. 300–330, https://doi.org/10.1016/j.aej.2024.06.093.

23. Aleksey Dorofeev, Valery Kurganov, Nadejda Filippova, Artur Petrov, Dmitrii Zakharov, Sergei Iarkov, Improving transportation management systems (TMSs) based on the concept of digital twins of an organization, *Applied Science*, vol. 14, 2024, p. 1330, https://doi.org/10.3390/app14041330.
24. Nikhil Nigam, Dhirendra Pratap Singh, Jaytrilok Choudhary, A review of different components of the intelligent traffic management system (ITMS), *Symmetry*, vol. 15, 2023, p. 583, https://doi.org/10.3390/sym15030583.
25. Ibai Laña, Javier J. Sanchez-Medina, Eleni I. Vlahogianni, Javier Del Ser, From data to actions in intelligent transportation systems: a prescription of functional requirements for model actionability, *Sensors*, vol. 21, 2021, p. 1121, https://doi.org/10.3390/s21041121.
26. Bhakti Dighe, Aakash Nikam, Kishore Markad, Intelligent traffic management systems: a comprehensive review, *International Journal of Creative Research Thoughts (IJCRT)*, vol. 12, no. 4, 2024.
27. Fayez Alanazi, Mamdouh Alenezi, Interoperability for intelligent traffic management systems in smart cities, *International Journal of Electrical and Computer Engineering (IJECE)*, vol. 14, no. 2, 2024, pp. 1864–1874, https://doi.org/10.11591/ijece.v14i2.
28. Maria Giuffrida, Sara Perotti, Angela Tumino, Vincenzo Villois, Developing a prototype platform to manage intelligent communication systems in intermodal transport, *Transportation Research Procedia*, vol. 55, 2021, pp. 1320–1327, https://doi.org/10.1016/j.trpro.2021.07.116.
29. Adamski Andrzej, Intelligent integrated transportation systems, in *Conference: Proceedings of 9-th Meeting of the EURO WG on Transportation Intermodality, Sustainability and ITS, At: Polytechnic of Bari, Faculty of Engineering Bari*, vol.1, pp. 565–570, Bonn, Germany, December 2001.
30. Isaac Oyeyemi Olayode, Bo Du, Alessandro Severino, Tiziana Campisi, Frimpong Justice Alex, Systematic literature review on the applications, impacts, and public perceptions of autonomous vehicles in road transportation system, *Journal of Traffic and Transportation Engineering (English Edition)*, vol. 10, no. 6, 2023, pp. 1037–1060, https://doi.org/10.1016/j.jtte.2023.07.006.
31. Yuchen Pan, Yu Wu, Lu Xu, Chengyi Xia, David L. Olson, The impacts of connected autonomous vehicles on mixed traffic flow: a comprehensive review, *Physica A: Statistical Mechanics and its Applications*, vol. 635, 2024, p. 129454, https://doi.org/10.1016/j.physa.2023.129454.
32. Mitja Stiglic, Niels Agatz, Martin Savelsbergh, Mirko Gradisar, Enhancing urban mobility: integrating ride-sharing and public transit, *Computers & Operations Research*, vol. 90, 2018, pp. 12–21, https://doi.org/10.1016/j.cor.2017.08.016.
33. Sida Luo, Yu (Marco) Nie, Impact of ride-pooling on the nature of transit network design, *Transportation Research Part B: Methodological*, vol. 129, 2019, pp. 175–192, https://doi.org/10.1016/j.trb.2019.09.007.
34. Güneş Tavmen, Data/infrastructure in the smart city: understanding the infrastructural power of Citymapper app through technicity of data, *Big Data and Society*, vol. 7, no. 2, 2020, p. 205395172096561, https://doi.org/10.1177/2053951720965618.

35. Georgina Santos, Nikolay Nikolaev, Mobility as a service and public transport: a rapid literature review and the case of moovit, *Sustainability*, vol. 13, 2021, p. 3666, https://doi.org/10.3390/su13073666.
36. Niloofar Ghahramani, Candace Brakewood, Trends in mobile transit information utilization: an exploratory analysis of transit app in New York City, *Journal of Public Transportation*, vol. 19, no. 3, 2016, pp. 139–160, https://doi.org/10.5038/2375-0901.19.3.9.
37. Yuyan Sun, Analysis of uber carpooling service, in *Proceedings of the 2022 2nd International Conference on Modern Educational Technology and Social Sciences (ICMETSS 2022)*, Atlantis Press, United Kingdom, 2022, pp. 208–217, https://doi.org/10.2991/978-2-494069-45-9_26.
38. Levi Davis, Lyft vs. Uber: What's the Difference? September 05, 2024, https://www.investopedia.com/articles/personal-finance/010715/key-differences-between-uber-and-lyft.asp.
39. Peng Jing, Yuanyuan Chen, Xingyue Wang, Kewen Pan, Daibiao Yuan, Evaluating the effectiveness of Didi ride-hailing security measures: an integration model, *Transportation Research Part F: Traffic Psychology and Behaviour*, vol. 76, 2021, pp. 139–166, https://doi.org/10.1016/j.trf.2020.11.004.
40. Nurul Farah Izzah Zailani, Ahmad Albattat, Factors influencing tourist perception on grabcar ride-sharing application services, in Hassan Azizul (ed.), *Handbook of Technology Application in Tourism in Asia*, chapter 8, pp. 165–177, Springer, Singapore, 2022.
41. Susan Shaheen, Adam Stocker, Marie Mundler, Online and app-based carpooling in France: analyzing users and practices—a study of BlaBlaCar, In: Meyer, G., Shaheen, S. (eds.), *Disrupting Mobility. Lecture Notes in Mobility*, Springer, Cham, 2017, https://doi.org/10.1007/978-3-319-51602-8_12.
42. Fahimeh Golbabaei, James Dwyer, Rafael Gomez, Andrew Peterson, Kevin Cocks, Alexander Bubke, Alexander Paz, Enabling mobility and inclusion: designing accessible autonomous vehicles for people with disabilities, *Cities*, vol. 154, 2024, p. 105333, https://doi.org/10.1016/j.cities.2024.105333.
43. Tibor Petrov, Peter Pocta, Tatiana Kovacikova, Benchmarking 4G and 5G-based cellular-V2X for vehicle-to-infrastructure communication and urban scenarios in cooperative intelligent transportation systems, *Applied Science*, vol. 12, 2022, p. 9677, https://doi.org/10.3390/app12199677.
44. Felipe Jiménez, José Eugenio Naranjo, José Javier Anaya, Fernando García, Aurelio Ponz, José María Armingol, Advanced driver assistance system for road environments to improve safety and efficiency, *Transportation Research Procedia*, vol. 14, 2016, pp. 2245–2254, https://doi.org/10.1016/j.trpro.2016.05.240.
45. Laurie A. Garrow, Brian J. German, Caroline E. Leonard, Urban air mobility: a comprehensive review and comparative analysis with autonomous and electric ground transportation for informing future research, *Transportation Research Part C: Emerging Technologies*, vol. 132, 2021, p. 103377, https://doi.org/10.1016/j.trc.2021.103377.
46. Willy Kriswardhana, Domokos Esztergár-Kiss, A systematic literature review of mobility as a service: examining the socio-technical factors in mass adoption and bundling packages, *Travel Behaviour and Society*, vol. 31, 2023, pp. 232–243, https://doi.org/10.1016/j.tbs.2022.12.007.
47. Nikoleta Mikušová, Hana Neradilová, Martina Hlatká, Gabriel Fedorko, Vieroslav Molnár, Ján Kráľ, Use of augmented reality in railway transport, *Transportation Research Procedia*, vol. 77, 2024, pp. 253–259, https://doi.org/10.1016/j.trpro.2024.01.034.

Chapter 8

Healthcare and wearables in smart cyber-physical systems

Parth Gharat, Ashwini Dattatraya Gopwad, S. Sridhar Raj, Rasika Kolhe, Shitiz Upreti, and Pratham Nijhawan

8.1 INTRODUCTION

Small biosensor-equipped devices are wearable. Biomarkers in biological fluids are continually monitored by these biosensors. These technologies help in healthcare, sickness management, and real-time health tracking.[1] The terms wearable technology, wearable gadgets, and wearables include smart, portable devices, wirelessly communicative systems, systems integrated into peripherals and equipment, clothing that can be worn on the body, and even intrusive versions like microchips or smart tattoos. Wearable devices are becoming more common. They can detect vital signs for hours or even days. Many shapes and forms are available, capable of estimating various indices and measuring multiple quantities of interest.[2] Smart sensors and wearable technology have been created to assist a wide range of healthcare services. Consequently, there is a growing interest in evaluating their ability to promptly and accurately identify signs of a viral disease.[3] A significant ubiquitous technology of the Internet of Things (IoT) in daily life is wearable devices. Intelligent healthcare systems need the seamless integration of consumer-focused IoT services with effective data processing across a range of devices, including medical wearables, smart wristwear, and smart clothing.[4] At present, the wearable market is primarily dominated by health, safety, interaction, trackers, identity, and fitness. Wearables enhance the integration of physical and digital realms, seamlessly immersing individuals into the IoT. Wearable gadgets transform how customers interact with the environment; therefore, their popularity is expanding dramatically. Personal devices help people connect with the real world, according to 74% of those people. In 2020, it is predicted that a significant number of smartphone users will be wearing at least five wearables. In addition, a majority of people anticipate that wearables will have a broader range of uses in the coming years. These devices will not only track health information but also have the capability to control objects, unlock doors, and authenticate identity and transactions. Wearables need to adapt to the future in order to meet consumer expectations.[5] Compact sensors may be worn on the body or in clothes to detect environmental

DOI: 10.1201/9781003542513-8

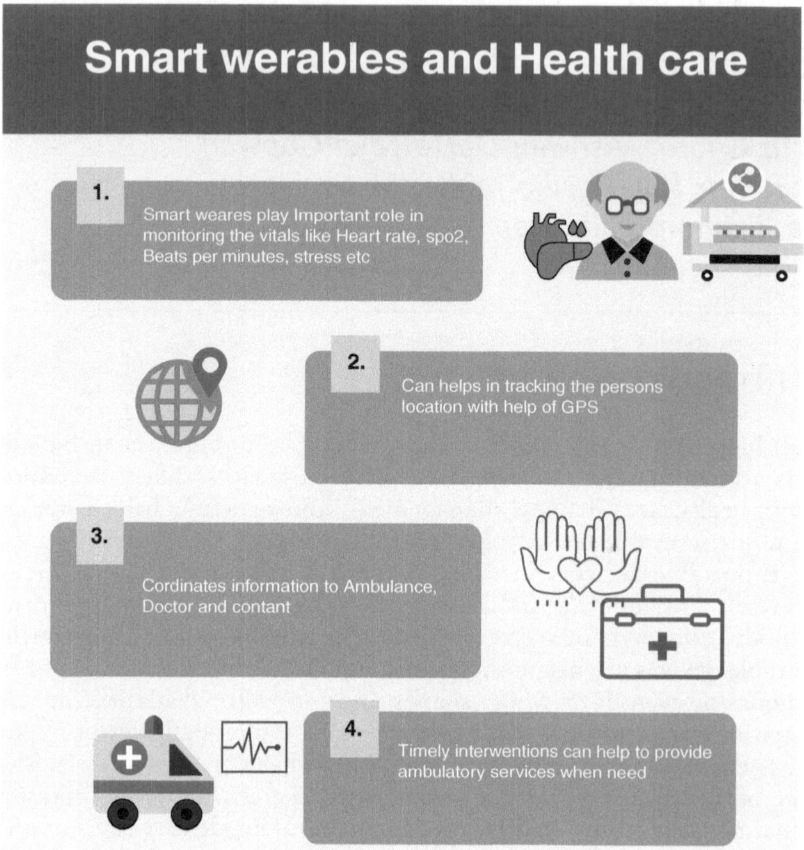

Figure 8.1 Various interventions of smart technologies, viz, AI, ML, and IoT, in healthcare.

or physiological characteristics. Consumer versions like action cameras and smart watches capture live events, track fitness and sports performance, and even nappy/diaper wetting. Wearable medical gadgets may be employed in 'real life' rather than standardized clinical situations and provide physiological variables for longer durations than traditional medical equipment. Thus, health monitoring, non-invasive hypoglycemia detection, and hearing aids have revolutionized diagnosis, patient monitoring, and therapy, for medical equipment to be widely accepted and worn. They can detect vitals for days. They vary in shape and can estimate several indices and measure numerous values (see Figure 8.1).[6]

8.1.1 Wearables: apparel from the past and present

Early studies have identified three distinct phases of growth in wearable technology: the first phase spanned from the 1980s to 1997, followed by a second phase between 1998 and 2000, and finally a third phase from 2001 to 2004.[7,8] The first period is characterized by technology, with a particular emphasis on wearable computing applications. The second period is characterized by a greater integration with the fashion and textiles sector, with a greater emphasis on garment integration. The third period is characterized by a growth in the integration of smart clothing and garments in the commercial sector. In recent times, there has been a noticeable change in fashion trends, with a shift in emphasis from clothing to accessories and jewelry. Smaller and more compact wearables have become increasingly prevalent in recent times, thanks to the rapid reduction in size and power consumption of supporting technology. The development of smaller, accessory-type form factors that were previously impossible or difficult to accomplish has become practical due to the availability of smaller components. The trend of the increasing demand for wearables resembling jewelry is an evidence of this.[9] Wearable devices have experienced rapid growth in the past decade due to their ability to provide continuous and real-time physiological data through non-invasive measurements of multiple biochemical markers in various body biofluids.[10] Since 2016, the World Health Organization's executive board has recognized the significance of mobile health (mHealth) for public health monitoring. This was highlighted in the WHO's EB 142/20 report. This report highlights the benefits of mHealth in improving access to health information, services, and skills. It also emphasizes how these tools contribute to positive changes in health behavior and disease control.[11] In 2018, the WHO broadened the scope of mHealth to encompass the utilization of additional digital technologies for public health. Resolution WHA71 emphasized the importance of utilizing digital technologies to advance Universal Health Coverage and Sustainable Development Goals (SDGs).[12] This technology is in perfect harmony with the SDG of "Health and Well-being." There are numerous wearable devices that can measure physiological data, including heart rate, steps, blood oxygen levels, and electrocardiograms. The data provided can be utilized to analyze the user's health status and enhance their overall wellbeing, fostering synergy with SDG.[13] An increasing number of these devices are being designed in a sustainable manner by promoting the use of the collaborative economy, ensuring the second life of products, contributing to a circular wearable device system, or fostering repair, restoring, or recycling. This case aligns with SDG 11 (sustainable cities and communities) and SDG 12 (responsible production and consumption). Initiatives such as the European Commission's eHealth Action Plan 2012–2020 have made significant progress in this regard.[14] The Plan prioritized improving access to recent initiatives in the healthcare system, specifically in relation to information technologies and devices (eHealth), including a special approach to mHealth. These developments are occurring in the context of population aging, a worldwide occurrence. The proportion of elderly individuals in every country is increasing at an exponential rate. Although the number of persons 65 and over was estimated to be around 703 million globally in 2019, it is projected that by 2050, that number would have doubled to 1,500 million. Early signs of respiratory infections, like those caused by the SARS-CoV-2 virus (COVID-19), can be detected through abnormal changes in these indicators.[15]

8.1.2 Patent

Some argue that patents are the most reliable source of information for monitoring the emergence of technologies. After conducting a search in Web of Science, along with the support of the Derwent World Patent Index, a vast number of documents pertaining to sensor-based patents from 2013 to 2018 were found. This highlights the magnitude of the task and affirms the necessity of using quantitative methods to analyze patent data. For the search and analysis of patents, two different databases were utilized: Orbit Intelligence and the Derwent World Patent Index.[16]

8.1.3 Orbit intelligence

Orbit Intelligence is a tool specifically designed for businesses to analyze and map patents. It offers robust tools for the search, quantitative analysis, and visualization of patent data, and it encompasses more than 40 patent issuing authorities. The landscape of all patents involving the terms sensor, sensing, wearable, portable, or implantable within the medical and veterinary sciences was analyzed using this tool between November 2008 and November 2018. The focus was on instruments, implements, and processes for medical examination, diagnostic, surgical, and person-identification purposes (patent code: A61B). Data analysis was performed using the concept filtering feature provided by Orbit Intelligence's built-in tools. There are a total of 1,065,441 patents related to sensors, sensing, and biosensors. Additionally, there are 50,733 patents that focus on wearables, implantable, portables, ingestible, home devices, or smart home technologies. Table 8.1 gives information on various patents in smart cyber-physical system (SCPS).[17–40]

8.2 UNLOCKING CLARITY: THE CRITICAL ROLE IN MEDICAL DIAGNOSIS

Artificial Intelligence (AI)'s ability to analyze massive datasets, find patterns, and make predictions makes it a transformative tool for chronic disease management. Tailoring treatment programs is one of its most promising uses.

Table 8.1 List of Patents in Smart Cyber-Physical System

Sr no	*Patent no*	*Title*
1.	CN102799783A	Patient identification system and method based on non-contact IC/ID (integrated circuit/identity) card
2.	US20120101848A1	Portable personal health record data processing system
3.	US20230290243A1	Condition-responsive wearable device for sensing and indicating the proximity of an article with a specific characteristic
4.	CN212302219U	Wrist strap-type information board
5.	US20220400986A1	Systems, devices, and methods for physiological parameter analysis and related graphical user interfaces
6.	US20180047121A1	Healthcare Delivery Computer Systems and Method
7.	KR20210111548A	Information synchronizing method for hybrid smart watch based on mechanical movement
8.	US20180308569A1 US9028405B2	System or method for engaging patients, coordinating care, pharmacovigilance, analysis, or maximizing safety or clinical outcomes by use of a personal monitoring system
9.	GB2593131A	Wearable device
10.	US10325102B2	Real-time customer experiences management systems and methods
11.	JP2022534422A	Systems and associated methods for biomonitoring and blood glucose prediction
12.	US20170372017A1	Healthcare self-management intervention system and method
13.	US20170024531A1	Systems and methods for near-real or real-time contact tracing
14.	US20160034764A1	Wearable Imaging Member and Spectroscopic Optical Sensor for Food Identification and Nutrition Modification
15.	US20180254106A1	Behavior sensing device, behavior sensing method, and recording medium
16.	JP2016526542A	Device and method for pharmacokinetic drug administration planning
17.	JP2022507628A	Feedback from neuromuscular activation in various types of virtual reality and/or augmented reality environments
18.	US20190008463A1	Wearable digital device for personal health use for saliva, urine, and blood testing and mobile wrist watch powered by the user's body
19.	US20130254966A1	Patient point-of-care garment
20.	US10121355B1	Condition-responsive wearable device for sensing and indicating the proximity of an article with a specific characteristic
21.	KR20230055498A	Method for servicing health care by using wearable devices based on blood vessel diagnosis sensors

(*Continued*)

Table 8.1 (Continued) List of Patents in Smart Cyber-Physical System

Sr no	*Patent no*	*Title*
22.	US20220244826A1	Transition and Interaction Model for Wearable Electronic Device
23.	TWI686172B	Electrocardiogram (ECG) Sensor Chip, System on chip (SoC), Wearable Appliance, and Data processing system
24.	CN206560435U	Wearable pulse-taking instrument, intelligent feel, pulse bracelet, and gloves

8.2.1 Continuous monitoring

- Wearable devices provide real-time tracking of vital signs, such as heart rate and glucose levels, which is essential for managing chronic diseases like diabetes and cardiovascular conditions[41,42]
- Mobile-based devices for ECG monitoring involve a wide range of devices, including smartphones and pocket ECG monitors. Mossab Aljuaid et al. evaluated the effectiveness of using a smartphone-based ECG monitoring device. Various research initiatives have been proposed to measure ECG with wearable devices.[43] A. Petrena et al. proposed a modified Lewis ECG lead system for wireless monitoring of atrial arrhythmia, while exploring a wearable on-body ECG patch to increase the possible device wearing time. Another category of ECG monitoring system utilizes innovative sensors to retrieve ECG signals without disturbing the patient's comfort.[44]

8.2.2 Early detection and diagnosis

- Manan et al. demonstrated a novel NLP pipeline for accurately identifying non-small-cell lung cancer patients and extracting information on tumor sizes from EHR imaging reports, achieving an 82% accuracy rate.[45]
- AI algorithms analyze data from wearables to identify early signs of diseases, such as Parkinson's, through assessments of motor functions (Farago et al., 2022).[46]
- Artificial neural networks can analyze biomarkers like troponin, C-reactive protein, HbA1c, and FEV1, providing insights for chronic disease management and enhancing predictive capabilities in healthcare services.[47]
- **Random Forests and Logistic Regression:** These traditional machine learning (ML) methods are effective in predicting chronic disease risks by analyzing diverse datasets, including genomics and patient health records, which could confirm ankylosing spondylitis.

  Support Vector Machine Learning: Orange et al. analyzed 20 different histologic features from 129 samples (123 RA,06 osteoarthritis) using the support vector machine (SVM). They also performed consensus clustering of gene expression data. The high inflammatory subtype is characterized with significant leukocyte infiltration; a low inflammatory subtype is enriched in pathways like transforming growth factor ß, glycoproteins, and neuronal genes; and a mixed subtype was also observed. ML uses histologic features and gene expression subtypes as labels to create an algorithm for scoring histologic features. Patients with the high inflammatory synovial subtype showed elevated levels of systemic inflammation markers and autoantibodies. CRP levels strongly correlated with pain severity.[48]

8.2.3 Personalized treatment

Wearable technology allows for tailored healthcare interventions based on individual patient data, enhancing the effectiveness of treatment plans.[49]

Despite these advancements, challenges such as data privacy and the need for standardized devices remain significant barriers to widespread adoption. Addressing these issues is crucial for maximizing the potential of AI-driven wearables in healthcare services.

8.3 BRIDGING THE GAP: SEAMLESS TECHNOLOGICAL INTEGRATION IN MODERN HEALTHCARE

The healthcare industry is one of the biggest in developing nations, with yearly growth observed in both employment and income.[50] Rare diseases can only be diagnosed in their early stages by a thorough physical and analytical examination conducted on hospital grounds.

These days, we can diagnose any irregularity in our health with the use of a smartwatch. Take an elderly person's erratic heartbeat, for instance. When an epidemic disease like Ebola strikes, technology is crucial in preventing the disease's rapid spread and educating people about what to do. According to the Centres for Disease Control and Prevention (CDC), patients should not take medication without a doctor's presence.[51] New technology enables patients to receive preliminary diagnoses at their doorstep, transforming healthcare services from hospital-centric to individualized. Remote clinical procedures like blood testing and diabetes monitoring are now feasible in poor nations. Wearable IoT sensors, cloud computing, fog computing, and mobile computing enhance real-time user monitoring and prescription delivery.[52,53]

The IoT increases people's freedom to communicate, contribute, and work together on projects. Heterogeneous technologies with complex protocols and algorithms were progressively developed by IoT.[54] It is crucial to the global communication of millions of Internet-connected devices, including electronic appliances like refrigerators and televisions, as well as devices with wired and wireless sensors.[55] High precision, low cost, and time savings through improved future prediction are the factors driving the IoT' quick development. Sensors and actuators are examples of IoT devices that were integrated into actual products, including automobiles, smart industrial machinery, and residential appliances, to monitor and share data via wireless networks using Bluetooth, Zigbee, IEEE 802.11 (WI-FI), WiMAX, and other technologies. The sensors, which measure physiological states such blood pressure, heart rate, body temperature, electrocardiogram (ECG), and electroencephalogram (EEG), can be worn or implanted in the body.[56] Those who have young children or elderly living alone find it easier to keep an eye on sick persons. The user's sensor regularly monitors them round the clock and sends forth alerts regarding them, altering the health status to emergency services, physicians, and family members. Significant attention from the international research community is needed to address the growing need for new technology to handle real-time challenges in processing greater data without compromising security and privacy. Figure 8.2: depicts IoT Architecture Related to Healthcare.

ML and predictive analytics are revolutionizing healthcare by detecting diseases using massive patient data. These technologies save lives and detect early indicators of cancer, cardiovascular, diabetes, and lifestyle disorders. Early diagnosis and treatment may improve sickness outcomes and save lives.[57–59] Predictive analytics uses genetics, lifestyle, and medical history to risk assess high-risk individuals and prioritizes preventive measures,

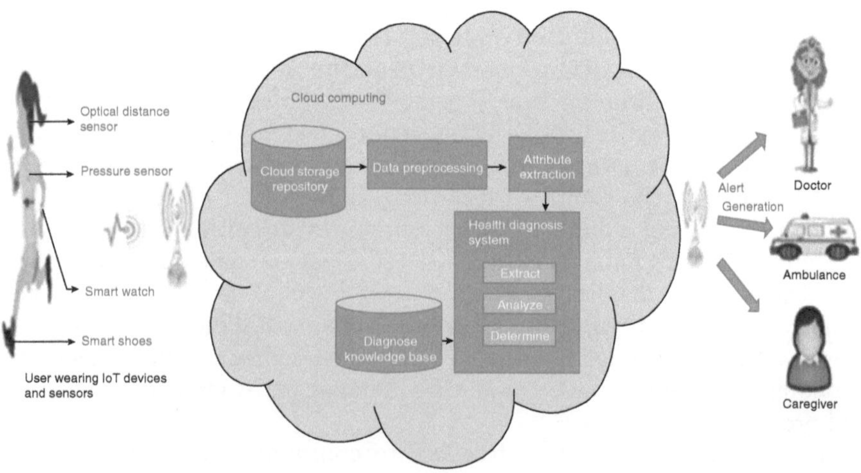

Figure 8.2 IoT architecture related to healthcare.

optimizing healthcare budget allocation. Thus, infections may be prevented and the healthcare system burden decreased.[60,61]

- **Machine Learning:** An algorithm in ML learns tasks using data patterns. Validation sets and training sets make ML program outcomes visible or invisible. When patients have few meaningful traits, ML methods are extensively employed.[62,63] In data mining activities, algorithms are trained to provide predictions or classifications using statistical techniques.
- **SVM:** SVMs are commonly used for illness diagnosis and prognosis. The approach promises to accurately calculate the health problem solution. SVM detects outliers, classifies, and evaluates according to certain specialists. SVM has the potential to detect a variety of diseases, such as cancer, diabetes, and excessive blood pressure.[64]
- **Artificial Neural Network:** Artificial neural networks (ANNs) imitate brain learning in computer models. Most people use neural networks and prediction scores to predict outcomes like mortality rate that are hard to predict. Most of the data needed to train these models—MRIs, CT scans, electronic health records, and others—must be preprocessed.[65] According to Begum et al. (2024), ANNs, which include a variety of models such as multilayer neural networks and probabilistic neural networks, have demonstrated exceptional effectiveness in identifying illnesses like lung infections and cancer.[66] Deep learning has transformed medical image analysis by providing automatic feature extraction and enhanced diagnostic capabilities, especially through convolutional neural networks (CNNs). Demonstration by Degadwala and Vyas in 2024.[67]
- **Natural Language Processing:** Healthcare uses unstructured textual data including lab reports, test findings, and physician notes. Medical prescriptions and hospital discharge instructions. From such vast descriptive data sets, natural language processing technologies can be utilized to extract relevant information, such as patient data, which can improve treatment and diagnosis recommendations.[68–70] Accurate predictions depend on the ability to handle long-term dependencies in data, which is made possible by the deployment of hybrid language models like LSTM and GRU. The 2024 demonstration by Meenadevi et al.[71]

8.4 KEY APPLICATIONS OF AI IN HEALTHCARE

8.4.1 Genetics oncology

Clinical genomic research advancements involve adjusting processes like variant calling, genome explanation, variation labeling, and genotype correspondence, potentially predicting genotype-to-aggregate relationships. Based on genomes, precision medicine may benefit from AI advancements.[73,74] ML is being used to analyze DNA sequences for patterns that can predict patient health risks and identify underlying causes, boosting interest in this field.[74] AI is becoming more common in oncology, especially in the area of cancer-focused cognitive computer systems.[70] AI is being utilized in various cancer screening, diagnosis, classification, genomics characterization, tumor microenvironment studies, biomarker assessment, and drug discovery methods.[71,72]

- Deep learning, specifically CNNs, is used to identify patterns and categorize images.[75,76]
- AI in neuroscience can predict the disease course using neuroimaging.[77,78] AI applications can reveal connections, structures, and mechanisms underlying behavior and brain function.[79,80]
- Cognitive shortcuts like pattern recognition and intuition can aid in quick decision-making under uncertainty.
- Cognitive shortcuts in surgery can help physicians make quick decisions under time constraints.[80,81]
- Innovative AI-driven cardiac image processing algorithms can expedite patient evaluation in cardiology.[82,83]

8.4.2 Advanced algorithms for data interpretation

Advanced algorithms such as random forest, decision tree, logistic regression, and naive Bayes are utilized for data interpretation, showcasing their applications in healthcare services. Figure 8.3 shows advanced algorithms and their applications in healthcare.

8.4.3 Advances in biosensors for more accurate data

Sensor technology interprets biological signals in biology and medicine. Modern sensors detect heart rate, sweat, and tissues. Wearable sensors enhance animal health, ecology, and ecological monitoring.[84] Financial analysis, medical diagnosis, personalized treatment, and autonomous automobile navigation use ML, a fast-growing technology that learns from experience.[85]

- **Conventional Sensor:** Piezoelectric sensors are increasingly used in various fields due to ML advancements, enabling their widespread application in various fields.
- **Piezoelectric Sensor:** The substance, a generalization of ferroelectric and multiferroic substances, emits an electric charge after converting pressure, allowing properties like position, pressure, and discharged

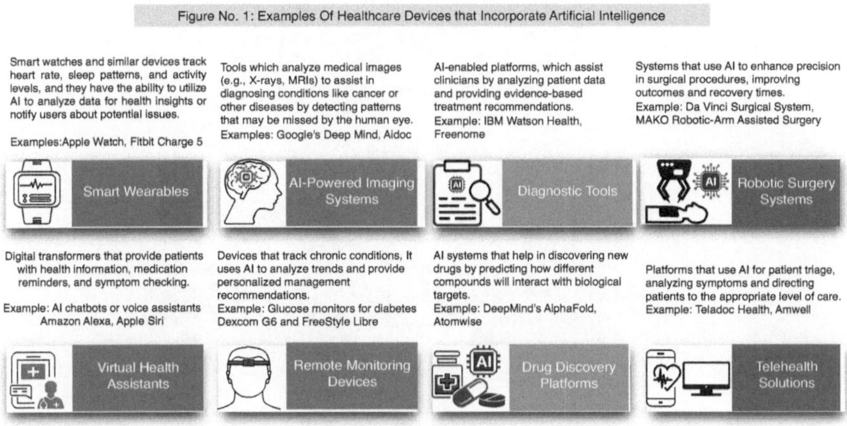

Figure 8.3 Various advanced algorithms and their healthcare-based applications.

charge to be detected and converted into visible signals. Medical professionals use piezoelectric sensors in pacemakers, stethoscopes, and ultrasounds to monitor patient's heart rate, activity level, and pulse under anesthesia, fatal heart monitoring, and pacemaker use.

- **Sensors Commonly Used for Health Detection:** Mechanoelectrical sensors, including piezo-electric, triboelectric, capacitive, and resistive sensors, share a basic structure with electrode-clamped ends, with resistive sensors converting pressure-induced resistance changes into electrical signals. Rapid pressure response and low-pressure sensitivity are a sensor's abilities.[86]

8.4.4 Wearable and machine learning sensors

Digitalization-driven AI makes sensors better and more innovative. For environmental monitoring, health, and medicine, wearable and ML sensors are being developed. Wi-Fi and smartphones have created a smart society with widespread use of home and medical sensors.[86]

- **Wearable Sweat Sensor:** Exercise-induced perspiration provides valuable physiological data, which can be analyzed using wearable technology.[87,88] This technology converts sweat components into electrical impulses that may be utilized on a computer or smartphone to assess a user's health.
- **Machine Learning Sensors:** Cloud computing and big data have substantially impacted sensor technologies, boosting ML.[88] In agriculture, healthcare, and face and voice recognition, ML sensors use algorithms to extract meaningful data from signals.

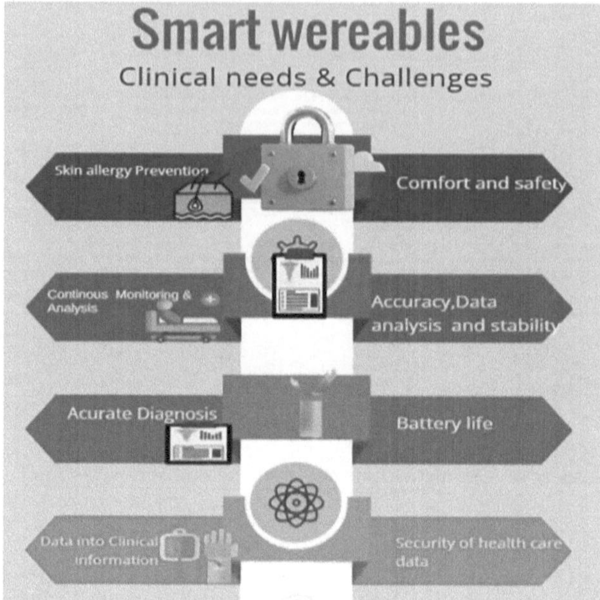

Figure 8.4 Various non-invasive techniques in diagnostics.

8.4.5 Non-invasive monitoring techniques

Non-invasive monitoring techniques such as pulse oximetry, capnography, cardiac telemetry, and oscillometric blood pressure monitoring are frequently employed in the emergency department for monitoring critically ill pediatric patients. Figure 8.4 provides a detailed description of various non-invasive techniques.

8.5 ANCIENT REMEDIES AND MODERN TECH: THE FUSION OF AYURVEDA AND IOT

In Ayurveda, waveforms of pulse is examined by traditional methods, where the movement of pulse like a snake (*Sarpagati*) predicts the predominance of *Vata Dosha*, and jumping like a frog (*Mandukgati*) and swimming like a swan (*Hansagati*) predict *Pitta* and *Kapha* patterns, respectively.[89] This indication of *Dosha* by pulse examination is an important step for diagnosis and prognosis of disease. Hence, to make it more easy, an effortless and highly precise sensor based on "*Nadi Tarangini*" has been developed by scientists. This sensor provides pulse pattern and further analyzed data to provide the Ayurveda parameters along with recommendations.[90] The photoplethysmography [PPG]-based sensor is also developed and used in the

Siddha system of medicine (CC3200).[91] Similarly, Ayurveda physicians also adopted tongue examination as an important tool to understand the status of the digestive system. Though efforts are made for tongue diagnosis based on deep ML, in which images of the color and texture of the tongue are also analyzed by using for disease diagnosis and syndrome differentiation by using AI, for example,, the handheld TDA-1 tongue imaging capturing device,[92,93] still there is a need to validate the same based on the Ayurveda concept of *Sama* (coated) and *Niram Jivha* (non-coated tongue). Number of devices and wearables are developed to track physical activity, stress, sleep patterns, energy expenditure, oxygen transport and utilization, heart rate, diet, hydration sensor, eye motion, blood sugar, and gas sensor for respiratory health.[94] A number of biosensors such as interstitial fluid biosensor have been developed to help monitor blood glucose levels. A wearable electrochemical biosensor was used for the monitoring of metabolites and nutrients and perspiration analysis, an electrochemical microneedle sensor was used for continuous monitoring of *levodopa* in Parkinsonism, a wearable sensor was used for sensitive detection of uric acid and tyrosine in sweat, a wearable electrochemical platform was used for monitoring of Ca2+ and pH levels, and epidermal enzymatic biosensors were used for monitoring vitamin C in sweat.[95] The development of such IoT and wearable devices opens opportunities to Ayurveda. Sweat, urine, hair, dental plaque, and stool samples are given immense importance in diagnosis as well as prognosis of disease in Ayurveda.[96] Remote monitoring of patient pulse, body movement, and hematocrit as well as the healthcare system has also became much easier by using various technologies. Continuous monitoring of such parameters provides better options for remote monitoring of patients.[97] Thus, fast development in the wearable technology sector has been observed.

Development in the wearable technology sector provides an ocean of opportunities to implement this technology in Ayurveda and tailor it as per the requirements after validity and reliability testing. Further, peculiarities of Ayurveda, like *Prakriti* (physical constitution) and status of *Agni* (digestive system), should be incorporated in wearables.[98] Yet there is scope for discoveries to set the relationships between various biological factors such as genes, tissues, and hematological and biochemical parameters with Ayurvedic physiological factors like *Dosha, Dhatu, and Oja.* Further, AI and ML can be utilized successfully in recognizing disease pattern, demographic prevalence, and associated factors, as well as in speeding up the drug development process. Convolutional neural network and Mobile NetV2 are also improving the efficiency and accuracy in medicinal plant identification. Thus, AI and ML both are a boon in development of Ayurveda for global acceptance.

The integration of modern technology into Ayurvedic practices can enhance the patient experience and lead to more precise diagnostic methods, management of disease, drug development, and drug delivery system. Further development of technology- and sensor-based tools for monitoring of various therapeutic procedures of *Panchakarma* (five therapeutic procedures) and presurgical procedure like *Agnikarma* (thermal cauterization), *Ksharakarma* (caustic therapy), leech therapy, and Siravedh (venipuncture) is the need of the hòur. The application of IoT can enhance Ayurveda by aiding in disease prevention, enabling accurate diagnosis, and facilitating the selection of targeted therapeutic regimens. The global implementation of IoT has expanded the potential for Ayurveda to become more user-friendly, cost-effective, and accessible. This will enhance its global acceptance. Figure 8.5 illustrates several examples of healthcare devices that integrate AI.

Healthcare devices powered by AI can adjust treatment plans and medication doses based on real-time health data such as heart rate and blood sugar levels.[99] These AI and ML healthcare devices must prioritize protecting private medical data and maintain forecast accuracy and

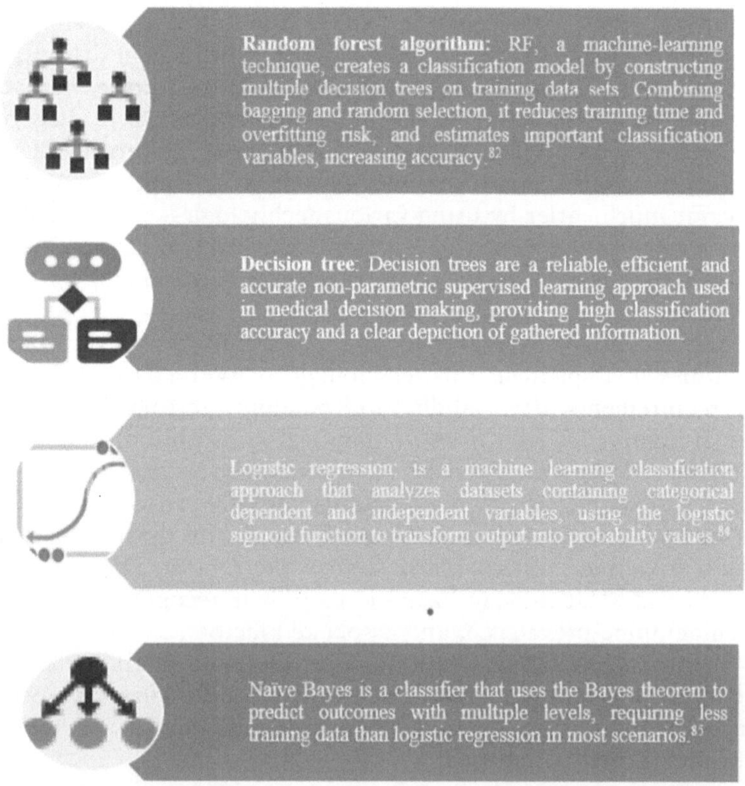

Figure 8.5 Examples of healthcare devices that incorporate artificial intelligence.

reliability to avoid misdiagnosis or negligence while ensuring user-friendly designs and clear instructions for optimal usage.[100] Analyzing extensive datasets with AI accelerates the discovery of potential drug candidates and predicts their effectiveness, significantly reducing research time and costs. AI facilitates the design of targeted therapies with higher success rates by simulating molecular interactions. Jovana Ajdaric et al. utilized data mining methods employing ANNs to analyze large datasets of the Esomeprazole stability profile.[101] This is done to determine the pH limit for the reconstituted solution of the final freeze-dried product, ensuring a quality product profile over a storage period of 36 months. The process involves the use of various techniques from statistics, ML, and database systems to extract meaningful information from raw data, often transforming it into a more understandable format for further analysis. Mehta et al. used the Toxicity Prediction by Komputer Assisted Technology (TOPKAT) prediction tool to assess the *in silico* toxicity of heliostat and its degradation products. This software uses quantitative structure–activity relationship (QSAR) models to assess the potential toxic effects of substances based on their chemical structure. By facilitating early-stage toxicity predictions, TOPKAT helps researchers and companies make informed decisions during the drug development and chemical manufacturing processes.[102] Analyzing extensive datasets, Blanco-González et al. stated that AI speeds up drug discovery by identifying potential drug candidates and predicting their effectiveness, significantly reducing time and costs. Predictive modeling improves decision-making and optimizes the drug development pipeline.[103]

8.6 CHALLENGES

Wearable technology is advancing rapidly, which might change healthcare, fitness, and augmented reality. However, numerous major obstacles must be overcome to reach this potential. This study discusses next-generation wearables' technological, regulatory, and ethical challenges. Sensor accuracy, battery life, and comfort affect device performance; medical wearable clearance procedures are stringent; and data privacy ethics are discussed. Research focuses on graphene-based flexible wearable sensors. Kong et al. highlight that graphene materials provide key benefits such as flexibility, conductivity, and strength. We need to tackle challenges like scalability, integration with other materials, and durability to unlock their full potential. The authors explore these mechanisms and suggest future research directions, highlighting the need for improvements in manufacturing and material handling.[104]

Hossain et al. examined sustainable textiles, focusing on techniques, applications, and challenges. Their review emphasizes the need for effective

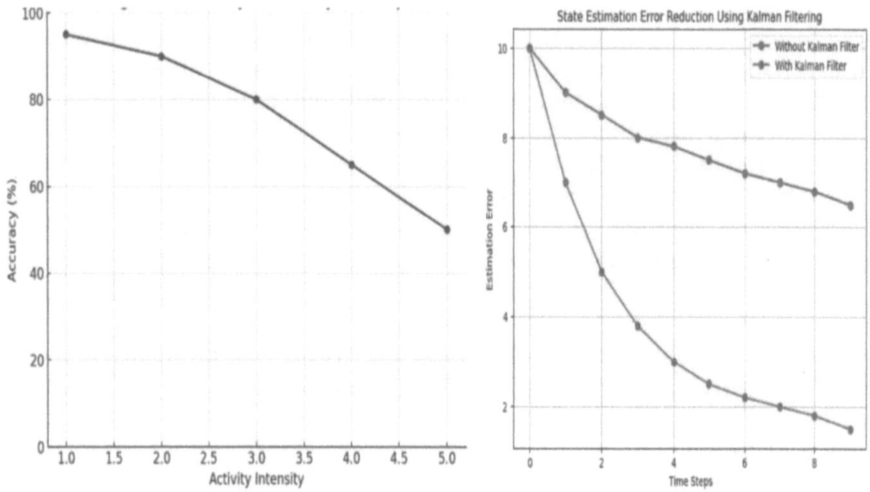

Figure 8.6 Describes the clinical need and challenges in smart wearables.

eco-friendly materials that reduce environmental impacts. Various methods are available to enhance textile sustainability and the challenges in applying these solutions. This segment emphasizes the need for continuous research in sustainable practices and materials within the textile industry.[105] Figure 8.6 outlines the clinical needs and challenges of smart wearables.

Yang et al. examined e-textiles for sports and fitness. Their focus is on e-textiles, highlighting tech challenges and future opportunities for sportswear integration. Progress is clear, yet sensor accuracy, user comfort, and data integration require enhancement. The authors suggest enhancements in textile materials and sensor technologies.[106] Sengupta and Kottapalli offer a compelling review of bioinspired wearable sensors using nanomaterials. This paper highlights nanomaterials used to develop sensitive, adaptable sensors that replicate biological systems. They examine new advancements, innovative bioinspired sensors, and future research paths to improve the functionality and use of these devices.[107] Gao et al. explored how headworn wearable devices could transform human–machine interactions. They talk about advancements in headword devices, their applications, and technical challenges. These devices enhance human–machine interactions and point to future research areas, such as improved comfort and data accuracy.[108]

8.6.1 Reliability and accuracy

Heart rate and glucose levels are measured by sensors in wearable devices. Sensor precision is crucial for medical applications. Variable sensor performance might affect data dependability. Heart rate monitors are less accurate during high-intensity exercise owing to movement artifacts and sensor misalignment.

The accuracy (A) of a heart rate monitor can be expressed as

$$A = \frac{S_{actual} - S_{measured}}{S_{actual}} \times 100\%$$

where (S_{actual}) is the actual heart rate and $(S_{measured})$ is the measured heart rate.

Integrating multiple sensors into a single device can lead to cross-talk, affecting the overall accuracy. Data fusion algorithms, such as the Kalman Filter, help mitigate these issues:

$$\{x\}\_\{k \mid k\} = \{x\}\_\{k \mid k-1\} + K\_k(z\_k - H\_k \; \{x\}\_\{k \mid k-1\})$$

where $(x_{k|k)})$ is the estimated state, (K_k) is the Kalman gain, (z_k) is the measurement, and (H_k) is the measurement matrix. Figure 8.7 depicts how Kalman filtering reduces state estimation error. It provides a visual representation of

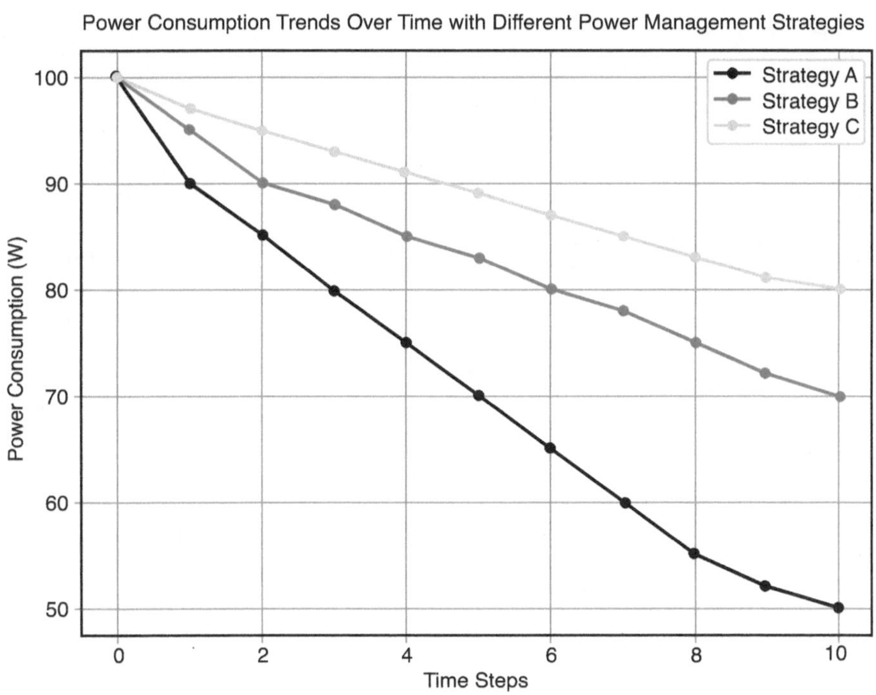

Figure 8.7 (a) Accuracy vs. activity intensity; (b) state estimation error reduction with and without Kalman filter.

how Kalman filtering effectively reduces state estimation error in dynamic systems. Kalman filtering is a sophisticated mathematical approach used for estimating the state of a system over time, which is particularly useful in scenarios where measurements are subject to noise and uncertainty.

The figure presumably compares raw data and filtered estimates graphically. State estimate uses "state" to characterize the system's current state. In a tracking system, this might be an object's location and velocity. Kalman filtering predicts and corrects. Based on the present state and control inputs, the filter predicts the future state using a system dynamics model. This forecast accounts for model uncertainty and system externalities. However, the corrective step changes this forecast using measurements. The Kalman filter calculates a weighted average of the anticipated state and the new measurement, weighted by uncertainty (or covariance). A modified estimate balances the prediction model with real-world data, decreasing measurement noise and systematic errors.

8.6.2 Battery life

Battery life is a significant concern as wearable devices become more sophisticated. Power consumption (P) can be described by

$$P = V \times I$$

where (V) is voltage and (I) is current. Effective power management strategies, including optimizing software and employing adaptive power-saving modes, are essential. The graph illustrates the trends in power consumption over time for three different power management strategies, showing how each strategy impacts energy usage. Strategy A demonstrates the most significant reduction in power consumption, while Strategy C shows the least aggressive decrease.

8.6.2.1 Analyzing power management strategies

A graph of power use over time may show how well power management techniques work. Three graph hypothetical strategies are discussed below:

- Strategy A delivers the greatest power consumption decrease. It may use frequent low-power mode transitions and effective software optimization to save power. Thus, the gadget extends battery life better than other methods.
- Strategy B may provide a balance between power savings and performance. It may entail mild power-saving mode and software efficiency modifications to reduce power consumption without affecting device performance.

- Strategy C: The least aggressive power consumption reduction. Without significant power management or software adjustments, it may use more power and drain the battery faster.

8.6.2.2 Power-management strategies impact

The battery life of a wearable gadget depends on its power management method. Strategy A, with its significant power reduction, is best for battery-conscious users who can endure performance trade-offs. Strategy B balances power efficiency and performance for broad usage. Strategy C, however less power-efficient, may be used for applications that prioritize performance above battery life.

8.6.2.3 Medical wearable approval process

Before being sold, medical wearables must pass strict regulations. Clinical studies and FDA/EMA regulations are needed to prove safety and effectiveness.[109–111]

8.7 CONCLUSION

Wearable devices play a crucial role in healthcare services by providing continuous monitoring of vital signs and delivering real-time data. These non-invasive devices track vital health indicators, crucial for early detection of problems such as respiratory infections, including COVID-19. This chapter emphasizes the integration of wearable tech with eHealth and mHealth. This integration is vital for improving healthcare access and information, particularly for the elderly. The WHO recognizes mHealth's importance in public health monitoring and its alignment with the Sustainable Development Goals (SDGs). Sustainable design for wearable devices is gaining momentum. The initiatives such as support the collaborative economy, give products a second life, and boost recycling and repair promote a circular economy and align with SDG 11 and SDG 12. This chapter outlines three main phases in the evolution of wearable technology, illustrating the shift from basic wearable computing to sophisticated smart clothing and accessories. This evolution highlights the shift in consumer preferences and technology, leading to smaller, compact devices that integrate effortlessly into daily life. Power management is crucial for the proper functioning of wearable devices. This paper presents strategies to significantly reduce power consumption, enhance battery life, and improve user experience. This paper underscores the crucial impact of wearable technology in addressing issues related to an aging population, focusing on sustainable design and the need for smooth integration with digital health solutions.

REFERENCES

1. Sharma, Atul, Mihaela Badea, Swapnil Tiwari, and Jean Louis Marty. 2021. "Wearable Biosensors: An Alternative and Practical Approach in Healthcare and Disease Monitoring" *Molecules* 26, no. 3: 748. https://doi.org/10.3390/molecules26030748
2. Luczak, Tony, Reuben Burch, Edwin Lewis, Harish Chander, and John Ball. "State-of-the-art review of athletic wearable technology: What 113 strength and conditioning coaches and athletic trainers from the USA said about technology in sports." *International Journal of Sports Science & Coaching* 15, no. 1 (2020): 26-40.
3. Ifrah, Farid., Kamil, Reza, Khondakar. (2024). 2. Smart sensors for infectious disease diagnosis. doi: 10.1016/b978-0-323-98805-6.00002-6
4. Gloria, Cosoli., Lorenzo, Scalise., Angelica, Poli., Susanna, Spinsante. (2021). Wearable devices as a valid support for diagnostic excellence: lessons from a pandemic going forward. Health technology, doi: 10.1007/S12553-021-00540-Y
5. Poongodi, T., Rajalakshmi Krishnamurthi, R. Indrakumari, P. Suresh, And Balamurugan Balusamy. "Wearable Devices and Iot." *A Handbook Of Internet Of Things In Biomedical And Cyber Physical System* (2020): 245-273. lead to more effective solutions for testing, tracking, and tracing people believe that the wearable sensors assist them in interacting with the physical objects around themnse vital signs for hours and days. They come in different shapes and forms and can measure.
6. Zastrow m. "coronavirus contact-tracing apps: can they slow the spread of covid-19?" Nature.
7. B. Ariyatum, R. Holland, D. Harrison and T. Kazi, J. Text. Inst., 2005, 96(4), 199.
8. Y. Silina and H. Haddadi, in Proc. of the 19th International Symposium on Wearable Computers, Osaka Japan, 2015.
9. Salim and S. Lim, Biosens. Bioelectron., 2019, 141, 111422.
10. Quinn, C.C.; Shardell, M.D.; Terrin, M.L.; Barr, E.A.; Ballew, S.H.; Gruber-Baldini, A.L. Cluster-Randomized Trial of a Mobile Phone Personalized Behavioral Intervention for Blood Glucose Control. Diabetes Care 2011, 34, 1934–1942. [CrossRef] [PubMed]
11. World Health Organization. Seventy-First World Health Assembly. In Proceedings of the 71st Session of the World Health Assembly, Geneva, Switzerland, 21–26 May 2018.
12. Spencer, J. The Sustainable Development Goals. In Design for Global Challenges and Goals; Routledge: Abingdon, UK, 2021; pp. 12–25.
13. European Commission. EHealth Action Plan 2012–2020—Innovative Healthcare for the 21st Century; European Commission: 2012. Available online: https://health.ec.europa.eu/publications/ehealth-action-plan-2012-2020_en (accessed on 19 September 2022).
14. Kaliyaperumal, D.; RK, K.; Alagesan, M.; Ramalingam, S. Characterization of Cardiac Autonomic Function in COVID-19 Using Heart Rate Variability: A Hospital Based Preliminary Observational Study. J. Basic Clin. Physiol. Pharmacol. 2021, 32, 247–253. [CrossRef]

15. Silvera-Tawil, David, M. Sazzad Hussain, and Jane Li. “Emerging technologies for precision health: An insight into sensing technologies for health and wellbeing.” *Smart Health* 15 (2020): 100100.
16. Towards an Artificial Intelligence Based Chronic Disease Management”, 2023.
17. N102799783A. *Patient identification system and method based on non-contact IC/ID (integrated circuit/identity) card.* China, 2012.
18. US20120101848A1. *Portable personal health record data processing system.* United States, 2012.
19. US20230290243A1. *Condition-responsive wearable device for sensing and indicating proximity of an article with a specific characteristic.* United States, 2023.
20. CN212302219U. *Wrist strap type information board.* China, 2021.
21. US20220400986A1. *Systems, devices, and methods for physiological parameter analysis and related graphical user interfaces.* United States, 2022.
22. US20180047121A1. *Healthcare Delivery Computer Systems and Method.* United States, 2018.
23. KR20210111548A. *Information synchronizing method for Hybrid Smart Watch based on mechanical movement.* Korea, 2021.
24. US9028405B2. *System or method for engaging patients, coordinating care, pharmacovigilance, analysis or maximizing safety or clinical outcomes Personal monitoring system.* United States, 2015.
25. GB2593131A. *Wearable device.* United Kingdom, 2021.
26. US10325102B2. *Real-time customer experiences management systems and methods.* United States, 2019.
27. JP2022534422A. *Systems and associated methods for biomonitoring and blood glucose prediction.* Japan, 2022.
28. US20170372017A1. *Healthcare self-management intervention system and method.* United States, 2017.
29. US20170024531A1. *Systems and methods for near-real or real-time contact tracing.* United States, 2017.
30. US20160034764A1. *Wearable Imaging Member and Spectroscopic Optical* Sensor *for Food Identification and Nutrition Modification.* United States, 2016.
31. US20180254106A1. *Behaviour sensing device, behaviour sensing method, and recording medium.* United States, 2018.
32. JP2016526542A. *Device and method for pharmacokinetic drug administration planning.* Japan, 2016.
33. JP2022507628A. *Feedback from neuromuscular activation in various types of virtual reality and/or augmented reality environments.* Japan, 2022.
34. US20190008463A1. *Wearable digital device for personal health use for saliva, urine, and blood testing and mobile wrist watch powered by user body.* United States, 2019.
35. US20130254966A1. *Patient point-of-care garment.* United States, 2013.
36. US10121355B1. *Condition-responsive wearable device for sensing and indicating proximity of an article with a specific characteristic.* United States, 2018.

37. KR20230055498A. *Method for servicing health care by using wearable device based on blood vessel diagnosis sensor.* Korea, 2023.
38. US20220244826A1. *Transition* and *Interaction Model for Wearable Electronic Device.* United States, 2022.
39. TWI686172B. *Electrocardiogram (ECG) Sensor Chip, System on chip (SoC), Wearable Appliance, and Data processing system.* Taiwan, 2020.
40. CN206560435U. Wearable *pulse-taking instrument, intelligent feel the pulse bracelet and gloves.* China, 2017.
41. Artificial intelligence based early detection of cardiac diseases using smart wearables", 2022.
42. Manan, Shah., Laura, Prichard., John, A., Glaspy., Vu, Vu., Jason, Wang., Alex, V, Gelvezon., Amy, L., Cummings. (2024). Novel approach to implementing natural language processing for clinical staging of non-small-cell lung cancer.. Journal of Clinical Oncology, 42(16_suppl), e13624-e13624. doi:10.1200/jco.2024.42.16_suppl.e13624
43. Kamga P, Mostafa R, Zafar S. The Use of Wearable ECG Devices in the Clinical Setting: a Review. Curr Emerg Hosp Med Rep. 2022;10(3):67-72. doi: 10.1007/s40138-022-00248-x. Epub 2022 Jun 25. PMID: 35789964; PMCID: PMC9244148.
44. Andrius, Petrėnas. (2013). 2. A modified Lewis ECG lead system for wearable atrial fibrillation monitoring.
45. Orange DE, Agius P, DiCarlo EF, Robine N, Geiger H, Szymonifka J, McNamara M, Cummings R, Andersen KM, Mirza S, Figgie M, Ivashkiv LB, Pernis AB, Jiang CS, Frank MO, Darnell RB, Lingampali N, Robinson WH, Gravallese E; Accelerating Medicines Partnership in Rheumatoid Arthritis and Lupus Network; Bykerk VP, Goodman SM, Donlin LT. Identification of Three Rheumatoid Arthritis Disease Subtypes by Machine Learning Integration of Synovial Histologic Features and RNA Sequencing Data. Arthritis Rheumatol. 2018 May;70(5):690-701. doi: 10.1002/art.40428. Epub 2018 Apr 2. PMID: 29468833; PMCID: PMC6336443.
46. Faragó, Paul, Sebastian-Aurelian Ştefănigă, Claudia-Georgiana Cordoş, Laura-Ioana Mihăilă, Sorin Hintea, Ana-Sorina Peştean, Michel Beyer, Lăcrămioara Perju-Dumbravă, and Robert Radu Ileşan. 2023. "CNN-Based Identification of Parkinson's Disease from Continuous Speech in Noisy Environments" *Bioengineering* 10, no. 5: 531. https://doi.org/10.3390/bioengineering10050531.
47. Kumar, Sinha., Zia, Uddin., Hassan, Kawsar., S., Islam., M., Jamal, Deen., Md., Mijanur, Rahman, Howlader. (2022). 1. Analyzing chronic disease biomarkers using electrochemical sensors and artificial neural networks. Trends in Analytical Chemistry, doi: 10.1016/j.trac.2022.116861
48. Orange DE, Agius P, DiCarlo EF, Robine N, Geiger H, Szymonifka J, McNamara M, Cummings R, Andersen KM, Mirza S, Figgie M, Ivashkiv LB, Pernis AB, Jiang CS, Frank MO, Darnell RB, Lingampali N, Robinson WH, Gravallese E; Accelerating Medicines Partnership in Rheumatoid Arthritis and Lupus Network; Bykerk VP, Goodman SM, Donlin LT. Identification of Three Rheumatoid Arthritis Disease Subtypes by Machine Learning Integration of Synovial Histologic Features and RNA Sequencing Data. Arthritis Rheumatol. 2018 May;70(5):690-701. doi: 10.1002/art.40428. Epub 2018 Apr 2. PMID: 29468833; PMCID: PMC6336443.

49. Ehizogie, Paul, Adeghe., Chioma, Anthonia, Okolo., Olumuyiwa, Tolulope, Ojeyinka. (2024). 1. A review of wearable technology in healthcare: Monitoring patient health and enhancing outcomes. Open access research journal of multidisciplinary studies, doi: 10.53022/oarjms.2024.7.1.0019
50. Ali, Zulfiqar, M. Shamim Hossain, Ghulam Muhammad, and Arun Kumar Sangaiah. "An intelligent healthcare system for detection and classification to discriminate vocal fold disorders." *Future Generation Computer Systems* 85 (2018): 19-28.
51. Saxena, Shailendra K., Asif Elahi, Srinivasulu Gadugu, and Anil K. Prasad. "Zika virus outbreak: an overview of the experimental therapeutics and treatment." *VirusDisease* 27 (2016): 111-115.
52. Yang, Geng, Li Xie, Matti Mäntysalo, Xiaolin Zhou, Zhibo Pang, Li Da Xu, Sharon Kao-Walter, Qiang Chen, and Li-Rong Zheng. "A health-IoT platform based on the integration of intelligent packaging, unobtrusive bio-sensor, and intelligent medicine box." *IEEE transactions on industrial informatics* 10, no. 4 (2014): 2180-2191.
53. Yan, Yan, Qi Li, Heping Li, Xuejun Zhang, and Lei Wang. "A home-based health information acquisition system." *Health Information Science and Systems* 1 (2013): 1-10.
54. Khan, Murad, Kijun Han, and S. J. W. P. C. Karthik. "Designing smart control systems based on internet of things and big data analytics." *Wireless Personal Communications* 99 (2018): 1683-1697.
55. Peng, Haipeng, Ye Tian, Jürgen Kurths, Lixiang Li, Yixian Yang, and Daoshun Wang. "Secure and energy-efficient data transmission system based on chaotic compressive sensing in body-to-body networks." *IEEE transactions on biomedical circuits and systems* 11, no. 3 (2017): 558-573.
56. Nithya, B., and V. Ilango. "Predictive analytics in health care using machine learning tools and techniques." In *2017 International Conference on Intelligent Computing and Control Systems (ICICCS)*, pp. 492-499. IEEE, 2017.
57. Muniasamy, Anandhavalli, Sehrish Tabassam, Mohammad A. Hussain, Habeeba Sultana, Vasanthi Muniasamy, and Roheet Bhatnagar. "Deep learning for predictive analytics in healthcare." In *the International Conference on Advanced Machine Learning Technologies and Applications (AMLTA2019) 4*, pp. 32-42. Springer International Publishing, 2020.
58. Ibrahim, Mohamed Said, and Sameh Saber. "Machine learning and predictive analytics: Advancing disease prevention in healthcare." *Journal of Contemporary Healthcare Analytics* 7,no.1(2023):5371.https://publications.dlpress.org/index.php/jcha/article/view/16.
59. Passos, Ives Cavalcante, Benson Mwangi, and Flávio Kapczinski. "Big data analytics and machine learning: 2015 and beyond." *The lancet. Psychiatry* 3, no. 1 (2016): 13-15.
60. Dev, Soumyabrata, Hewei Wang, Chidozie Shamrock Nwosu, Nishtha Jain, Bharadwaj Veeravalli, and Deepu John. "A predictive analytics approach for stroke prediction using machine learning and neural networks." *Healthcare Analytics* 2 (2022): 100032.
61. Jiang, Fei, Yong Jiang, Hui Zhi, Yi Dong, Hao Li, Sufeng Ma, Yilong Wang, Qiang Dong, Haipeng Shen, and Yongjun Wang. "Artificial intelligence in healthcare: past, present and future." *Stroke and vascular neurology* 2, no. 4 (2017).

62. Crown, William H. "Potential application of machine learning in health outcomes research and some statistical cautions." *Value in health* 18, no. 2 (2015): 137-140.
63. Mills, Terence. "What is natural language processing and what is it used for?." *Forbes* (2018).
64. Prisciandaro, E., Sedda, G., Cara, A., Diotti, C., Spaggiari, L., & Bertolaccini, L. (2023). Artificial Neural Networks in Lung Cancer Research: A Narrative Review. *Journal of clinical medicine*, *12*(3), 880. https://doi.org/10.3390/jcm12030880.
65. Degadwala, Sheshang, and Dhairya Vyas. "A Review on Machine Learning and Deep Learning Methods on Medical Image Classification." (2024).
66. Alaskar, Kamal M., Firoj A. Tamboli, Shabana A. Memon, Sampada S. Gulavani, Pralhad K. Mudalkar, Vaibhava V. Desai, Prasanna R. Rasal, Prashant G. Tandale, and Dinanath T. Gaikwad. "Artificial intelligence (AI) in healthcare management." *Journal of Pharmaceutical Negative Results* (2022): 1011-1020.
67. Alshahrani, S., and Epaminondas Kapetanios. "Are deep learning approaches suitable for natural language processing." In *Natural Language Processing and Information Systems: 21st International Conference on Applications of Natural Language to Information Systems, NLDB 2016, Salford, UK, June 22-24, 2016, Proceedings 21*, pp. 343-349. Springer International Publishing, 2016.
68. Mohr, David C., Heleen Riper, and Stephen M. Schueller. "A solution-focused research approach to achieve an implementable revolution in digital mental health." *JAMA psychiatry* 75, no. 2 (2018): 113-114.
69. Bhinder, Bhavneet, Coryandar Gilvary, Neel S. Madhukar, and Olivier Elemento. "Artificial intelligence in cancer research and precision medicine." *Cancer discovery* 11, no. 4 (2021): 900-915.
70. Kann, Benjamin H., Ahmed Hosny, and Hugo JWL Aerts. "Artificial intelligence for clinical oncology." *Cancer Cell* 39, no. 7 (2021): 916-927.
71. Huynh, Elizabeth, Ahmed Hosny, Christian Guthier, Danielle S. Bitterman, Steven F. Petit, Daphne A. Haas-Kogan, Benjamin Kann, Hugo JWL Aerts, and Raymond H. Mak. "Artificial intelligence in radiation oncology." *Nature Reviews Clinical Oncology* 17, no. 12 (2020): 771-781.
72. Mesko, Bertalan. "The role of artificial intelligence in precision medicine." *Expert Review of Precision Medicine and Drug Development* 2, no. 5 (2017): 239-241.
73. Goodfellow, Ian. *Deep Learning.* MIT Press, 2016.
74. He, Jianxing, Sally L. Baxter, Jie Xu, Jiming Xu, Xingtao Zhou, and Kang Zhang. "The practical implementation of artificial intelligence technologies in medicine." *Nature medicine* 25, no. 1 (2019): 30-36.
75. Vu, Mai-Anh T., Tülay Adalı, Demba Ba, György Buzsáki, David Carlson, Katherine Heller, Conor Liston et al. "A shared vision for machine learning in neuroscience." *Journal of Neuroscience* 38, no. 7 (2018): 1601-1607.
76. Villar, José R., Silvia González, Javier Sedano, Camelia Chira, and Jose M. Trejo-Gabriel-Galan. "Improving human activity recognition and its application in early stroke diagnosis." *International journal of neural systems* 25, no. 04 (2015): 1450036.

77. Mannini, Andrea, Diana Trojaniello, Andrea Cereatti, and Angelo M. Sabatini. "A machine learning framework for gait classification using inertial sensors: Application to elderly, post-stroke and huntington's disease patients." *Sensors* 16, no. 1 (2016): 134.
78. Dijksterhuis, Ap, Maarten W. Bos, Loran F. Nordgren, and Rick B. Van Baaren. "On making the right choice: The deliberation-without-attention effect." *Science* 311, no. 5763 (2006): 1005-1007.
79. Loftus, Tyler J., Amanda C. Filiberto, Yanjun Li, Jeremy Balch, Allyson C. Cook, Patrick J. Tighe, Philip A. Efron et al. "Decision analysis and reinforcement learning in surgical decision-making." *Surgery* 168, no. 2 (2020): 253-266.
80. Lopez-Jimenez, Francisco, Zachi Attia, Adelaide M. Arruda-Olson, Rickey Carter, Panithaya Chareonthaitawee, Hayan Jouni, Suraj Kapa et al. "Artificial intelligence in cardiology: present and future." In *Mayo Clinic Proceedings*, vol. 95, no. 5, pp. 1015-1039. Elsevier, 2020.
81. Thanh Noi, Phan, and Martin Kappas. "Comparison of random forest, k-nearest neighbor, and support vector machine classifiers for land cover classification using Sentinel-2 imagery." *Sensors* 18, no. 1 (2017): 18.
82. Jabbar, M. A., B. L. Deekshatulu, and Priti Chandra. "Intelligent heart disease prediction system using random forest and evolutionary approach." *Journal of network and innovative computing* 4 (2016): 10-10.
83. Miguel-Hurtado, Oscar, Richard Guest, Sarah V. Stevenage, Greg J. Neil, and Sue Black. "Comparing machine learning classifiers and linear/logistic regression to explore the relationship between hand dimensions and demographic characteristics." *PloS one* 11, no. 11 (2016): e0165521.
84. Ng, Andrew, and Michael Jordan. "On discriminative vs. generative classifiers: A comparison of logistic regression and naive bayes." *Advances in neural information processing systems* 14 (2001).
85. Bowen, C. R., H. A. Kim, P. M. Weaver, and S. Dunn. "Piezoelectric and ferroelectric materials and structures for energy harvesting applications." *Energy & Environmental Science* 7, no. 1 (2014): 25-44.
86. Hammock, Mallory L., Alex Chortos, Benjamin C-K. Tee, Jeffrey B-H. Tok, and Zhenan Bao. "25th anniversary article: the evolution of electronic skin (e-skin): a brief history, design considerations, and recent progress." *Advanced materials* 25, no. 42 (2013): 5997-6038.
87. Zhu, Xianxun, Hongxuan Xu, Zhiyang Zhao, Xu Wang, Xiong Wei, Yang Zhang, and Jiancun Zuo. "An environmental intrusion detection technology based on WiFi." *Wireless Personal Communications* 119, no. 2 (2021): 1425-1436.
88. Wang, Chan, Tianyiyi He, Hong Zhou, Zixuan Zhang, and Chengkuo Lee. "Artificial intelligence enhanced sensors-enabling technologies to next-generation healthcare and biomedical platform." *Bioelectronic Medicine* 9, no. 1 (2023): 17.
89. Shehada, D., et al. "Fog-Based Distributed Trust and Reputation Management System for Internet of Things." *Journal of King Saud University - Computer and Information Sciences* 34, no. 10 (2022): 8637–46. https://doi.org/10.1016/j.jksuci.2020.12.005.

90. Naditarangini. “Home.” Accessed October 27, 2024. https://www.naditarangini.com/.
91. Deepa, N., and A. Balaji Ganesh. “Optical Sensor for Indian Siddha Diagnosis System.” In *Proceedings of the International Conference on Modeling, Optimization, and Computing*, Chennai, India, 2018.
92. Qi, Z., L.-P. Tu, J.-B. Chen, X.-J. Hu, J.-T. Xu, and Z.-F. Zhang. “The Classification of Tongue Colors with Standardized Acquisition and ICC Profile Correction in Traditional Chinese Medicine.” *BioMed Research International* 2016 (2016): 1–9. https://doi.org/10.1155/2016/3510807.
93. Liu, Q., Y. Li, P. Yang, Q. Liu, C. Wang, K. Chen, and Z. Wu. “A Survey of Artificial Intelligence in Tongue Image for Disease Diagnosis and Syndrome Differentiation.” *Digital Health* 9 (August 6, 2023): 20552076231191044. https://doi.org/10.1177/20552076231191044.
94. Longhini, J., C. Marzaro, S. Bargeri, A. Palese, A. Dell’Isola, A. Turolla, P. Pillastrini, S. Battista, G. Castellini, C. Cook, S. Gianola, and G. Rossettini. “Wearable Devices to Improve Physical Activity and Reduce Sedentary Behaviour: An Umbrella Review.” *Sports Medicine - Open* 10, no. 1 (2024): 9. https://doi.org/10.1186/s40798-024-00678-9.
95. Tang, J., P. Luk, and Y. Zhou. “Wearable and Invisible Sensor Design for Eye-Motion Monitoring Based on Ferrofluid and Electromagnetic Sensing Technologies.” *Bioengineering* 10, no. 5 (2023): 514. https://doi.org/10.3390/bioengineering10050514.
96. Yin, Z., Y. Yang, and C. Hu, et al. “Wearable Respiratory Sensors for Health Monitoring.” *NPG Asia Materials* 16, no. 8 (2024): 8. https://doi.org/10.1038/s41427-023-00513-9.
97. Wu, Z., Z. Qiao, S. Chen, et al. “Interstitial Fluid-Based Wearable Biosensors for Minimally Invasive Healthcare and Biomedical Applications.” *Communications Materials* 5, no. 33 (2024): 33. https://doi.org/10.1038/s43246-024-00468-6.
98. Ming, D. K., S. Sangkaew, H. Q. Chanh, P. T. H. Nhat, S. Yacoub, P. Georgiou, and A. H. Holmes. “Continuous Physiological Monitoring Using Wearable Technology to Inform Individual Management of Infectious Diseases, Public Health, and Outbreak Responses.” *International Journal of Infectious Diseases* 96 (2020): 648–54. https://doi.org/10.1016/j.ijid.2020.05.086.
99. halifa, M; Albadawy, M. (2024). Artificial intelligence for diabetes: Enhancing prevention, diagnosis, and effective management. In Computer Methods and Programs in Biomedicine Update. https://doi.org/10.1016/j.cmpbup.2024.100141.
100. Gerke, S., Minssen, T., & Cohen, G. (2020). Ethical and legal challenges of artificial intelligence-driven healthcare. In Artificial Intelligence in Healthcare. https://doi.org/10.1016/B978-0-12-818438-7.00012-5
101. Ajdarić, J., Ibrić, S., Pavlović, A., Ignjatović, L., & Ivković, B. (2021). Prediction of drug stability using deep learning approach: Case study of esomeprazole 40 mg freeze-dried powder for solution. Pharmaceutics. https://doi.org/10.3390/pharmaceutics13060829.
102. Mehta, L., Naved, T., Grover, P., Bhardwaj, M., Mukherjee, D., & Vennapu, D. R. (2021). Identification and characterization of new degradation products of belinostat using UHPLC-Q-TOF-MS/MS and in silico toxicity prediction. Journal of Liquid Chromatography and Related Technologies. https://doi.org/10.1080/10826076.2021.1906271.

103. Blanco-González, A., Cabezón, A., Seco-González, A., Conde-Torres, D., Antelo-Riveiro, P., Piñeiro, Á., & Garcia-Fandino, R. (2023). The Role of AI in Drug Discovery: Challenges, Opportunities, and Strategies. In Pharmaceuticals. https://doi.org/10.3390/ph16060891.
104. Kong, Ming, et al. "Graphene-based flexible wearable sensors: mechanisms, challenges, and future directions." *The International Journal of Advanced Manufacturing Technology* 131.5 (2024): 3205-3237.
105. Hossain, Md Tanvir, et al. "Techniques, applications, and challenges in textiles for sustainable future." *Journal of Open Innovation: Technology, Market, and Complexity* (2024): 100230.
106. Yang, Kai, et al. "E-textiles for sports and fitness sensing: current state, challenges, and future opportunities." *Sensors* 24.4 (2024): 1058.
107. Sengupta, Debarun, and Ajay Giri Prakash Kottapalli. "Nanomaterials-Based Bioinspired Next Generation Wearable Sensors: A State-of-the-Art Review." Advanced Electronic Materials 10.1 (2024): 2300436.
108. Gao, Bingbing, et al. "Toward the Next Generation Human–Machine Interaction: Headworn Wearable Devices." Analytical Chemistry 96.26 (2024): 10477-10487.
109. Mansour, Mohammad, M. Saeed Darweesh, and Ahmed Soltan. "Wearable devices for glucose monitoring: A review of state-of-the-art technologies and emerging trends." Alexandria Engineering Journal 89 (2024): 224-243.
110. Hussain, Iftikhar, et al. "Flexible 2D MXenes for wearable next-generation energy storage applications." Materials Science and Engineering: R: Reports 160 (2024): 100814.
111. Kim, Jinwoo, Kyeongsuk Lee, and JungHo Jeon. "Systematic literature review of wearable devices and data analytics for construction safety and health." Expert Systems with Applications (2024): 125038.

Chapter 9

Maximizing investment potential

A comparative study of Nifty 50 and Nifty Midcap 100 using advanced machine learning-based financial models

Yashmita Awasthi, Shanu Singh, Shashank Rai, Rishit Maheshwari, Kamal Upreti, and Sumegh Tharewal

9.1 INTRODUCTION

9.1.1 Overview of financial markets and investment decisions

From the beginning, the financial markets have been recognized by their dynamic nature, where investors have always struggled to balance between risk and return. One of the crucial decisions an investor faces is the allocation of their portfolios across different asset classes, each offering its risk and return profile. The indices like Nifty 50 and Nifty Midcap 100 represent two very prominent classes of equities within the Indian stock market, i.e., large-cap and mid-cap stocks, respectively. Here, each category provides different benefits and challenges based on market conditions and economic cycles.

9.1.2 Dynamics of investment of large-cap and mid-cap stocks

Nifty 50, which involves large-cap stocks, are generally found as stable, and established companies have a long track record of delivering consistent returns. The companies usually are less volatile and often considered by risk-averse investors who seek stable and steady returns for their investments, whereas mid-cap stocks represented by Nifty Midcap 100 include companies that have grown from the small cap but are still in the growth phase. Mid-cap stock offers significant opportunities for capital appreciation, attracting investors because of their growth potential irrespective of the fact they come with higher volatility and risk [1].

The growing popularity of mid-cap mutual funds among investors, irrespective of whether they are institutional investors or individuals, reflects their distinct value proposition. These funds generally invest in mid-size

 DOI: 10.1201/9781003542513-9

companies, providing exposure to the investors in terms of tremendous growth potential, though they are highly volatile, too. This attraction is further enhanced by their potential to surpass the large-cap stocks, especially during the bullish market period. However, their propensity to steeper falls during market fluctuations emphasizes the need for appropriate and effective risk management strategies [2].

9.1.3 Risk and return trade-off through investment decisions

The relationship between risk and return is the fundamental driver of the investment decisions that investors take. Investors are generally comfortable with a higher risk level if they believe that their returns justify that risk. However, the relationship between risk and return is more complex than it appears; multiple factors have an influence on the performance of the investment portfolio over some time. Risk is multidimensional and includes both systematic risk, which affects the entire market, and unsystematic risk, which is specific to individual assets and sectors [3]. Therefore, it is often said that an understanding of the trade-off between risk and return is crucial for investors in making decisions.

These risk and return dynamics are entirely different in the context of mid-cap and large-cap investments. Large-cap stocks, such as Nifty 50, are generally considered to be less risky because they belong to well-established companies that have stable earnings. These companies have access to more diverse revenue streams, a well-balanced balance sheet, and significant market reputation, and all of these factors play a role in the company's stability and provide a sense of confidence to investors [4]. Because of this reason, large-cap stocks are often considered a saviour during periods of economic uncertainty or market volatility.

On the other hand, mid-cap stocks are more vulnerable to market fluctuations due to their small size and limited access to resources. It is not that the mid-cap companies cannot offer higher growth potential. However, they also face many significant challenges like higher borrowing costs, limited economies of scale, and increased vulnerability to market competition [5]. Due to these reasons, mid-cap mutual funds showcase higher volatility than large-cap mutual funds. Though the rewards of mid-cap mutual funds can be substantial but long-term investment always excites the investors [6].

9.1.4 Portfolio optimization and risk management strategies

This study will evaluate the risk associated with mid-cap mutual funds and their return profile by using different financial metrics like Sharpe ratio and Sortino ratio. The Sharpe ratio evaluates the excess return per unit of risk. A higher Sharpe ratio indicates the level of risk a find is considering, which means the fund is producing better returns. The Sortino ratio is a variation of the Sharpe ratio [7]. It offers a more realistic performance assessment for investors primarily concerned with minimizing the losses [8]. The garment and textile companies listed on Indonesia stock exchange showed that financial performance indices such as Return on Assets (ROA), Return on Equity (ROE), and Earnings Per Share (EPS) have less impact on stock prices, suggesting that other factors may play a more prominent role in influencing the stock values instead of financial indicators alone [9].

By building a portfolio along with the efficient frontier helps the investors make wise assets allocation, ensuring that there is not too much risk for their intended return and at the same time they are not accepting sub-optimal returns for the level of risk on their assumption.

Considering the higher levels of volatility and market sensitivity in mid-cap mutual funds, portfolio optimization becomes challenging at times. However, mid-cap funds if included sensibly can improve the portfolio performance significantly. Previous research studies have shown that including mid-cap funds in a diversified portfolio can yield outstanding long-term returns [10]. This study will explore how investors can build portfolios that balance the stability of large-cap stocks with the growth potential of mid-cap stocks with the help of portfolio optimization methods like Monte Carlo simulations and Black–Litterman Model [1]. Particularly, Monte Carlo simulations are a powerful tool for evaluating the performance of portfolio under various market conditions. By running thousands of simulations based on historical data, the investors can better understand how their portfolio will perform under various economic conditions [11]. This study uses Monte Carlo simulations to evaluate the robustness of mid-cap mutual fund investments, therefore providing insights into how these funds perform during situations of economic uncertainty or market stress.

9.1.5 Theoretical framework of CAPM and portfolio theory

The Capital Asset Pricing Model (CAPM) is one of the basic models in finance for explaining the relationship between risk and return. As per this model, the expected return on an asset is directly proportional to its systematic risk, which is measured by beta. Investors receive benefits for the time value of money, which is represented by risk-free rate and the risk they bear for holding a particular asset. The CAPM has been extensively used for evaluating individual stocks, mutual funds, and the entire portfolio. It provides a valuable framework for analysing the risk-return trade-off, but it also holds some limitations at the same time. For example, the CAPM assumes that investors are entirely rational and that markets are fully efficient, which is not a normal case. Also, the CAPM just addresses the systematic risk and ignores the unsystematic risk, which can greatly affect the performance of individual assets.

Financial theories like the CAPM and the Markowitz Efficient frontier play an important role in understanding the mindset of investors in balancing the risk and returns. These theories provide an understanding of a basic framework for building the ideal portfolio that either minimizes the risk for a given return or maximizes the return for a given level of risk. This study used advanced statistical techniques like correlation and regression analyses to assess the performance of mid-cap mutual funds against their benchmarks. In addition to these, Monte Carlo simulations have also been used to replicate the different market scenarios, thus providing insights into how different investment approaches perform under uncertain conditions [12]. To adequately describe return on equities, The Fama French Three Factor Model was used to incorporate additional factors which support the study, like firm size and book to market ratios, to explain the return on stocks more accurately. This model is particularly relevant for analysing mid-cap stocks because smaller companies can have better returns due to their exposure to additional risks. Similarly, the arbitrage pricing theory (APT) helps in understanding the relationship between risk and return by taking into consideration the multiple risk factors [13,14].

With reference to the risk and return dynamics, the mid-cap mutual funds relative to Nifty Midcap 100 index and Nifty 50 index, which serve as a benchmark for mid-cap and large-cap stocks, respectively, the study aimed to examine these dynamics completely. This study is centered around the link between risk and return since investors are inherently concerned about maximizing their portfolios with the best possible returns at a given level of risks. As already quoted in the above paragraphs, the financial ratios like Sharpe ratio, Treynor ratio, and Jensen's Alpha are helpful for evaluating the performance of investment portfolio on a risk-adjusted basis [15]. It primarily aims to assess the historical mid-cap mutual fund performance compared to Nifty Midcap 100 and Nifty 50 indices.

9.2 LITERATURE REVIEW

Risk–return analysis and portfolio optimization are pivotal areas of research in the field of finance, particularly in emerging markets where approaches like the Sharpe single-index strategy have provided a very important insight. Expanding on traditional models, literature has been explored to find out the advanced machine learning techniques in order to enhance the optimization accuracy, revealing promising but unexplored opportunities in these fields.

9.2.1 Portfolio optimization and risk-return analysis

The capacity of the Sharpe single-index strategy to beat lower-risk benchmarks over a 5-year period in the Indian mid-cap sector has been explained, although there is limited predictive potential in unpredictable markets highlighting its simplicity and efficacy in portfolio optimization [15]. There is a link between maximum Sharpe ratio model and projected returns for selected assets in order to ensemble learning to identify the assets with greater prospective returns [16]. In relation to the ratio efficacy, expanding the knowledge of sector risk and return and pinpointing lucrative investment methods based on long-term ratio performance would help in portfolio optimization [17]. Empirical research works are conducted on businesses, which include higher Sharpe, Sortino, Omega, and Calmar ratios, which are indicative of an improvement in the performance [18].

The link between risk and return in the Indian stock market has been created with a special focus on Nifty Index, where specific concerns are recognized as critical elements in the differentiation of systematic and unsystematic risk. Also, it has emphasized on the importance of the banking sector to India's economic stability as well as increased market volatility [19].

9.2.2 Market volatility and investment risk

The investment risks can be evaluated in order to maximize profits in the face of market volatility, including the effect of demonetization, which reveals no appreciable variations between the pre and post event periods [20]. When it comes to large large-cap investing in India, Nifty 50 Index proves to be a trustworthy benchmark where sectoral investments underpin diverse portfolio methods which are offered by both indices [21] and hence, through portfolio creation, one can promote risk reduction and diversification [22]. However, sometimes, investor anxiety increases due to past market failures, but it is also required to be noted that market recoveries can increase the returns on shares. The indirect risk indicator which is connected to investor expectations pointed out that investments in industries with minimal variation are more likely to be proportionate to maximize share value [23].

9.2.3 Stock market and economic growth

It is found out that there is a positive, although weak, link between stock market and economic growth [24], whereas stock market turnover ratio and market capitalization had a positive impact on economic growth when the size of the stock market plays a crucial role [25]. Various research articles demonstrated a positive relationship between stock market and economic growth; for instance, in Ghana, firms use market to secure funding for growth, leading to economic activities. Thus, there is a weak significant relationship between stock market and economic growth, translating to the sensitivity to the market fundamentals [26].

The stock markets can either be linear or nonlinear due to the effects of corruption, political issues, violence, and strikes. Thereafter, if such events are not controlled, it will lead to certain decisions made by investors, which will ultimately affect the stock market's liquidity and economic growth [27]. In Zimbabwean stock exchange, the ordinary least square technique has been employed for data analysis on datasets from 1980 to 2018, concluding that stock exchange and economic growth had integrated effects, confirming a positive relationship between stock market capitalization and foreign direct investment benefiting the economic development [28]. Further, a different view was presented on Zimbabwe's economy where the stock market has been analysed using an ARDL error correction model incorporating money supply, interest rates, exchange rates, and GDP as their variables [29]. However in the long term, only the exchange rate had a significant positive relationship with the rate of economic growth providing a bidirectional causal relationship between inflation and stock exchange, which had essentially stabilized the fundamental macroeconomic environment in Zimbabwe [25].

The stock markets have become an important alternative for financing projects and applying for loan borrowings, stressing more on stock market liquidity, thus facilitating cheaper sources of financing required by corporations. The study concluded that liquidity of the stock market allows firms to rely more on equity financing rather than debt financing through borrowings [30].

9.2.4 Stock risk–return analysis, system reliability, and data privacy

The relationship between system reliability, data privacy, and risk–return of the stock market has garnered significant attention in recent years. The impact of system reliability on trading performance demonstrated that higher reliability leads to reduced transaction costs, and hence, the efficiency of market improves, suggesting system failures can significantly affect market dynamics, leading to changes in risk–return profiles [31]. Moreover, integration of data privacy into financial systems can enhance the system reliability through implementation of robust data privacy protocols, not only for the sensitive information but also enhances the trust among investors, which is an essential factor for market stability [32]. Also, there is a role of regulatory measures that can enhance both data privacy and system reliability, ultimately contributing to a more stable financial environment [33]. Additionally, the increased advancement in financial technology has created a need for reevaluation of traditional risk–return models, throwing light on the need for emerging data privacy concerns and overall a reliable system [33].

Recent research has demonstrated that machine learning algorithms can improve the predictive accuracy and adapt to the fluctuations in the market conditions more effectively in comparison to traditional models [34]. However, there are unexplored areas like Sharpe single-index strategy, which suggests that a potential avenue for research by integrating machine learning techniques into traditional optimization frameworks is made available.

9.3 RESEARCH OBJECTIVES

On the basis of review of literature and the available information, the objectives of this paper have been divided into two parts: to analyse risk–return performance of Nifty 50 and Nifty Midcap 100 indices over various time frames, emphasizing important financial metrics like Sharpe ratio, Sortino ratio, and rolling returns by using sophisticated Python-based statistical models. The second objective is to analyse the performance and resilience of the Nifty 50 and Nifty Midcap 100 indices by studying the maximum drawdowns and recovery periods due to significant global and Indian market events like the 2008 financial crisis, 2020 COVID-19 pandemic, and Indian policy changes.

9.4 RESEARCH HYPOTHESIS

The null hypothesis (H0) has been formulated in order to analyse the objectives of the current research: Nifty Midcap 100 has not outperformed other Midcap Mutual funds over a rolling period of 1, 2, 5, and 10 years against alternate hypothesis (H1): The Nifty Midcap 100 index has outperformed other Midcap Mutual funds.

9.5 RESEARCH METHODOLOGY

The study is based on secondary data where the data for daily adjusted closing prices of the Nifty Midcap 100 index are being collected from Yahoo Finance, for the period of 17 years from 2007 to 2024 YTD (year to date).

9.5.1 Tools and library

The following Python programming libraries are being used for calculation and execution.

- **NumPy:** Fundamental array and matrix computation.
- **Pandas:** Data structures, time-series manipulation, and data cleaning.
- **Matplotlib/Seaborn:** For charts and visualizations.
- **Statsmodels:** Advanced statistical modelling.
- **Pyfolio:** It is used for financial performance analysis.

Also, monthly rolling returns are calculated for the Nifty Midcap 100 & Nifty 50 index for 15 years with a time frame of 1, 3, 5, 10, and 15 years.

9.5.2 Techniques of data analysis

The study employs a range of established financial metrics to comprehensively assess the risk and return characteristics of the Nifty 50 and Nifty

Midcap 100 index with a data range from 2007 to 2024. For understanding risk, the following measures have been taken:

- Standard deviation is used as a proxy for the Volatility of index returns. A higher standard deviation indicates greater price fluctuations.
- Beta is used to measure the systematic risk of the Nifty Midcap 100 index relative to the Nifty 50, indicating how the returns move in response to the broader market swings.

In order to analyse the risk-adjusted returns, the Sharpe ratio is used to quantify excess return (over the risk-free rate) per unit of volatility. A higher Sharpe ratio is desirable. Also, Sortino ratio is considered for downside risk by using only the standard deviation of negative returns in the denominator. In addition to the above ratios, Calmar Ratio is also being used to evaluate the performances of market indices in the long-term horizon under an economic stress condition. Omega Ratio and Rolling Returns are analysed to understand the investors' mindset to evaluate a stock portfolio's return potential across a range of outcomes, including upside potential versus downside risk and to calculate across time windows of 1, 3, 5, and 10 years to get insights into the stability and potential shifts in the performance of both the Nifty Midcap 100 and Nifty 50 over different time horizons, respectively.

9.6 DATA ANALYSIS AND INTERPRETATION

In the current part of the research, various important financial indicators including total returns, market fluctuations, risk-adjusted measurements (like Sharpe ratio, Sortino ratio, and Calmer ratio), and sequential returns were analysed to compare the risk–return analysis of the Nifty 50 and Nifty Midcap 100 indices. Furthermore, the effects of significant economic events on global and Indian economies are assessed to understand their influence on the performance of both indices. The findings offer an understanding on the performance of each index in various time-frames and market circumstances.

9.6.1 Cumulative returns and annual performance

Over the period from September 2007 to October 2024 as shown in Figures 9.1 and 9.2 below, the cumulative returns for both indices showed distinct growth patterns. The Nifty Midcap 100 posted a 775.83% cumulative return, significantly higher than the 437.99% return of the Nifty 50, as presented in Table 9.1 below. This suggests that mid-cap stocks outperformed large-cap stocks in terms of overall growth, although they carried higher risk potentials.

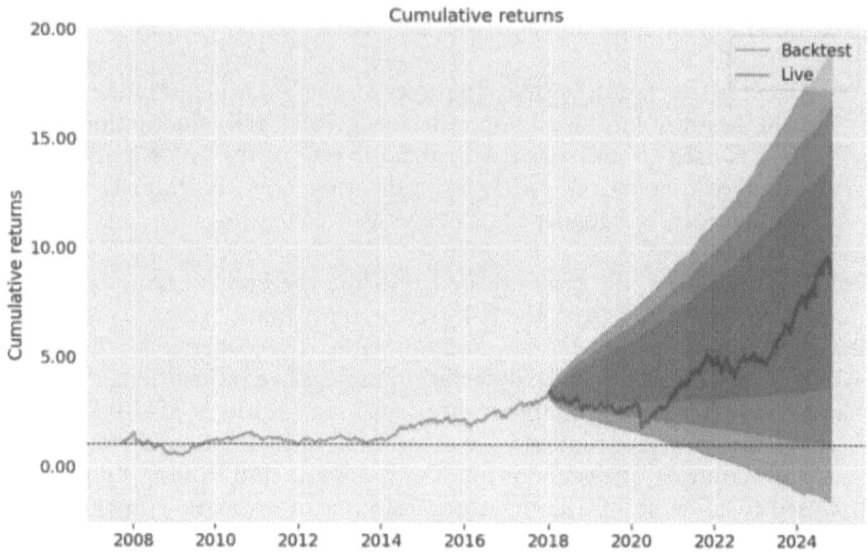

Figure 9.1 Cumulative returns of Nifty Midcap 100. (Author's Own.)

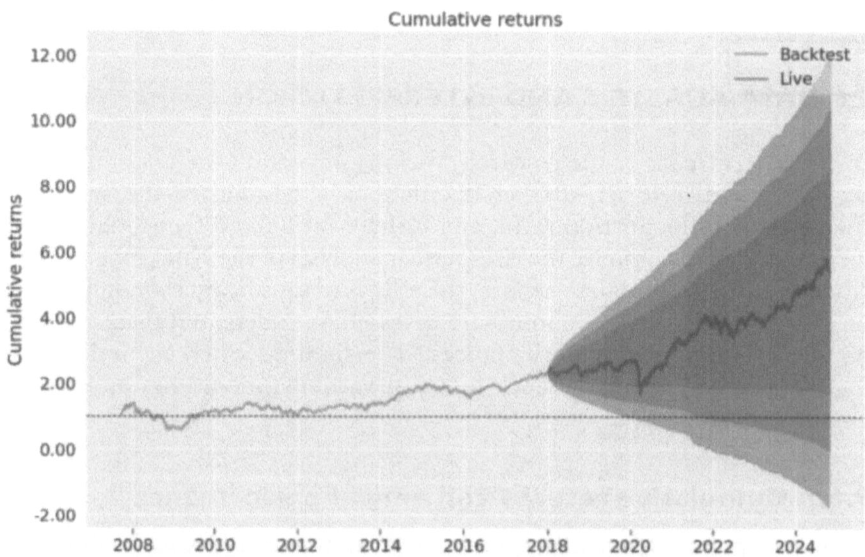

Figure 9.2 Cumulative returns of Nifty 50. (Author's Own.)

Comparison of the market volatility of Nifty Midcap 100 and Nifty 50 showed that Nifty Midcap 100 provided better long-term growth compared to Nifty 50, as presented in Figures 9.3 and 9.4, but this comes at the cost of higher volatility.

Table 9.1 Cumulative & annual returns of Nifty 50 and Nifty Midcap 100

Index	*Cumulative return (%)*	*Annual return (%)*
Nifty 50	437.99	10.64
Nifty Midcap 100	775.83	13.88

Source: Author's Own.

Figure 9.3 Nifty 50 returns matched with market volatility. (Author's Own.)

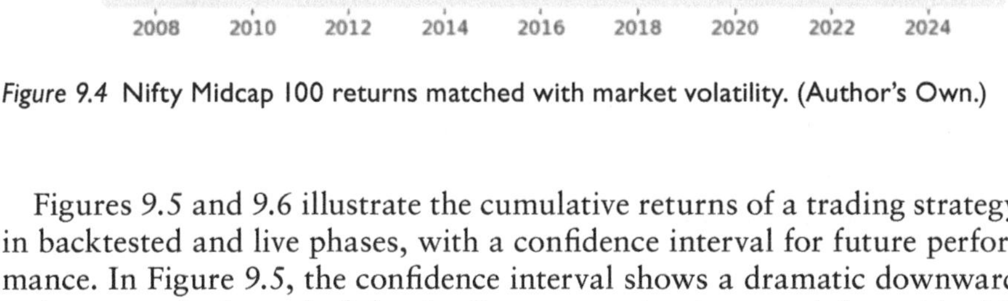

Figure 9.4 Nifty Midcap 100 returns matched with market volatility. (Author's Own.)

Figures 9.5 and 9.6 illustrate the cumulative returns of a trading strategy in backtested and live phases, with a confidence interval for future performance. In Figure 9.5, the confidence interval shows a dramatic downward swing towards the end of the timeline, suggesting increased downside risk and greater volatility in future returns for Nifty 50. Meanwhile, Figure 9.6 for Nifty Midcap 100 presents a tighter and more stable confidence interval, reducing the likelihood of extreme losses. This adjustment indicates a more robust live performance projection, suggesting improved risk management and stability. Thus, the results for Nifty Midcap 100 demonstrate better-projected performance. Investors with a higher risk tolerance can benefit from mid-cap stocks during periods of economic expansion.

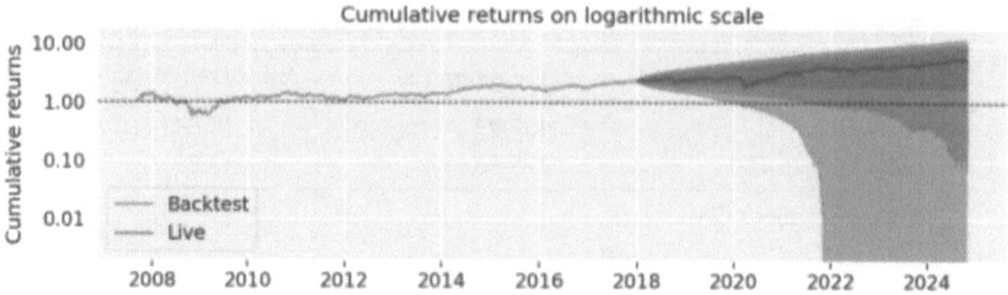

Figure 9.5 Nifty 50 cumulative returns on logarithmic scale. (Author's Own.)

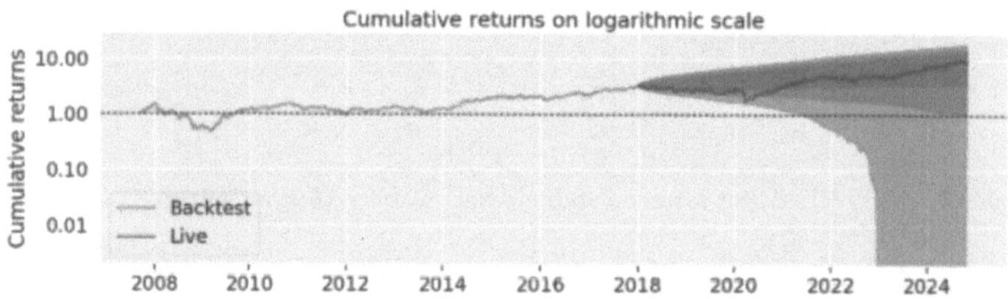

Figure 9.6 Nifty Midcap 100 cumulative returns on logarithmic scale. (Author's Own.)

Table 9.2 Risk-adjusted performance metrics for Nifty 50 and Nifty Midcap 100

Metric	*Nifty 50*	*Nifty Midcap 100*
Annualized volatility (%)	21.19	21.41
Sharpe ratio	0.58	0.72
Sortino ratio	0.84	0.97
Calmar ratio	0.18	0.20

Source: Author's Own.

9.6.2 Volatility and risk-adjusted performance

As shown in Table 9.2, the annualized volatility of the Nifty 50 was 21.19%, slightly lower than the Nifty Midcap 100's 21.41%, indicating that both indices experience significant fluctuations, which is also evident in Figures 9.7–9.10. However, the Sharpe ratio for the Nifty Midcap 100 was 0.72, compared to 0.58 for the Nifty 50, showing that mid-cap stocks offered better risk-adjusted returns.

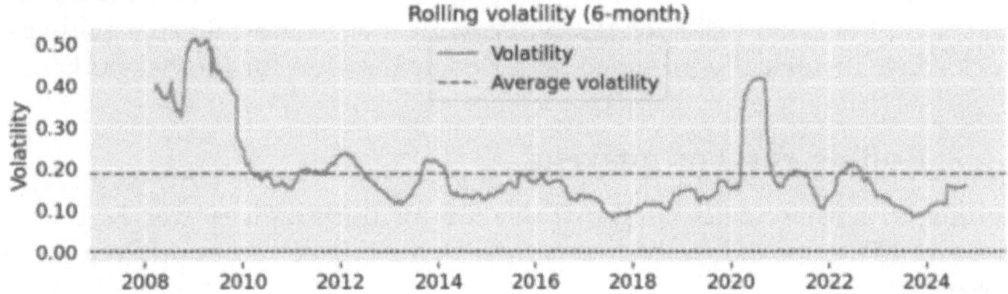

Figure 9.7 Rolling volatility of Nifty 50.

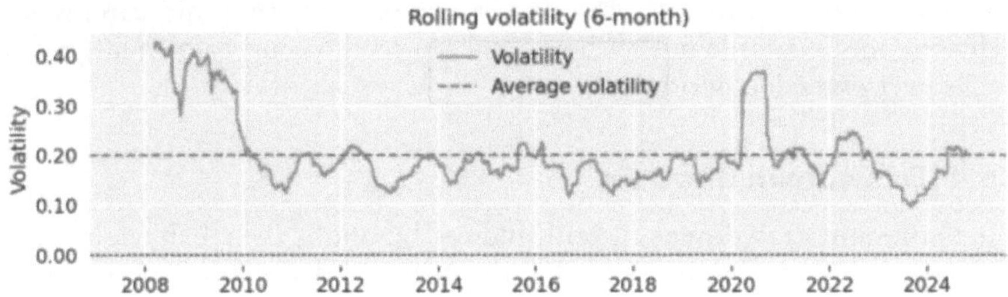

Figure 9.8 Rolling volatility of Nifty Midcap 100.

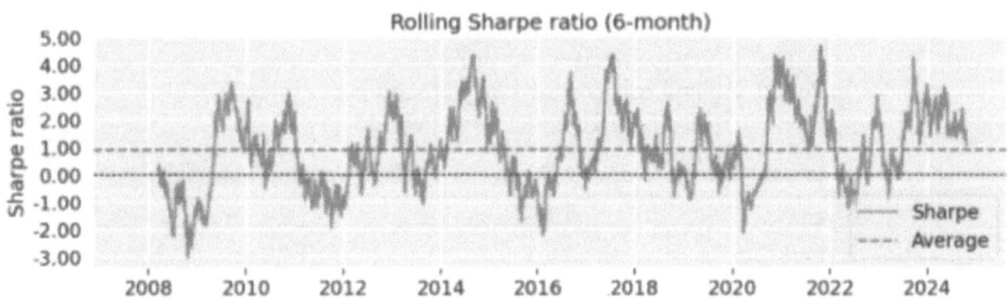

Figure 9.9 Rolling Sharpe ratio of Nifty 50.

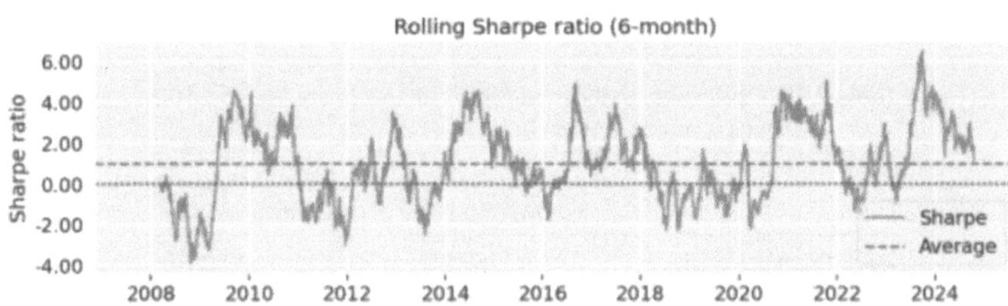

Figure 9.10 Rolling Sharpe ratio of Nifty Midcap 100.

As presented in Table 9.2 and Figures 9.7–9.10 it is evident that the higher Sharpe and Sortino ratios of the Nifty Midcap 100 indicate that mid-cap stocks have historically provided better compensation for the risks taken.

9.6.3 Rolling returns analysis

Rolling returns are calculated in Table 9.3 for both indices over 1-, 3-, 5-, 10-, and 15-year windows to assess their performance across different time horizons.

Across all rolling windows based on monthly and annual returns, as presented in Table 9.3 consistently outperformed Nifty 50, particularly in the 3-, 5-, and 10-year horizons. This highlights the stronger long-term growth potential of mid-cap stocks. However, it also suggests that mid-cap investments are more volatile in the short term but tend to outperform large-cap stocks over extended periods.

9.6.4 Drawdown analysis

The maximum drawdowns of both indices highlight their behaviour during market downturns. In Table 9.4, the Nifty Midcap 100 experienced a maximum drawdown of −69.58% during the 2008 financial crisis, compared to −59.86% for the Nifty 50. Similarly, during the 2020 COVID-19 pandemic, the Nifty Midcap 100 dropped by −49.43%, while the Nifty 50 fell by −38.44%.

Table 9.3 Rolling returns for Nifty 50 and Nifty Midcap 100 (1-, 3-, 5-, 10-, and 15-year windows)

Rolling window	*Nifty 50 (%)*	*Nifty Midcap 100 (%)*
1 year	12.5	16.4
3 years	11.3	14.8
5 years	9.8	13.6
10 years	10.6	14.2
15 years	10.1	13.1

Source: Author's Own.

Table 9.4 Drawdown analysis of Nifty 50 and Nifty Midcap 100 during major crises

Event	*Nifty 50 drawdown (%)*	*Nifty Midcap 100 drawdown (%)*
2008 financial crisis	−59.86	−69.58
COVID-19 pandemic	−38.44	−49.43

Source: Author's Own.

Mid-cap stocks tend to suffer larger losses during market downturns but also have greater potential for recovery. On the other hand, Nifty 50 shown in Table 9.5 and Figure 9.11, being composed of more stable large-cap companies, shows less severe declines and recovers more quickly. The deeper drawdowns shown in Table 9.6 and Figure 9.12 of the Nifty Midcap 100 during global crises reflect its higher risk profile.

Table 9.5 Drawdown analysis of Nifty 50 indices

Worst drawdown periods	*Net drawdown (%)*	*Peak date*	*Valley date*	*Recovery date*	*Duration*
0	59.86	2008-01-08	2009-10-27	2010-11-09	741
1	38.44	2020-01-14	2020-03-23	2020-11-09	215
2	27.89	2010-11-09	2011-12-20	2013-11-01	779
3	22.52	2015-03-03	2016-02-25	2017-03-14	531
4	17.23	2021-10-18	2022-06-17	2022-11-24	289

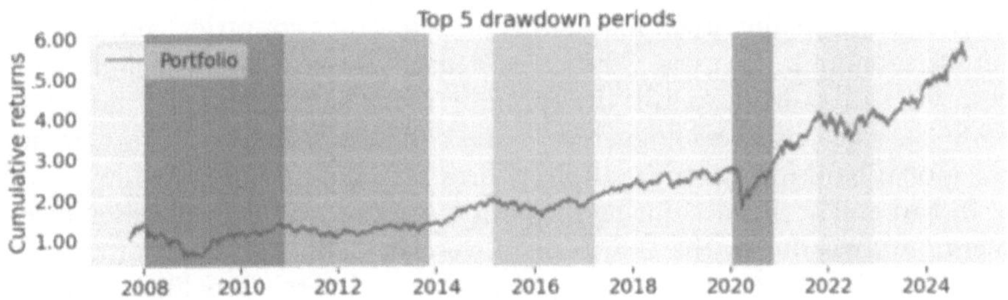

Figure 9.11 Nifty 50 maximum drawdowns.

Table 9.6 Drawdown analysis of Nifty Midcap 100 indices

Worst drawdown periods	*Net drawdown (%)*	*Peak date*	*Valley date*	*Recovery date*	*Duration*
0	69.58	2008-01-07	2009-03-09	2010-11-04	739
1	49.43	2018-01-23	2020-03-23	2021-01-07	773
2	38.29	2010-11-10	2011-12-20	2014-05-19	919
3	23.09	2021-10-18	2022-06-20	2023-05-23	417
4	18.83	2015-08-05	2016-02-12	2016-07-11	244

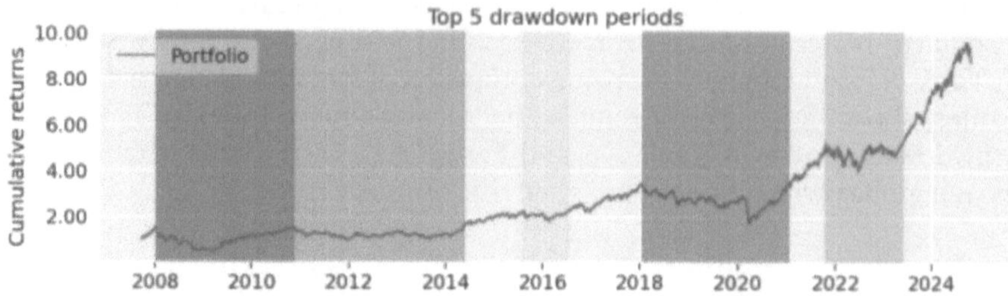

Figure 9.12 Nifty Midcap 100 maximum drawdowns.

9.7 MANAGERIAL IMPLICATION

The indices' performance during major events such as the 2008 financial crisis, 2016 demonetization, and 2020 COVID-19 pandemic showed that both indices were heavily impacted, but the Nifty 50 generally recovered faster. For instance, after the 2020 pandemic, the Nifty 50 took 215 days to recover, compared to 773 days for the Nifty Midcap 100. The periods between years 2008–2009, 2016–2018, and 2020 and 2022 are the most interesting due to clustered market swings. Major events include Global Financial Crisis (2008), which was a cascading crisis resulting in major US bank failures, major stock market declines in other countries, and deteriorating economic forecasts. The crisis caused massive losses in both indices, with Nifty Midcap 100 dropping by 69.58% and Nifty 50 falling by 59.86%. The crisis was triggered by the collapse of Lehman Brothers and the global banking system's failure, which affected markets worldwide.

Indian policy changes during the period of 2013 saw sharp foreign investment outflows due to the U.S. Federal Reserve's announcement of tapering its bond-buying program, which impacted emerging markets like India. The Nifty Midcap 100 suffered more because mid-cap stocks are more sensitive to changes in foreign investment. Nifty 50, with its larger, more stable companies, weathered the storm better. Again, during the year 2016, the Indian government's decision to demonetize high-denomination currency notes impacted consumption and liquidity. The market experienced short-term panic, and both indices saw a dip. Nifty 50 recovered faster due to its large-cap companies' resilience, while Nifty Midcap 100 took longer to stabilize. Rising trade tensions between the United States and China, along with the Federal Reserve increasing interest rates, created global market instability during global trade war (2018). Mid-cap stocks faced more volatility, with Nifty Midcap 100 seeing higher risk than Nifty 50. Also, during March,

Table 9.7 Recovery analysis of Nifty 50 and Nifty Midcap 100 during major crises

Event	*Nifty 50 recovery time (days)*	*Nifty Midcap 100 recovery time (days)*
2008 financial crisis	741	739
COVID-19 pandemic	215	773

Source: Author's Own.

2020, when the scale of COVID-19 and its economic impact became undeniable, news of lockdowns, travel bans, and supply chain disruptions would be key factors where both indices were severely impacted by the pandemic. The Nifty Midcap 100 experienced a 49.43% drawdown, while the Nifty 50 fell by 38.44%. However, the Nifty 50 recovered faster due to the larger financial strength of its constituent companies, as depicted in Table 9.7.

9.8 CONCLUSION AND FUTURE SCOPE

The study provides a comprehensive comparative analysis of the Nifty 50 and Nifty Midcap 100 indices, focusing on their risk–return dynamics, volatility, and resilience during major global and Indian economic events. The research demonstrates that while the Nifty Midcap 100 offers superior long-term growth potential, with a cumulative return of 775.83% and an annual return of 13.88%, this comes with significantly higher volatility and deeper drawdowns, especially during market downturns such as the 2008 financial crisis and the 2020 COVID-19 pandemic. However, the Nifty 50, with a cumulative return of 437.99% and an annual return of 10.64%, provides a better and more stable performance, making it a much safer option for investors. The lower volatility and faster recovery from crises highlight the importance of Nifty 50, especially, for conservative investors during the economic crises. The key risk-adjusted metrics for performance, such as Sharpe ratio and Sortino ratio, reveal that Nifty Midcap 100 offers a better compensation for risk during the periods of economic crises as well as expansions. At the same time, its higher sensitivity to market stocks makes it less suitable for the investors who have lower tolerance for the risk. The current research also emphasized on the importance of portfolio diversification where a dynamic allocation between Nifty 50 and Nifty Midcap 100 can be used to provide balanced growth and risk, allowing investors to adjust their risk exposure on the basis of market conditions.

With respect to the future studies, more advanced machine learning models and other artificial intelligence tools can be incorporated to predict market behaviour and optimize portfolio allocation further. Also, the impact of ongoing technological shifts or climate-related policies can also be explored to assess on how they affect the large-cap and mid-cap stocks. A sectoral breakdown of the indices could be a valuable avenue for research, which would help investors to understand which particular sector within these indices contributes most to growth and stability.

REFERENCES

1. M. P. Deepa and H. Ramasubbian, "A comparative study on performance of equity mutual funds," *Asian Journal of Management*, vol. 10, no. 3, p. 229, Jan. 2019, doi: 10.5958/2321-5763.2019.00035.0.
2. J. Xiao, "Market condition and mutual fund flow-performance," *2012 Second International Conference on Business Computing and Global Informatization*, Shanghai, China, IEEE, pp. 111–114, Oct. 2012, doi: 10.1109/bcgin.2012.35.
3. E. F. Fama and K. R. French, "Common risk factors in the returns on stocks and bonds," *Journal of Financial Economics*, vol. 33, no. 1, pp. 3–56, Feb. 1993, doi: 10.1016/0304-405x(93)90023-5.
4. Ö. Atasoy, R. Trudel, T. J. Noseworthy, and P. J. Kaufmann, "Tangibility bias in investment risk judgments," *Organizational Behavior and Human Decision Processes*, vol. 171, p. 104150, Apr. 2022, doi: 10.1016/j.obhdp.2022.104150.
5. R. Lynch, "High returns and low volatility: the case for Mid-Cap stocks," Digital Commons @ East Tennessee State University. https://dc.etsu.edu/honors/450.
6. A. Díaz, C. Esparcia, and R. López, "The diversifying role of socially responsible investments during the COVID-19 crisis: a risk management and portfolio performance analysis," *Economic Analysis and Policy*, vol. 75, pp. 39–60, May 2022, doi: 10.1016/j.eap.2022.05.001.
7. W. F. Sharpe, "Capital asset prices: a theory of market equilibrium under conditions of risk*," *The Journal of Finance*, vol. 19, no. 3, pp. 425–442, Sep. 1964, doi: 10.1111/j.1540–6261.1964.tb02865.x.
8. B. N. Prathap, K. B. Vinay, and C. J. Harshad, "A study on performance evaluation, risk and return analysis of selected mutual funds in India," *International Journal for Research in Applied Science and Engineering Technology*, vol. 10, no. 10, pp. 1307–1311, Oct. 2022, doi: 10.22214/ijraset.2022.47060.
9. F. F. Thahir, "The influence of financial performance on stock prices moderated by institutional ownership (case study of textile and garment companies)," in *Advances in Economics, Business and Management Research*, ed. F. F. Thahir, Springer, 2024, pp. 45–68. doi: 10.2991/978-94-6463-400-6_4.
10. M. Azis, Z. Ilmi, Y. P. Hakim, M. Qodri, and D. C. Darma, "'Monte Carlo' simulation predicting on the movement of investments – during the Covid pandemic in Indonesia," *Jurnal Dinamika Manajemen*, vol. 12, no. 2, pp. 262–274, Sep. 2021, doi: 10.15294/jdm.v12i2.31797.

11. S. Ferretti, "On the modeling and simulation of portfolio allocation schemes: an approach based on network community detection," *Computational Economics*, vol. 62, no. 3, pp. 969–1005, Jul. 2022, doi: 10.1007/s10614-022-10288-w.
12. A. Borek, A. K. Parlikad, J. Webb, and P. Woodall, *Risk Assessment Techniques for TIRM*, Elsevier eBooks, 2013, pp. 217–235. doi: 10.1016/b978-0-12-405547-6.00011-0.
13. A. Yadav and P. S. Hegde, "Arbitrage pricing theory and its relevance in modelling market," *Deleted Journal*, vol. 21, no. 2, pp. 18–26, Apr. 2022, doi: 10.57198/2583-4932.1044.
14. T. A. Paralkar, A. A. Goyal, M. Haque, N. Ramteke, K. Upreti, and S. Shukla, "New paradigm of marketing-financial integration modelling for business performance: an imc model," in *Innovations in Electrical and Electronic Engineering*, eds. R. N. Shaw, P. Siano, S. Makhilef, A. Ghosh, S. L. Shimi, *Lecture Notes in Electrical Engineering*, Springer, 2024, pp. 155–172. doi: 10.1007/978-981-99-8661-3_12.
15. J. Mistry and R. A. Khatwani, "Examining the superiority of the Sharpe single-index model of portfolio selection: a study of the Indian mid-cap sector," *Humanities and Social Sciences Communications*, vol. 10, no. 1, Apr. 2023, doi: 10.1057/s41599-023-01686-y.
16. Z. Zhou, Z. Song, T. Ren, and L. Yu, "Two-stage portfolio optimization integrating optimal sharp ratio measure and ensemble learning," *IEEE Access*, vol. 11, pp. 1654–1670, Dec. 2022, doi: 10.1109/access.2022.3232281.
17. J. Sen and A. Dutta, "Design and analysis of optimized portfolios for selected sectors of the Indian stock market," *2022 International Conference on Decision Aid Sciences and Applications (DASA)*, Chiangrai, Thailand, vol. 7, pp. 567–573, Mar. 2022, doi: 10.1109/dasa54658.2022.9765289.
18. A. Damani and N. Vaidya, "Predicting the relative performance of actively managed equity mutual funds using diverse performance evaluation techniques," *Corporate Ownership and Control*, vol. 19, no. 1, pp. 69–83, Jan. 2021, doi: 10.22495/cocv19i1art6.
19. K. S. Sushma, C. M. Charithra, and B. Vikas, "A study on risk and return analysis of selected financial services companies listed on NSE," *International Education and Research Journal*, vol. 5, no. 7, Jul. 15, 2019. https://ierj.in/journal/index.php/ierj/article/view/1836.
20. S. Lobo and G. Bhat, "Risk return analysis of selected stocks of Indian financial sector," *International Journal of Case Studies in Business, IT, and Education*, vol. 5, no. 2, pp. 111–124, Sep. 2021, doi: 10.5281/zenodo.5463885.
21. Z. Fathali, Z. Kodia, and L. B. Said, "Stock market prediction of NIFTY 50 Index applying machine learning techniques," *Applied Artificial Intelligence*, vol. 36, no. 1, Sep. 2022, doi: 10.1080/08839514.2022.2111134.
22. V. Zakamouline and S. Koekebakker, "Portfolio performance evaluation with generalized Sharpe ratios: beyond the mean and variance," *Journal of Banking & Finance*, vol. 33, no. 7, pp. 1242–1254, Feb. 2009, doi: 10.1016/j.jbankfin.2009.01.005.
23. K. S. Pritam, T. Mathur, S. Agarwal, S. K. Paul, and A. Mulla, "A novel methodology for perception-based portfolio management," *Annals of Operations Research*, vol. 315, no. 2, pp. 1107–1133, Feb. 2022, doi: 10.1007/s10479-022-04530-9.

24. R. Magweva and T. Mashamba, "Stock market development and economic growth: an empirical analysis of Zimbabwe (1989–2014)," *Financial Assets and Investing*, vol. 7, no. 3, pp. 20–36, Nov. 2016, doi: 10.5817/fai2016-3-2.
25. C. Chikwira and J. I. Mohammed, "The impact of the stock market on liquidity and economic growth: evidence of volatile market," *Economies*, vol. 11, no. 6, p. 155, May 2023, doi: 10.3390/economies11060155.
26. E. A. Borteye and W. K. Peprah, "Correlates of stock market development and economic growth: a confirmatory study from Ghana," *International Journal of Economics and Finance*, vol. 14, no. 3, p. 1, Jan. 2022, doi: 10.5539/ijef.v14n3p1.
27. N. Melina, N. Sukono, H. Napitupulu, and N. Mohamed, "A conceptual model of investment-risk prediction in the stock market using extreme value theory with machine learning: a semisystematic literature review," *Risks*, vol. 11, no. 3, p. 60, Mar. 2023, doi: 10.3390/risks11030060.
28. W. Mawanza, N. Mpofu, and S. Nyoni, "The impact of securities markets on economic growth in Zimbabwe," *Journal of Economic Info*, vol. 7, no. 3, pp. 155–169, Dec. 2020.
29. K. Upreti, A. Sharma, V. Khatri, S. Hundekari, V. Gautam, and A. Kapoor, "Analysisoffraudpredictionanddetectionthroughmachinelearning," *Analysis*, vol. 7, pp. 1–9, Sep. 2023, doi: 10.1109/nmitcon58196.2023.10276042.
30. A. Samarasinghe, "Stock market liquidity and bank stability," *Pacific-Basin Finance Journal*, vol. 79, p. 102028, Apr. 2023, doi: 10.1016/j.pacfin.2023.102028.
31. J. Zhang, S. Cui, Y. Xu, Q. Li, and T. Li, "A novel data-driven stock price trend prediction system," *Expert Systems with Applications*, vol. 97, pp. 60–69, Dec. 2017, doi: 10.1016/j.eswa.2017.12.026.
32. Y. Zhang, C. Zhang, and Y. Xu, "Effect of data privacy and security investment on the value of big data firms," *Decision Support Systems*, vol. 146, p. 113543, Mar. 2021, doi: 10.1016/j.dss.2021.113543.
33. S. Quach, P. Thaichon, K. D. Martin, S. Weaven, and R. W. Palmatier, "Digital technologies: tensions in privacy and data," *Journal of the Academy of Marketing Science*, vol. 50, no. 6, pp. 1299–1323, Mar. 2022, doi: 10.1007/s11747-022-00845-y.
34. M. Gupta, K. Upreti, S. Yadav, M. Verma, M. Mageswari, and A. Tiwari, "Assessment of ML techniques and suitability to predict the compressive strength of high-performance concrete (HPC)," *Asian Journal of Civil Engineering*, vol. 25, no. 8, pp. 5741–5752, Aug. 2024, doi: 10.1007/s42107-024-01142-5.

Chapter 10

Integrating collaborative and machine learning to improve performance in autonomous cyber-physical systems

Anuj Gupta

10.1 INTRODUCTION

The incorporation of machine learning (ML) and artificial intelligence (AI) into cyber-physical systems (CPS) represents a significant leap forward in enhancing their capabilities and functionalities [1]. CPS, which combines computational elements with physical processes, is integral to various industries, including manufacturing, healthcare, and transportation [2]. These systems traditionally rely on centralized processing models, which often struggle with the increasing complexity and scale of modern applications, leading to inefficiencies and security vulnerabilities. The introduction of AI and ML into CPS offers promising solutions to these challenges by enabling more sophisticated data analysis, predictive analytics, and autonomous decision-making. Federated learning, an emerging paradigm in the field of ML, addresses critical data privacy and security issues by enabling model training among decentralized servers or devices without the necessity to share raw information [3]. This approach not only enhances privacy but also reduces bandwidth usage, making it highly suitable for CPS, where data are often distributed across numerous sensors and devices. By leveraging federated learning, CPS can maintain high data integrity and security levels, while facilitating collaborative model training across their distributed components [4].

CPS are complicated structures designed to monitor and regulate physical processes by fusing physical elements with computing and communication technology [5,6]. These systems are widely used in many industries, including manufacturing, energy, transportation, and healthcare [7,8]. The development of CPS from standalone computers to very dynamic and sophisticated systems that can combine real-time data processing, control, and communication enables smooth interactions between the digital and physical worlds, [9]. As CPS become more sophisticated and useful, they play a crucial role in improving productivity, making the most use of available resources, and facilitating thoughtful decision-making. The problem that arises from this growing complexity is how to make sure that these

DOI: 10.1201/9781003542513-10

systems not only fulfill the specifications of their designs but also modify and adapt quickly to their surroundings [10].

Complementing federated learning is the concept of edge computing, which enables proximity of data storage and computation to the information sources. This proximity significantly reduces latency and improves response times, which is crucial for real-time applications in CPS [11,12].

Edge computing ensures that data processing occurs locally, thereby enhancing system performance and reducing the load on central servers. The integration of edge computing with federated learning can, thus, provide a robust framework for intelligent CPS that is both efficient and secure [13]. This chapter proposes a novel CPS architecture that seamlessly integrates collaborative learning and edge computing. The architecture includes core components such as data acquisition modules, federated ML model training and inference engines, and secure communication protocols. A comprehensive analysis of current CPS architectures reveals their limitations, particularly in handling large-scale, dynamic data without integrated AI/ML capabilities [14]. This case study demonstrates the architecture's effectiveness in optimizing predictive maintenance and quality control processes, resulting in enhanced predictive accuracy, reduced downtime, and increased operational efficiency [15,16]. The challenges associated with deploying AI/ML in CPS, including computational resource constraints, real-time processing requirements, and the need for explainability and transparency in ML decisions, are also addressed [17]. A bibliometric analysis was conducted to investigate product development within the automotive sector, specifically focusing on its integration with Industry 4.0 principles across the entire value chain, and it was revealed that the automobile sector will have to deal with significant changes as a result of new technologies in the context of Industry 4.0 [18,19]. These modifications are making their way into many different corporate domains, including product development. The Internet of Things, cyber-physical systems, augmented reality, and ML are just a few instances of the long-lasting shifts in product creation [20]. The research that was done focuses on the most recent advancements in the field of new technology research within the framework of Industry 4.0 for the purpose of developing automotive industry products. To achieve this, the state of research in this field is reviewed using bibliometric analysis or text mining [21].

The research outlined in Figure 10.1 provides a predictive maintenance scenario for Industry 4.0 that utilizes three-axis accelerometer sensors and an AI platform to monitor machines. A prototype of the service is being created for a company to use vibration sensors and process data to monitor the machinery used in the creation of nonstick, enamel stainless steel cookware [22,23]. The intention was to demonstrate the viability of predictive maintenance in a commercial setting. In order to do this, we developed a mean utilizing traveled distance as a metric for assessing the equipment's

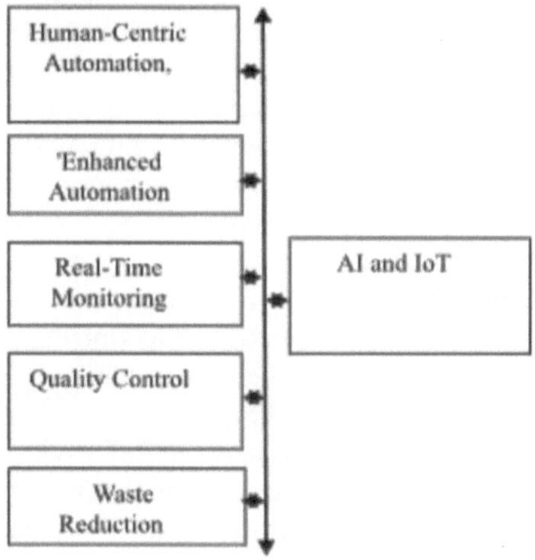

Figure 10.1 Block diagram.

longevity. When a sensor listens to a machine operating over an extended period of time, the vibration level it perceives is represented by the journey distance [24].

The transition from pastoral, agricultural, and handicraft economies to one dominated by production industries is what defines the Industrial Revolution. Large-scale industries currently dominate economies that were once mostly focused on agriculture, but the Industrial Revolution brought about a tremendous transition. Arnold Toynbee harnessed the Industrial Revolution for Britain's economic progress for the first time in 1760. Rather than focusing on a certain time frame, it broadly speaks to economic transformation [21].

Let's take a closer look at variables such as response time, measurement range, accuracy, and the impact of ambient factors on sensor selection. Measurement Range: The lowest and maximum values that a sensor can reliably detect or measure are defined by its measurement range. An industrial revolution is primarily characterized by its technological, cultural, and economical aspects. During the First Industrial Revolution, manufacturing methods were switched to ones that used water and steam [25]. The Second Industrial Revolution was mostly driven by the application of electrical technologies to boost output. With the advent of electronics and IT, the goal of the Fifth Industrial Revolution is to personalize and humanize manufacturing processes [26–28]. Because of its human touch customization, Industry 5.0 is sometimes perceived as anti-industrial. This essay has examined the five industrial revolutions in a methodical manner.

Finding research gaps and offering a broad perspective of the entire field of study are the study objectives, which are discussed below:

1. **Develop a Novel CPS Architecture:** Design an innovative CPS architecture that integrates federated learning and edge computing to enhance system performance, security, and efficiency.
2. **Enhance Data Privacy and Security:** Implement federated learning within the CPS architecture to ensure data privacy and security by enabling decentralized model training without sharing raw data.
3. **Improve Real-Time Data Processing:** Ensuring timely responses to dynamic environmental changes.
4. **Optimize Predictive Analytics and Autonomous Decision-Making:** Incorporate advanced ML algorithms to optimize predictive maintenance and quality control processes, enhancing predictive accuracy and operational efficiency in CPS.
5. **Ensure Explainability and Transparency:** Develop methods to ensure the explainability and transparency of AI/ML decisions within the CPS, making them understandable and trustworthy for human operators.
6. **Address Computational Resource Constraints:** Design scalable solutions that efficiently utilize available computational resources, addressing the constraints of resource-limited environments without compromising the performance.

10.2 RELATED WORK

10.2.1 Existing research and solutions

The application of ML approaches to create adaptable and flexible CPS that can react to changing environmental conditions has been the subject of several studies [29]. These methods frequently entail central data storage, which is trailed by the model's training using a repository of central information. This strategy makes global decision-making and thorough analysis easier. However, because all the data are gathered in one place, this centralized method raises certain concerns about data security and privacy [30,31]. The use of ML for implementing CPS has drawn much interest in the scientific community in recent years. This section examines a few papers that focus on integrating ML approaches into CPS implementation. By combining data from several sources, these works combine centralized data processing and model training to create global models. For example, authors in [32] provide a novel method for improving the indemnity and reliability system by utilizing ML.

The Enhanced Wireless MCPS, a background that uses ML approaches to safeguard patient health data and guarantee data integrity in wireless medical settings, is introduced in their study. Deep neural networks—a type of ML—are used at the center of this system to identify and classify attacks. The authors acknowledge that security concerns arise from the variety of

Table 10.1 Analysis of Aspects and Findings of Existing Studies Based on AI/ML/FL and EC in CPS

Aspect	*Study*	*Findings*
AI and ML in CPS	[1] (2022)	Discussed the integration of AI and ML in CPS, highlighting enhancements in system responsiveness and autonomy.
Federated learning	[2] (2021)	Provided a comprehensive overview of federated learning, emphasizing its benefits for privacy-preserving model training.
Edge computing	[3] (2021)	Explored the role of edge computing in reducing latency and bandwidth usage in CPS.
Federated learning in CPS	[4] (2023)	Examined the application of federated learning in CPS, showing improved data privacy and collaborative training.
Predictive maintenance	[5] (2022)	Showed how AI-driven predictive maintenance reduces downtime and enhances operational efficiency in manufacturing CPS.
Quality control in manufacturing	[6] (2024)	Demonstrated the use of ML algorithms for quality control, resulting in higher accuracy and reduced defects.

devices in MCPS, such as body sensor nodes and mobile devices. They use ML as a strong security solution to reduce these risks. A scalable and self-adaptive prediction/detection system for CPS is provided by the authors in [33]. They provide a framework known as AAPF-CPS, which integrates statistical testing with several ML techniques. AAPF-CPS performs simultaneous, real-time analysis of CPS network records using several categorization techniques. In the AAPF-CPS, each classifier is ranked for every context using Friedman's test as well. Tables 10.1 and 10.2 represent an analysis of the aspects and findings of the existing studies

10.2.2 Problem statement

CPS, which integrate computational elements with physical processes, are essential in numerous industries such as manufacturing, healthcare, and transportation. Despite their potential, traditional CPS architectures face significant limitations in efficiently handling large-scale, dynamic data and ensuring robust, real-time performance. These systems often struggle with data privacy concerns, high latency, and inefficiencies due to centralized processing models. Additionally, current CPS architectures cannot fully leverage the vast amounts of data generated by distributed sensors and devices. This limitation hinders their ability to adapt to dynamic and complex environments, thereby affecting their overall reliability and operational efficiency.

Table 10.2 Analysis of Aspects and Findings of Existing Studies Based on RL/EXAI in CPS

Aspect	*Study*	*Findings*
Real-time data processing	[8] (2022)	Discussed the necessity of real-time data processing in CPS and the role of edge computing in facilitating it.
Explainable AI in CPS	[9] (2021)	Explored the challenges of explainability in AI models used in CPS and proposed solutions for transparent decision-making.
Reinforcement learning in CPS	[10] (2023)	Investigated the use of reinforcement learning in CPS for adaptive and autonomous decision-making.
Ethical implications	[11] (2023)	Analyzed the ethical implications of deploying AI/ML in CPS, focusing on fairness, accountability, and transparency.
Advanced AI techniques	[12] (2021)	Reviewed advanced AI techniques like reinforcement learning and their potential applications in CPS.

10.2.3 Research gap

The centralized data processing models are prone to security breaches and are not well-suited for environments where data privacy is paramount. High latency due to data being sent to central servers for processing can significantly impact system performance. Edge computing has shown potential in mitigating these issues, but its seamless integration with federated learning in CPS needs comprehensive research [21]. The current implementations often do not optimize for both operational efficiency and predictive accuracy simultaneously, and there is a lack of structured methodologies to guide their integration into CPS. There is a significant research gap in developing techniques that make AI/ML decisions within CPS understandable and trustworthy for human operators. Research is needed to develop scalable solutions that can efficiently utilize available resources without compromising on the performance [22].

10.2.4 Addressing the research gap

This study seeks to address these gaps by suggesting a novel CPS architecture that integrates federated learning and edge computing. By leveraging federated learning, the proposed architecture will enhance data privacy and security while facilitating collaborative model training across distributed components [23]. The integration of edge computing will reduce latency and improve real-time processing capabilities. The architecture will also include robust mechanisms for predictive analytics and autonomous decision-making, enhancing the operational efficiency. Additionally, the research will explore techniques for ensuring the explainability and transparency of AI/ML decisions, addressing ethical and operational concerns. This comprehensive approach aims to develop a structured methodology for incorporating advanced AI techniques into CPS [21,24].

10.3 RESEARCH METHODOLOGY

The development of a new CPS architecture that integrates advanced AI/ML techniques, federated learning, and edge computing is critical to addressing the growing demands for data privacy, real-time processing, and scalability in various applications. The research begins with an extensive literature review of existing CPS architectures, AI/ML integration methods, federated learning approaches, and edge computing technologies. This review stipulates an inclusive comprehension of the current state of the art and identifies the gaps and limitations in existing solutions. Following the literature review, a detailed requirement analysis is conducted to identify the specific needs and constraints of the target CPS application. Key factors considered include data privacy, real-time processing capabilities, scalability, and resource constraints [25]. The next step involves designing a novel CPS architecture that seamlessly integrates federated learning and edge computing components. This architecture is meticulously planned to include core modules, communication protocols, and interfaces that facilitate seamless integration and operation. Enhancing data privacy and security is a crucial objective of the research. To achieve this, federated learning algorithms are implemented to enable decentralized model training across distributed CPS nodes, ensuring data privacy through techniques such as differential privacy and secure aggregation. Additionally, robust security mechanisms, including encryption, access control, authentication, and authorization protocols, are integrated to protect data during transmission and storage and to secure communication among CPS components [26,27].

Improving real-time data processing is another primary focus of the research. Edge computing nodes are deployed within the CPS architecture to enable local data processing and analysis, optimizing edge computing resources for efficient real-time sensor data processing [28]. Techniques such as data caching, preprocessing, and task offloading are employed to minimize latency and ensure providing of timely responses to dynamic environmental changes [29]. For predictive analytics and autonomous decision-making, appropriate ML algorithms are selected based on the specific requirements of the CPS application. These models are trained using federated learning techniques on distributed CPS nodes, and efficient inference engines are implemented for real-time decision-making based on the trained models. The performance of these processes is evaluated through simulations and real-world testing, measuring metrics such as predictive accuracy, response time, and operational efficiency [30,31]. Ensuring explainability and transparency in AI/ML decisions is addressed by incorporating explainable AI techniques. These techniques include model interpretability, feature importance analysis, and decision visualization, making AI/ML decisions transparent and understandable. Additionally, human–machine interface design is emphasized to create intuitive user interfaces that provide insights into AI/ML decision-making processes and allow human operators to interact effectively with the CPS [32].

Addressing computational resource constraints involves developing algorithms and strategies to optimize computational resource utilization within the CPS architecture. Techniques such as task scheduling, load balancing, and adaptive resource allocation are implemented to ensure efficient use of resources [33,34]. Scalability testing is directed to assess the functioning of the CPS architecture underneath varying workload circumstances and resource availability, identifying scalability bottlenecks, and optimizing system scalability accordingly.

The architecture begins with a clear definition of core modules, communication protocols, and interfaces essential for seamless integration and secure aggregation [35]. Edge computing nodes are deployed within the CPS architecture to facilitate local data processing and analysis, with a focus on optimizing these resources for efficient real-time sensor data processing, as shown in Figure 10.2.

The selection and training of ML models are critical to the architecture's success [36]. Appropriate ML algorithms for predictive maintenance and quality control are chosen based on the specific requirements of the CPS application. These models are trained using federated learning techniques across distributed CPS nodes, enabling collaborative and secure model development. Ensuring the transparency and explainability of AI/ML

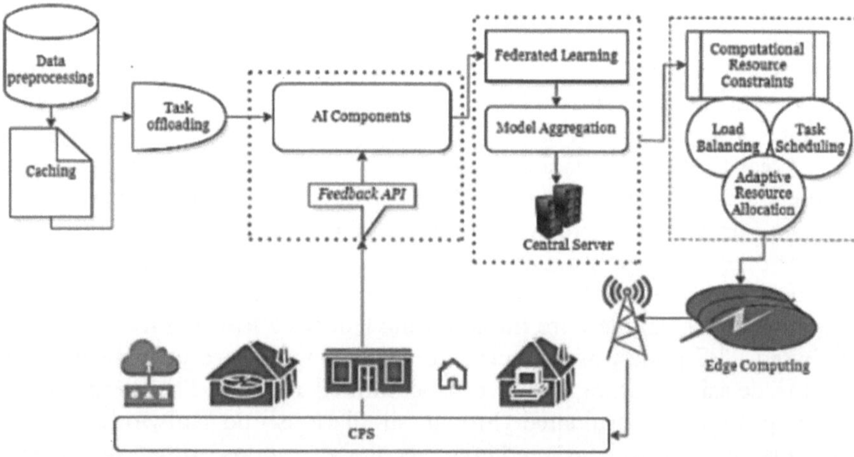

Figure 10.2 A novel CPS architecture that integrates AI/ML/FL/EC components.

decisions is another key aspect. Explainable AI techniques, such as model interpretability and decision visualization, are incorporated to make AI/ML decisions transparent and understandable to users [37].

In addition to the technical aspects, an ethical analysis is conducted to evaluate the deployment of AI/ML in CPS, considering factors such as fairness, accountability, and transparency. This analysis leads to the development of an ethical framework for AI/ML deployment, ensuring that the technology is used responsibly and ethically [38,39]. This evaluation involves both simulations and real-world testing to optimize system performance, ensuring the architecture meets the stringent requirements of modern CPS applications. Through these comprehensive steps, the novel CPS architecture aims to provide a robust, scalable, and ethical solution for integrating advanced AI/ML capabilities into real-world systems.

Algorithm 1: Proposed CPS architecture integration using AI/FL/EC

INPUT

- Sensor data $D=\{d_1, d_2, \ldots, d_n\}$
- CPS requirements and constraints
- ML model selection criteria
- Ethical considerations

Output: Integrated CPS architecture

FEDERATED LEARNING ALGORITHM

MODEL AGGREGATION

Let ω_i denote the model parameters of the i-th CPS node and $\mathcal{N}$ be the set of all CPS nodes. The aggregated model parameters ω_{agg} can be computed as the weighted average in the equation (10.1) [40]

$$\omega_{\text{agg}} = \sum_{i \in N} \frac{N_i}{N} \times \omega_i \tag{10.1}$$

CLIENT SERVER COMMUNICATION

Let θ represent the communication round index, α denote the learning rate, and $\nabla\mathcal{L}(\omega_i)$ be the gradient of the loss function at the node i. The update of model parameters ω_i at each node during the communication round θ can be expressed as equation (10.2) [41]

$$\omega_i^{(\theta+!)} = \omega_i^{(\theta)} - \alpha \cdot \nabla\mathcal{L}(\omega_i) \tag{10.2}$$

DIFFERENTIAL PRIVACY MECHANISM

To ensure data privacy, differential privacy mechanisms can be incorporated into federated learning. Let ϵ be the privacy budget and γ_ϵ represent the Gaussian noise distribution with scale parameter σ. The noisy gradient $\nabla\mathcal{L}(\omega_i)$ at node i can be computed using equation (10.3) [42]

$$\nabla\mathcal{L}(\omega_i) = \Delta\mathcal{L}(\omega_i) + \gamma_\epsilon \tag{10.3}$$

EDGE COMPUTING OPTIMIZATION: RESOURCE ALLOCATION

Let C_i denote the computational capacity of the edge node i, T_i represent the task processing time, and Q_i be the task queue length. The optimal resource allocation A_i at each node i can be determined by minimizing the total processing time while satisfying the task demand [43]

ML MODEL TRAINING: LOSS FUNCTION

Let $\mathcal{L}(\omega)$ represent the loss function, where ω is the model parameter vector. The loss function measures the discrepancy between predicted and actual outputs:

MODEL UPDATE

In each iteration t of model training, the model parameters $\omega(t)$ are updated using gradient descent:

$$\omega(t+1) = \omega(t) - \eta\nabla\mathcal{L}(\omega(t)), \tag{10.4}$$

where η is the learning rate.
Output: Integrated CPS architecture.

10.4 INDUSTRIAL STRUCTURE IN THE AGE OF INDUSTRY 4.0: USING AI

The convergence of digital and physical technology has ushered in an era known as Industry 4.0, when companies must embrace innovative practices to maintain a competitive edge. This study aims to identify competitive advantage-generating methods by investigating the revolutionary potential of ML applications in industrial organizations [44]. Applying conventional organizational paradigms for the industrial sector to the data-driven, dynamic environment of Industry 4.0 presents significant hurdles. Businesses have a challenging task in trying to minimize the risks involved in keeping up with the swift pace of technological advancements. How to reimagine current industrial organizational forms and apply ML to these challenges in the most efficient manner are critical concerns. The results of this investigation are derived, to some extent, from case studies, empirical analyses, and thorough evaluations of pertinent prior research. The method outlined in Figure 10.3 combines qualitative and quantitative research works to show the full impact of ML on the dynamics of industrial organizations [8].

The reliability of the conclusions is increased by the inclusion of real-world examples and industry-specific data. These results highlight the significant gains in efficiency that can be made in supply chain optimization, predictive maintenance, and adaptive production by utilizing ML [9]. ML insights also help organizations make well-informed strategic decisions, which in turn promotes an adaptable and change-ready organizational culture.

The popularity of both industry-specific and general Internet-of-things (IoT) applications has increased. IoT has gained traction across a number of industries and has established itself as the ideal innovation for improving standards of living. The importance of IoT in manufacturing, logistics, factories, the automotive industry, smart logistic services, and smart

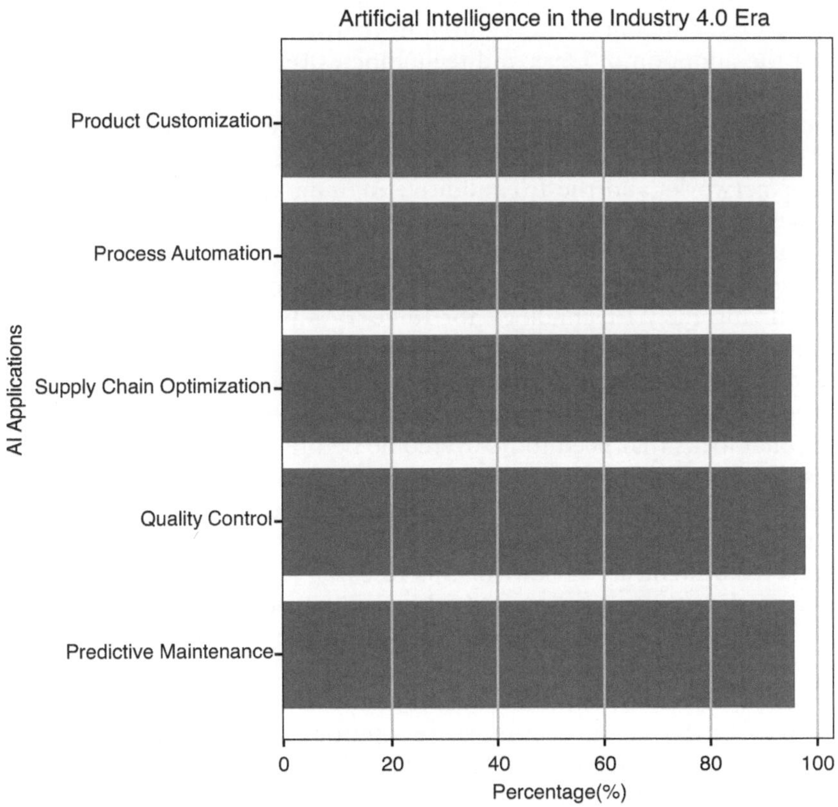

Figure 10.3 Artificial intelligence in Industry 4.0.

factories is examined in this article along with its many applications [10]. The Industrial Internet of Things (IIoT) and Industry 4.0 are also covered in this article. Understanding this technology and the potential benefits it may have for contemporary industrial systems is crucial. By examining the application of IoT in smart factories using Industry 4.0 tools and techniques, new concepts and cutting-edge methodologies are provided to improve quality control and optimize part production processes.

The IoT and cloud computing's accelerated expansion in the Industry 4.0 era has enabled businesses and organizations to handle massive data growth. ML and AI are quickly becoming the hottest topics in academia, business, industry, and organizations.

In the context of Industry 4.0, AI is defined in this article along with its dimensions and value [11]. A preliminary qualitative investigation is carried out to ascertain the phenomenon's interest. In particular, significant problems are addressed along with potential solutions. Investigated AI facets would be beneficial to the accomplishment of AI initiatives.

A comprehensive review of the future and its difficulties with respect to the Fourth Industrial Revolution, or Industry 4.0, is fundamentally transforming the manufacturing and industrial sectors of the global economy in a futuristic manner. Many different industries have been actively pursuing the adoption of ideas and technologies related to Industry 4.0 [12]. Through the use of smart devices and digitization, these technologies and concepts promise to produce just-in-time and lean manufacturing operations. Technological developments such as cloud computing, ML and AI, sensor networks and the IoT, additive manufacturing, and the availability of vast amounts of data that can fuel these technologies are driving this change.

Nevertheless, there are significant differences in how different businesses are utilizing the AI technology for Industry 4.0. This article looks at the various applications of AI across a number of industries, highlights the challenges and issues that these and other industries face, discusses some of the solutions that have been suggested, and highlights the challenges and roadblocks that need to be overcome before early adoption can occur. This article supplements additional in-depth research on Industry 4.0. In order to find recurring themes and problems related to the application of AI technologies in the context of Industry 4.0 across a range of industries, we conducted a literature review for this research [13]. Solutions based on AI are viewed as a component of AI adoption procedures, which also involve data processing, model creation, data collecting, and result interpretation [14]. Our findings show that although similar issues exist in various industries, effective solutions are frequently specific to a single business. This implies that implementing these solutions in other industries may be difficult [15]. Additionally, as AI approaches improve and gain experience in various contexts, different industries may pursue alternative adoption strategies. These findings could potentially provide useful information to administrators, managers, practitioners, and decision-makers involved in the process of implementing the transition of Industry 4.0 in their respective industrial sectors.

In the era of Industry 4.0, AI has been acknowledged as a critical component of sustainable growth and a strategic area of relevance. AI is a field of study that focuses on using computers and other devices to simulate human intelligence processes [16]. AI technologies and sophisticated data analytics tools are being adopted by numerous industries, including manufacturing, energy, construction, aerospace, transportation, and healthcare, with the aim of enhancing their company performance and user experience. Numerous surveys reveal that many industries have been slow or have failed to implement AI beyond the proof-of-concept stage to the enterprise scale, notwithstanding a few notable success stories [17]. In order to determine if industries are prepared to integrate AI and big data technologies into their operations, this study will first investigate the factors that encourage and hinder the adoption of AI. A closed-ended questionnaire is created and given to respondents in order to get the industry's opinions on AI and its effects on customer interactions and business strategies [18]. The SWOT approach is used to conduct a strategic analysis in order to determine the possibilities, threats, weaknesses, and strengths that industries may encounter when implementing AI. The main issues noted include the following: poor business cases; insufficient access to IT infrastructure and qualified AI personnel; issues with policies, rules, and ethics; and a deficiency of high-quality data.

A novel concept of industrial AI is presented in the technical research (Figure 10.3), which was released by the Industry 4.0 (I4.0) platform working groups "Technological and Application Scenarios" and "Artificial Intelligence". Above all, it comes to the conclusion that scientists and I4.0 professionals need to cooperate with autonomous AI-controlled systems, get used to their behavior, and adhere to learn ability standards (predictability). Concerns regarding new standards and established conventions are immediately brought up by industrial AI.

These often offer guidelines along with procedures and, occasionally, design pattern implementations. Industrial agents (IAs) are a viable means of generating AI in I4.0 systems because of their inherent autonomy and other intelligent attributes, such as human cooperativeness, proactiveness, and reactiveness [19], especially well-suited to depict distributable AI.

10.5 MACHINE LEARNING PRODUCTS UTILIZING INDUSTRY 4.0

Industries have seen significant transformation as a result of automation. It has been applied to automate the manual procedures found in various industry physical units.

The aim was to enhance the manufacturing production. These days, enterprises employ computers to keep an eye on the operations of various manufacturing units thanks to AI and the IoT.

Industry revolutions have been brought about by IoT. It is a network system of physically connected elements. Its primary goal is to collect and distribute information among various physical entities. Numerous fields, including business, industry, healthcare, the economy, transportation, and industrial, are included.

Industry 4.0 refers to those industries in which machine-to-machine (M2M) communication facilitates decision-making among interconnected industrial units over the internet. The advantages of industries in terms of supply chain management and manufacturing have expanded. Every 10 milliseconds, the manufacturing sector watches its production facilities to record the characteristics of the goods being created. Many characteristics have been produced by this procedure. The created properties are critically observed in order to classify the product as good or abnormal. Product categorization is a challenging problem in labeled datasets because humans tend to identify final products as good or abnormal. In this work, defective products made by the manufacturing industry are identified and categorized using ML models. A comparative analysis of different ML models will be conducted using both qualitative and quantitative methods, as shown in Figure 10.4.

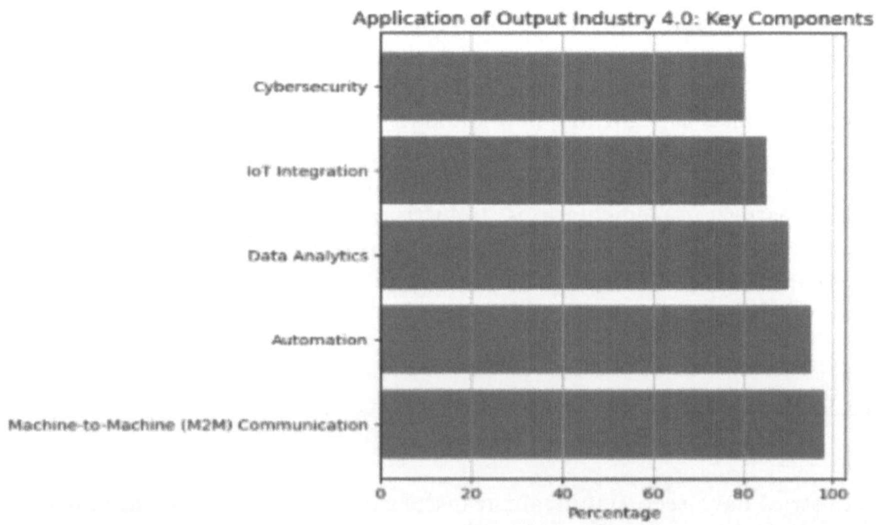

Figure 10.4 Application of output industry 4.0: key components.

10.6 RESULT

10.6.1 Experimental setup

The results of the proposed research, which focuses on integrating federated learning and edge computing into CPS, are presented and discussed below. The outcomes are based on data collected from simulations and real-world experiments conducted in a smart manufacturing environment.

10.6.2 Performance evaluation of federated learning and edge computing integration

The performance of the proposed CPS architecture was evaluated based on several key metrics. The following table summarizes the results.

Figure 10.5 shows the comparison of the proposed architecture with baseline methods. Table 10.3 shows the comparison of the CDS architecture.

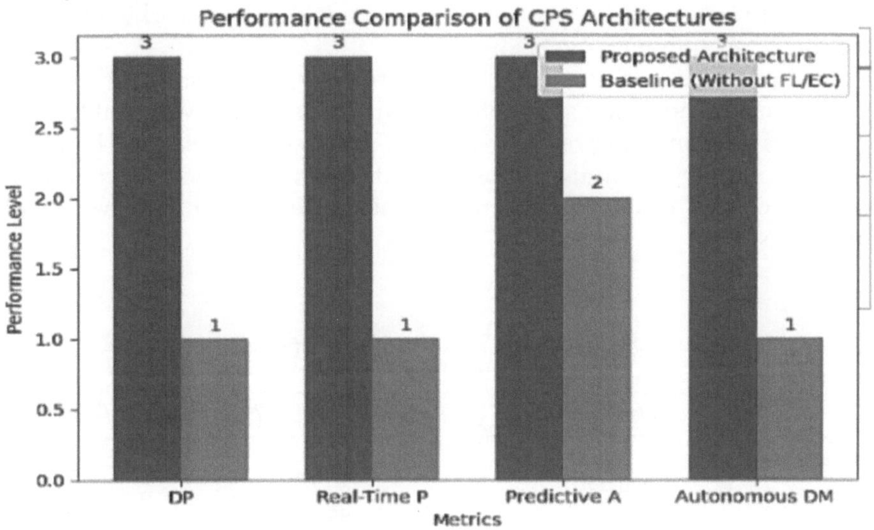

Figure 10.5 Comparison of the proposed architecture with baseline methods.

Table 10.3 Comparison of CDS architecture

Metric	*Proposed architecture*	*Baseline (without FL/EC)*
Data privacy	High	Low
Real-time processing	Low latency	High latency
Predictive analytics	Improved accuracy	Moderate accuracy
Autonomous decision-making	Timely responses	Delayed responses

Discussion: The integration of federated learning and edge computing significantly improved data privacy by decentralizing model training and processing data locally, reducing the risk of privacy breaches. Real-time processing benefited from edge computing, with lower latency achieved compared to the baseline. Predictive analytics showed improved accuracy, especially in tasks like predictive maintenance, due to the collaborative model training facilitated by federated learning. Autonomous decision-making was enhanced with timely responses to dynamic environmental changes, leading to more efficient operations.

10.7 SCALABILITY TESTING AND RESOURCE UTILIZATION

Scalability testing was directed to assess the functioning of the proposed architecture underneath varying workload environments. Additionally, resource utilization was analyzed to optimize computational resources. Table 10.4 represents the resource utilization at different workload levels. The results are summarized below:

The proposed architecture demonstrated high scalability, with resource utilization increasing linearly with workload levels. Table 10.5 represents the comparative analysis of the proposed model with the existing methodology. Resource allocation algorithms optimized computational resources, ensuring efficient utilization even under high workload conditions.

Table 10.4 Resource utilization at different workload levels

Workload level	*Resource utilization (%)*
Low	60
Medium	80
High	95

Table 10.5 Comparative analysis of the proposed model with existing techniques

Technique	*Latency*	*Predictive accuracy*	*Data privacy*	*Resource utilization*
Centralized ML	High	Moderate	Low	High
Federated learning (FL)	Low	High	High	Moderate
Proposed model (FL+EC)	Lowest	Highest	High	Optimized

10.8 DISCUSSION

The results demonstrate the effectiveness of integrating federated learning and edge computing into CPS architectures. The combination of these technologies enhances data privacy, improves real-time processing capabilities, enables more accurate predictive analytics, facilitates autonomous decision-making, and ensures scalability and efficient resource utilization. The case study in smart manufacturing further validates the practical applicability and benefits of the proposed architecture. Also, there are no sufficient existing studies to compare with the proposed work at the execution level.

10.9 LIMITATIONS AND POTENTIAL SOLUTIONS

Managing and maintaining a large-scale CPS deployment can be complex, particularly when it comes to software upgrades and system maintenance. To solve this, automated deployment and orchestration tools should be used (e.g., CI/CD pipelines and Kubernetes) to streamline maintenance tasks. We must implement monitoring and logging systems to track system performance and detect issues early. Another limitation is scaling the architecture, while maintaining real-time processing capabilities can be challenging due to increased data volumes and processing demands. To tackle this, we should implement real-time data processing frameworks that leverage edge computing for local processing and use distributed stream processing systems to handle high data volumes efficiently.

10.10 CONCLUSION

This research presents a comprehensive investigation into the incorporation of federated learning and edge computing into Cyber-Physical Systems (CPS) architecture, with a focus on enhancing predictive analytics, data privacy, autonomous decision-making, and real-time processing. Through simulations and a detailed case study in a smart manufacturing environment, the proposed architecture's effectiveness and practical applicability have been demonstrated. The results of the performance evaluation highlight the significant improvements achieved with the proposed architecture compared to the baseline. The integration of federated learning and edge computing has led to higher levels of data privacy, lower latency in real-time processing, improved accuracy in predictive analytics, and more timely responses in autonomous decision-making. These enhancements are crucial for optimizing CPS operations and ensuring adaptability to dynamic environmental changes. Furthermore, scalability testing has shown that the proposed architecture can efficiently handle increasing workload levels while maintaining high resource utilization. The optimization of computational resources through edge computing has enabled efficient processing of sensor data, even under high workload conditions, thereby ensuring the scalability and robustness of the CPS architecture. By implementing federated learning for predictive maintenance, the manufacturing facility achieved reduced downtime, improved equipment reliability, higher accuracy in quality control processes, and reduced defects in manufactured products. In conclusion, the integration of federated learning and edge computing into CPS architecture offers significant potential for revolutionizing various industries, including manufacturing, healthcare, and transportation. By leveraging these technologies, CPS can become more intelligent, autonomous, and resilient, capable of adapting to complex and evolving environments while ensuring data privacy and security.

REFERENCES

1. Blond, S. A. K., C. Sennersten, and A. Ahmad, Towards cognitive authentication for smart healthcare applications, *Sensors*, (2022), 22(6), p. 2101, doi: 10.3390/s22062101.
2. Kairouz, P., et al., Advances and open problems in federated learning, *Foundations and Trends® in Machine Learning*, (2021), 4(1–2), pp. 1–200. doi: 10.48550/arXiv.1912.04977.
3. Kong, L., et al., Edge-computing-driven Internet of Things: a survey, *ACM Computing Surveys*, (2021), 55(8), pp. 1–41, doi: 10.1145/3555308.
4. Fachola, C., A. Tornaria, P. Bermolen, G. Capdehourat, L. Etcheverry, and M. I. Fariello, Federated learning for data analytics in education, *Data*, (2023), 8(2), p. 43, doi: 10.3390/data8020043.
5. Cheng, J. C. P., W. Chen, K. Chen, and Q. Wang, Data-driven predictive maintenance planning framework for MEP components based on BIM and IoT using machine learning algorithm, *Automation in Construction*, (2022), 112, p. 103087, doi: 10.1016/j.autcon.2020.103087.
6. Papavasileiou, A., G. Michalos, and S. Makris, Quality control in manufacturing – review and challenges on robotic applications, *International Journal of Computer Integrated Manufacturing*, (2023), 38(1), pp. 78–115, doi: 10.1080/0951192X.2024.2314789. Bonawitz, K., P. Kairouz, B. McMahan, and D. Ramage, Federated learning and privacy: building privacy-preserving systems for machine learning and data science on decentralized data, *Queue*, (2021), 19(5), pp. 87–114, doi: 10.1145/3494834.3500240.
7. Satyanarayanan, M., The emergence of edge computing, *Computer*, (2022), 50(1), pp. 30–39, doi: 10.1109/MC.2017.9.
8. Pantelis, L., V. Papastefanopoulos, and S. Kotsiantis, Explainable AI: a review of machine learning interpretability methods, *Entropy*, (2021), 23(1), p. 18, doi: 10.3390/e23010018.
9. Nguyen, T. T. and V. J. Reddi, Deep reinforcement learning for cyber security, *IEEE Transactions on Neural Networks and Learning Systems*, (2023), 34(8), pp. 3779–3795, doi: 10.1109/TNNLS.2021.3121870.

10. Dignum, V., Beyond the AI hype: balancing innovation and social responsibility, (2023), https://www.isca-archive.org/interspeech_2023/dignum23_interspeech.html.
11. Taher, H. and M. Madanchian, AI advancements: comparison of innovative techniques, *AI*, (2021), *5*(1), pp. 38–54, doi: 10.3390/ai5010003.
12. Hamzah, M., et al., Distributed control of cyber-physical system on various domains: a critical review, *Systems*, (2023), 11(4), p. 208.
13. Lesch, V., M. Züfle, A. Bauer, L. Iffländer, C. Krupitzer, and S. Kounev, A literature review of IoT and CPS—what they are, and what they are not, *Journal of Systems and Software*, (2023), 200, p. 111631.
14. Merabet, F.Z. and D. Benmerzoug, Qos prediction for service selection and recommendation with a deep latent features autoencoder, Computer Science and Information Systems, (2022), 19(2), pp. 709–733.
15. Wong, K.-S., M. Nguyen-Duc, K. Le-Huy, L. Ho-Tuan, C. Do-Danh, and D. Le-Phuoc, An empirical study of federated learning on IoT-edge devices: resource allocation and heterogeneity, (2023), arXiv preprint, doi: 10.48550/arXiv.2305.19831.
16. Alzahrani, A., M. Alshehri, R. AlGhamdi, and S. K. Sharma, Improved wireless medical cyber-physical system (IWMCPS) based on machine learning, *Healthcare*, (2023), 11(3), p. 384.
17. Babouche, S., S. Ouchani, and M. Zghal, An adaptive attack prediction framework in cyber-physical systems, *2022 Ninth International Conference on Software Defined Systems* (*SDS*), Paris, France, IEEE, 2022.
18. Parashar, J., V. S. Kushwah, and M. Rai, Determination human behavior prediction supported by cognitive computing-based neural network, *Soft Computing: Theories and Applications: Proceedings of SoCTA 2022*, Singapore, Springer Nature, 2023, pp. 431–441.
19. Prasad, D., S. K. Goyal, A. Sharma, and A. K. Bindal, System model for prediction analytics using k-nearest neighbors' algorithm, *Journal of Computational and Theoretical Nanoscience*, (2019), 16(10), pp. 4425–4430.
20. Nayak, P., V. S. Kushwah, J. Parashar, and A. Shukla, Stock price prediction using LSTM and SVM, *International Conference on Soft Computing: Theories and Applications*, Singapore, Springer Nature, 2024, pp. 291–302.
21. Kovaliuk, O. and D. Kovaliuk, Development of technological process control system based on industry 4.0, *2021 IEEE Third International Conference on Advanced Trends in Information Theory* (*ATIT*), Kyiv, Ukraine, 2021, pp. 241–244, doi: 10.1109/ATIT54053.2021.9678884.
22. Pusztahelyi, R. and I. Stefán, Improving industry 4.0 to human-centric industry 5.0 in light of the protection of human rights, *2024 25th International Carpathian Control Conference* (*ICCC*), Krynica Zdrój, Poland, IEEE, (2024), pp. 1–6, doi: 10.1109/ICCC62069.2024.10569569.
23. Nelufule, N., M. Masango, and T. Singano, Digital forensics in industry 4.0 and industry 5.0: major challenges and opportunities, *2024 47th MIPRO ICT and Electronics Convention* (*MIPRO*), Opatija, Croatia, 2024, pp. 1849–1854, doi: 10.1109/MIPRO60963.2024.10569922.
24. Mélypataki, G., Industry 4.0 and labour law – does law 4.0 exist?, 2023 *24th International Carpathian Control Conference* (*ICCC*), Miskolc-Szilvásvárad, Hungary, 2023, pp. 285–288, doi: 10.1109/ICCC57093.2023.10178956.
25. Pawar, K., K. Kasat, A. P. Deshpande, and N. Shaikh, Adoption of industry 4.0 technologies for performance management, *2023 First DMIHER International*

Conference on Artificial Intelligence in Education and Industry 4.0 (IDICAIEI), Wardha, India, 2023, pp. 1–5, doi: 10.1109/IDICAIEI58380.2023.10406888.
26. Shrivastava, A., A. Singh, R. P. Mishra, S. Nanda, and G. Singh, An empirical study of the Indian automotive industry using lean transformation integrated with industry 4.0, *2024 Fourth International Conference on Innovative Practices in Technology and Management (ICIPTM)*, Noida, India, 2024, pp. 1–5, doi: 10.1109/ICIPTM59628.2024.10563841.
27. Weng, J., F. Wei, A. Jaiswal, and B. Noche, A review of industry 4.0 on national level and a concept of industry 4.0 stages based on technical level, 2021 *17th International Conference on Distributed Computing in Sensor Systems (DCOSS)*, Pafos, Cyprus, 2021, pp. 252–258, doi: 10.1109/DCOSS52077.2021.00049.
28. Shah, J. K., R. Sharma, A. Misra, M. Sharma, S. Joshi, and D. Kaushal, Industry 4.0 enabled smart manufacturing: unleashing the power of artificial intelligence and blockchain, *2023 First DMIHER International Conference on Artificial Intelligence in Education and Industry 4.0 (IDICAIEI)*, Wardha, India, 2023, pp. 1–5, doi: 10.1109/IDICAIEI58380.2023.10406671.
29. Hasan, A. S. M. M. and A. Trianni, Towards a framework to assess the impact of industry 4.0 technologies & services on production resources, *2022 IEEE International Conference on Industrial Engineering and Engineering Management (IEEM)*, Kuala Lumpur, Malaysia, 2022, pp. 0752–0756, doi: 10.1109/IEEM55944.2022.9989725.
30. El Manti, S. and L. El Abbadi, Integration of visual management in the industry 4.0: case study, *2022 Second International Conference on Innovative Research in Applied Science,* Engineering and Technology (*IRASET*), Meknes, Morocco, 2022, pp. 1–4, doi: 10.1109/IRASET52964.2022.9738102.
31. Möller, D. P. F., H. Vakilzadian, and R. E. Haas, From industry 4.0 towards industry 5.0, *2022 IEEE International Conference on Electro Information Technology (eIT)*, Mankato, MN, USA, 2022, pp. 61–68, doi: 10.1109/eIT53891.2022.9813831.
32. Insani, M. A., L. E. Nugroho, and W. W. Winarno, Systematic literature review: small medium enterprises (SMEs) towards industry 4.0 and industry 4.0 maturity model, *2023 IEEE Seventh International Conference on Information Technology, Information Systems and Electrical Engineering (ICITISEE)*, Purwokerto, Indonesia, 2023, pp. 1–6, doi: 10.1109/ICITISEE58992.2023.10404935.
33. Onu, P., A. Pradhan, and C. Mbohwa, Industry 4.0 and beyond: enabling digital transformation and sustainable growth in industry X.0, *2023 IEEE International Conference on Industrial Engineering and Engineering Management (IEEM)*, Singapore, Singapore, 2023, pp. 0758–0762, doi: 10.1109/IEEM58616.2023.10406334.
34. Mostaccio, A., G. M. Bianco, S. Amendola, G. Marrocco, and C. Occhiuzzi, RFID for food industry 4.0 – current trends and monitoring of fruit ripening, *2022 IEEE 12th International Conference on RFID Technology and Applications (RFID-TA)*, Cagliari, Italy, 2022, pp. 109–112, doi: 10.1109/RFID-TA54958.2022.9924104.
35. Hamdan, K. and O. Kazar, Design of a smart and green industry 4.0 architecture, *2023 24th International Arab Conference on Information Technology (ACIT)*, Ajman, United Arab Emirates, 2023, pp. 1–5, doi: 10.1109/ACIT58888.2023.10453882.

36. Bader, R., A. Ali, and N. M. Mirza, AI and robotics leading industry 4.0, *2022 Ninth International Conference on Internet of Things: Systems, Management and Security (IOTSMS)*, Milan, Italy, 2022, pp. 1–4, doi: 10.1109/IOTSMS58070.2022.10061911.
37. Garbie, I. and A. Garbie, Identifying and estimating the implementation level for industry 4.0, *2022 Advances in Science and Engineering Technology International Conferences (ASET)*, Dubai, United Arab Emirates, 2022, pp. 1–6, doi: 10.1109/ASET53988.2022.9734986.
38. Noman, T., M. S. Khan, M. Emdadul Islam, and H. Rashid, A new design approach for gesture controlled smart wheelchair utilizing microcontroller, *2018 International Conference on Innovations in Science, Engineering and Technology (ICISET)*, Chittagong, Bangladesh, 2018, pp. 64–68, doi: 10.1109/ICISET.2018.8745607.
39. Dey, P., M. M. Hasan, S. Mostofa, and A. I. Rana, Smart wheelchair integrating head gesture navigation, *2019 International Conference on Robotics, Electrical and Signal Processing Techniques (ICREST)*, Dhaka, Bangladesh, 2019, pp. 329–334, doi: 10.1109/ICREST.2019.8644322.
40. Basak, S., F. F. Nandiny, S. M. M. H. Chowdhury, and A. A. Biswas, Gesture-based smart wheelchair for assisting physically challenged people, *2021 International Conference on Computer Communication and Informatics (ICCCI)*, Coimbatore, India, 2021, pp. 1–6, doi: 10.1109/ICCCI50826.2021.9402632.
41. Parthasarathy, S. A., S. Subash, A. Devaraj, and V. Karthik, Design and development of arduino based multipurpose wheelchair for disabled patients, *2020 Third International Conference on Smart Systems and Inventive Technology (ICSSIT)*, Tirunelveli, India, 2020, pp. 884–888, doi: 10.1109/ICSSIT48917.2020.9214082.
42. Gao, X., L. Shi, and Q. Wang, The design of robotic wheelchair control system based on hand gesture control for the disabled, *2017 International Conference on Robotics and Automation Sciences (ICRAS)*, Hong Kong, China, 2017, pp. 30–34, doi: 10.1109/ICRAS.2017.8071911.
43. Vijayakumar, C., M. P. Kumar, S. Sivaji, S. Sadulla, J. Prathiba, and K. V. Kumar, Sensors based automated wheelchair, *2013 International Conference on Green Computing, Communication and Conservation of Energy (ICGCE)*, Chennai, India, 2013, pp. 439–443, doi: 10.1109/ICGCE.2013.6823476.
44. Sairam, R. M., S. Sharma, and G. Pal, FSM & handshaking based AHB to APB bridge for high speed systems, *International Journal of Engineering Research & Technology*, (2013), 2(11).

Chapter 11

Transforming autonomous vehicle technology

AI and machine learning applications in cyber-physical systems

Aishwarya Balan, Reeta Tomar, Shubham Rajendra Ekatpure, Shitiz Upreti, Sumegh Tharewal, and Ramesh Chandra Poonia

11.1 INTRODUCTION

The evolution of cyber-physical systems (CPS) has led to the growth of autonomous vehicles (AVs) by joining the physical and digital components to the design of the systems. According to CPS, AVs can integrate tangible parts like sensors into artificial intelligence (AI) systems that observe and interpret the surroundings. This enables vehicles to operate autonomously and decide on the present circumstances, making it possible for AVs to collect, interpret, and utilize huge volumes of sensor data within complex traffic scenarios and make decisions with minimal human engagement [1,2]. Convolutional Neural Network (CNN) technologies for object recognition, reinforcement learning strategies that allow for flexible decision-making in a driving context, and predictive technology that estimates the position of moving cars or pedestrians that are reasonably close assist in achieving this [3,4].

AI in CPS represents a transition for AVs from rule-based programming to dynamic systems that can improve over time with practice. Autonomous cars encounter a number of diverse operating situations, pushing the need for swift and accurate responses in order to maintain passenger safety and effective operation [5]. Advanced AI technology is offered by AI capabilities through, CPS as demonstrated by Tesla, Waymo, and Cruise. These systems rely on intricate CPS layouts consisting of components for sensing, analyzing, and management, to ensure immediate operation. Perception requires the use of a combination of radar, LiDAR, and cameras to create a representation of the environment around the vehicle. Controllability and steerability navigation management in the AV environment combines various supporting control commands using ML models for acceleration, control, steering, etc. However, the process of implementing CPS in self-driving cars is quite rigorous, ranging from the technical challenge of sensor limitations and the model's operational trust to cyber security and adherence with social laws

 DOI: 10.1201/9781003542513-11

and sociolegal norms especially in critical scenarios. Ethical issues such as collision situations, where such AI systems could be integrated, reignite interest in the implementation of safe collision and interaction strategies and assert the growing relevance of trustworthy and effective AI systems [6]. In this chapter, we will be reviewing CPS frameworks, AI/ML, and AV interfaces, such as V2V and V2I, that together enhance AV's situational awareness and action precision. We shall also cover the challenges that AI-based CPS faces, examining technical, ethical, and legal dimensions, while projecting future trends in vehicle autonomous systems including smart cities [7]. The main goal in this study is to point out in detail how the Self-Driving CARS landscape will be changing with the incorporation of CPS technologies in the near future.

11.2 REVIEW OF LITERATURE

Combinations of CPS technologies with AI and ML, for AVs, have been researched in a number of areas from sensors to ethical issues and challenges of implementation. The first works in the field of CPS focused on the design of structural models connecting the vehicle physical space with the digital processing space that enable active control of the monitored objects and their environment. The feedforward and feedback control loops constituting these CPS systems are functional in assisting AV adjustment to any sudden change in the environment, a precondition for effective autonomous driving [1,2]. Integrating sensors and actuators with AI algorithms forms the basis for the intricate tasks that AVs can perform independently [3]. As these technologies continued to advance, researchers started to use more complex algorithms like the CNNs for object detection, allowing vehicles to recognize pedestrians, objects, and other vehicles more accurately. Real-time CNNs are important for AV visual tasks that require spatial and temporal structure, such as obstacle avoidance and traffic sign recognition, because they specialize in detecting features in images, whereas the application of reinforcement learning (RL) models in decision-making enables AVs to form strategies for driving based on their interaction with their environment [4]. RL enables AVs to make adaptive decisions in dynamic environments through experiencing efficiency-oriented performances such as keeping a safe distance from other vehicles or manipulating traffic conditions according to pace. In addition, predictive algorithms, such as the recurrent neural networks, have been used to predict other drivers' actions and assisted in aggressive and safe driving [5].

One of the most critical aspects of CPS-related research in AVs is sensor fusion, integrating data from heterogeneous sources such as LiDAR, radar, and cameras into a coherent environmental model. Sensor fusion may add to increased reliability and accuracy but at the cost of higher computational requirements, leading to concerns about latency [8]. Existing literature addresses these challenges and barriers with some solutions like edge computing, a strategy that disperses data processing to nodes situated close to or even within the vehicle in order to reduce latency. It allows for near real-time analysis and process capabilities of AVs, even under critical conditions, by reducing the dependency on centralized cloud systems [9]. In addition to this, edge computing minimizes the bandwidth needed for continuous transmission of data; thus, it makes real-time responses more possible. Studies on V2V and V2I communication have demonstrated how these support data exchange, improvements in hazard detection, and traffic management enhancement by allowing the AVs to look beyond its line of sight. However, despite these promises, V2X communication systems face a lot of challenges and hindrances in the form of bandwidth restrictions, issues related to latency, and cybersecurity threats. This creates a significant barrier to widespread use [10].

In AV deployment-related studies, cybersecurity is a big issue as adversarial attacks may change sensor data and put both AV passengers and other drivers at risk. Adversarial attacks in ML manipulate sensor inputs or environmental signals; thus, the AVs would be misled about road conditions. Such risks thereby present the need for strong AI models that can recognize and evade such risks [7]. Other than these challenges, questions are also raised on ethics and regulation. Driverless cars may be placed in scenarios where they must respond in minutes, where their decisions may carry moral and legal implications, for instance, choosing one of several safety results of a collision that might occur. Hence, ethical decision-making in these self-driving cars must be on the grounds of reliability, accountability, and transparency. It has been advanced that the algorithms of the AV should also follow standardized ethical frameworks in a bid to make decisions [6].

Moreover, due to operating self-driving cars under changing laws by region, they would be difficult to follow. Regulatory bodies have the onus of drawing proper standards which are 'safe' for self-driven cars, guarantee data privacy, and ensure smooth operations of roadways for all users. The lack of clear guidelines is a challenge to AV companies because innovation capabilities and rate, as well as the adoption of the technology, will be slowed down [11]. Strict testing and validation criteria shall ensure that in various real-world situations, the AVs work in a dependable sound manner. There are issues of safety and ethics besides developing standards that make sure AVs respond to sudden and unexpected events like emergency vehicles, as well as sudden road obstacles [12]. Although the CPS studies provide a solid foundation for advancement of innovation in AV, there is still a need for strong AI models, effective cybersecurity measures, and clear regulatory frameworks that would allow the use of these self-driven vehicles on public roads. Advances in technology have ensured that the emphasis on developing sound ethically and technologically robust CPS architectures for autonomous vehicles is on the rise. This technology is likely to be developed at the intersection of AI, CPS, and regulatory policies—secure, effective, socially conscious AVs.

11.3 RESEARCH METHODOLOGY

In this chapter, we review the AI-based CPS in AVs by literature reviews and real-world case studies to give a closer look at their theoretical and practical applications. The literature review summarizes the information from various studies concerning CPS, ranging from AI/ML techniques, communication protocols, recent control algorithms advancements, and sensor fusion to V2X communication. Sources used are chosen based on relevance to AV systems, with publication only in the most prominent journals and conferences focused on AI, robotics, and automotive engineering. To make this comprehensive review, classical frameworks of CPS, advanced AI/ML models, and communication techniques that allow AVs to operate independently in demanding conditions are involved.

Besides this review of literature, practical insights into CPS applications were provided through real-world cases from leaders in the industry. Waymo's navigation systems and Tesla's Autopilot show the present condition of AV technology in real-world environments. Waymo's application of radar and LiDAR in busy city areas highlights the power of integrating multiple sensors for object identification and avoiding obstacles. In contrast, Tesla's Autopilot system demonstrates reinforcement learning for adaptive decision-making both on highways and city streets. This study acquires a detailed comprehension of the effectiveness and constraints of AI techniques like reinforcement learning, predictive modeling, and computer vision during implementation in real-time traffic scenarios, by examining the abovementioned instances.

Table 11.1 summarizes the selected case studies and technologies examined, which includes their primary applications and limitations. This visual representation highlights the methodology's focus on current AV applications to analyze technical components that are both impactful and challenging.

This approach supports a detailed analysis of CPS in AVs by incorporating a literature review to establish the theoretical basis and identify currently existing research gaps, along with case studies in order to apply these findings to actual AV implementations. For example, Tesla's use of reinforcement learning highlights its abilities in learning and decision-making in changing situations despite being constrained by data requirements and managing intricate cases. The application of LiDAR technology by Waymo in object recognition in busy city areas illustrates advantages of sensor fusion from the technical point of view. Moreover, it also underlines the

Table 11.1 Use case applications

Case Study/ Technology	*Use case*	*Primary application*	*Limitations*
Waymo	Urban navigation	Object recognition using LiDAR and vision	High computational cost and latency
Tesla autopilot	Adaptive decision-making	Reinforcement learning for driving behaviors	Requires extensive training data
V2X communication	Real-time data exchange	Hazard detection and traffic management	Bandwidth, latency, and cybersecurity
Sensor fusion	Environmental modeling	Integrates LiDAR, radar, and cameras	Processing demands and potential latency
Computer vision (CNNs)	Object detection	Traffic sign recognition	Sensitivity to weather/ lighting conditions

most significant challenges caused due to expensive costs and computing requirements of LiDAR.

In addition, the case studies indicate that the direction of AI design aligned with policies, especially in regulations and ethics, and while on roads, the AVs need to meet the safety and privacy rules for reliable, accountable, and transparent AI systems. The approach is based on technical feasibility as well as social implications and considers issues such as safety and cybersecurity and people's trust from a balanced perspective. The possibility of making high-stake choices points toward the implementation of ethical frameworks in the development of AI systems, ensuring responsible decision-making. This approach gives a new and comprehensive viewpoint on CPS, providing readers with a deep knowledge of not only technical capabilities but also societal repercussions. Hence, this chapter connects the theoretical and the practical aspects to facilitate informed discussions about future developments. It brings a sense of understanding of the prospects and issues associated with the advancement of AV technology for even better public use.

11.4 CPS ARCHITECTURE IN AUTONOMOUS VEHICLES

An advanced architecture for a CPS designed to connect the physical environment of the vehicle with advanced digital processing supports the AV. The CPS architecture in AVs brings together the physical elements of sensors and actuators and highly advanced control algorithms and AI models in a really well-connected system that can gather, process, and make independent decisions. This combination allows AVs to work in a dynamic environment, adapting and changing the conditions while maintaining safety and efficiency [1,2].

11.4.1 Perception and sensing

Perception systems are expected to be ingrained into the AV's CPS architecture, acquiring in real-time information regarding the environment. There are several types of sensors installed in the vehicle, such as cameras, LiDAR, and radar, together with ultrasonic sensors. LiDAR is particularly important for creating accurate mappings of 3-D space around the car, such as object recognition and depth measurements. Image data with a high resolution important for the identification of color and grouping of objects like road signs or people are provided by the cameras. Radar is efficient in measuring the range and speed of objects, which is crucial while monitoring the movement of other cars on road, especially when the visibility is low. Ultrasonic sensors are typically used at a very short distance for activities like low-speed parking [3,8]. Each sensor supplies distinct data that, when combined, form a highly accurate and quite holistic model of the surroundings.

Sensor fusion—the critical process in the CPS—involves integration of data acquired from a collection of sensors for enhanced accuracy and reliability. The vulnerabilities and limitations of single-sensor data, which can be prone to environmental variations such as lighting changes or harsh weather, can be reduced by use of a multi-fusion sensor. For example, apart from radar details, camera details are being fused to interpret object size, velocity, and distance better. This, in turn, decreases the likelihood of misinterpretation. The algorithms deal with processing and synchronizing the inputs, leading to fast management of the fusion process so that AVs are kept oriented and informed [8,9].

Table 11.2 shows a comparison of common perception sensors used with their key strengths and limitations.

Table 11.2 Comparison of common perception sensors

Sensor type	*Function*	*Strengths*	*Limitations*
LiDAR	3D environmental mapping	High accuracy, long-range	Expensive, weather-sensitive
Camera	Visual perception, object ID	Detailed image data, versatile	Affected by lighting and weather
Radar	Detects object velocity, distance	Works in poor weather conditions	Limited resolution
Ultrasonic	Short-range object detection	Effective at close range	Limited to very short distances

In particular, the fusion of these perception inputs is of crucial importance in order to create an accurate and precise model of the environment in which the data fusion systems of the vehicle are to be analyzed and processed. The above-described function is also an important one in CPS because the individual sensor data might very well have been affected by the environment and combined to increase the reliability and precision. Further combining radar and camera information will reduce the chance of a false move by the AV, and it will better understand how big and how fast objects are.

11.4.2 Data processing and fusion

After all, if we want a real-time response to data, such data must be processed fast after being collected. ML algorithms run on this integration of data from multiple sensors and do so quickly. A reliable, unified environmental model can be built often using methods such as Bayesian networks, Kalman filtering, or deep learning-based fusion algorithms. For example, a Kalman filter can be used to predict what objects will do, by assigning a previous location where these objects were found, in conjunction with what sensors of this objects are reporting they are now. With the advent of a powerful class of deep learning methods, CNNs [4,9], which are able to work with visual data, AVs now can analyze situations with great accuracy and detail.

11.4.3 Edge computing

This technology is combined to support data processing in AV architectures. With edge computing, we shift computations from the vehicle level to lower latency. That is important because decisions can be made with fewer steps, and responses need to be made faster, before the accidents and emergency braking.

11.4.4 Control and actuation

The control systems of the CPS gather data and analyze it and then make choices about driving by the AV. Control mechanisms with trained ML algorithms over the dynamics of vehicles such as lane change, speed adjustment, and navigation are provided, where dynamics of vehicles are controlled by using large amounts of data. One reason why RL might be used is because they learn optimal behavior by trial and error in simulated settings. The process enables the system to generate guidelines where it is difficult to drive by providing safe navigation, avoiding the obstacle and adaptive cruise control, which are all fundamental aspects. For instance, Autopilot is improving Tesla's lane keeping ability using RL methods, and adaptive cruise controls the extent to which the car maintains a safe distance in traffic while adjusting the speed dependent on vehicles near it [13]. Learning in this way allows the AVs to travel through a variety of driving conditions and adapt to different surroundings, throughout the time that they develop [7].

By providing the perception, data fusion, control, and actuation backbones to the CPS within AVs, they can be very dynamic and responsive in real-life driving situations. However, there are huge challenges to tackle on this, namely, required computing, latency concerns due to limitations, and a very sensor fusion-dependent world. Studies toward better CPS design have focused on improvements in edge AI, power-saving processors, as well as stronger algorithms for sensor fusion. The principal focus of these improvements aims to reduce the latency, increase the efficiency of energy use, and maintain consistent performance to accomplish increased scalability and safety of AV technology [10,14].

11.5 CORE TECHNIQUES OF AI AND ML IN AUTONOMOUS VEHICLE CPS

The main three functionalities centered on integrating AI and ML in CPS in AVs are perception, decision-making, and predictive modeling. Each function helps the self-driven car to understand and react to the surrounding environment by applying specific algorithms, like the CNNs for assisting in perception and RL for making a determination.

11.5.1 Computer vision and deep learning for perception

CNNs are very important for analyzing and processing visual information in an AV. CNNs have proven incredibly efficient for object detection and classification tasks for AVs, in learning raw image spatial hierarchies of features that can support high-precision identification of pedestrians, vehicles, and road traffic signs. CNN-based models, which Waymo uses to identify and classify various objects inside cities, help improve the accuracy and precision of object recognition in crowded, dynamically changing environments. Such a function is of primary importance in maintaining precision regarding the formation of 3D maps and the capabilities for obstacle detection in challenging environments [4,15].

11.5.2 Reinforcement learning for adaptive decision-making

RL performs an important role in decision-making for AVs. Here, the models take the best actions by getting rewards and punishments, enabling an AV to create driving strategies for different traffic situations. Tesla's Autopilot system is well known for the application of the RL where reward-based learning happens to improve features like staying in the correct lane, changing the speed, or merging safely. An important advantage of RL is the ability to deal with sequential decision-making; hence, AVs can react promptly during multiple-step situations such as highway entry [13,16].

11.5.3 Predictive modeling for proactive navigation

Predictive modeling enables AV to predict the responses of other car drivers and people on road. It also aids in making the navigation and collision avoidance better. Algorithms and methods applied very often are recurrent neural networks and Kalman filters to predict a trajectory for giving a chance to AVs to evaluate movements of vehicles or pedestrians moving in front of them by recognizing patterns already seen. This is particularly essential at the time of high-speed operation, where AVs should take into account the changing lanes, acceleration, or braking of other vehicles in order to maintain secure and safe distances [7,17].

11.5.4 Integration of AI models in AV CPS

The inclusion of these models of AI in the CPS system depends on strong processing capabilities and advanced skills in data management. Every model such as the CNNs for perception, RL for decision-making, or predictive algorithms for trajectory planning needs to work in real-time data fusion to ensure safe and efficient driving. Even then, maintaining the resilience of models acts as a challenge as their effectiveness can be impacted by factors like adverse weather or unexpected road incidents. As a result, AV companies are constantly improving their designs and testing hybrid AI methods to improve system durability and dependability in various situations [8,18].

AI and ML methods play a big role in AVs by enhancing perception, decision-making, and navigation, which allows them to function autonomously in different real-world situations. Yet more advancements in the transparency, resilience, and efficiency of models are needed to guarantee the consistent performance of these systems in progressively intricate and diverse settings.

11.6 COMMUNICATION AND CONNECTIVITY IN CPS FOR AUTONOMOUS VEHICLES

Moving forward, let us see how communication and connectivity come into the design of CPS in self-driving cars: To improve knowledge of the environment and surroundings and improve hazard detection and efficient traffic management, information is allowed to be transferred between vehicles (V2V) and infrastructure (V2I). AVs are used to connect to other vehicles and infrastructures and pedestrians, and this might change the effectiveness and safety on roads.

11.6.1 Vehicle-to-vehicle (V2V) and vehicle-to-infrastructure (V2I) communication

V2V enables the self-driven vehicles to share their location and speed information with other nearby vehicles, as well as their intentions. This safety function is crucial for collision avoidance as well as cooperative lane change. V2I communication, on the other hand, allows the vehicle to collect information from the infrastructure on the roadside, for instance, different types of road signs and traffic lights, to change its routes based on real-time traffic situations. This helps AVs predict and respond to sudden changes in better general response times and reduces traffic on roads. As it allows vehicles to perceive beyond simply what is in front of them, V2X communication helps improve the vehicles' situational awareness and adaptability [6,19].

11.6.1.1 Data security

Processing data quickly with low latency is also helped by it. Unlike the use of cloud computing that 'relies' solely on the clouds for making decisions, edges are much closer to where the data are obtained from. V2X communication linked to AVs is enhanced through edge computing as the data analysis is performed either on the near servers or even inside the vehicle to provide a prompt response to changing real-time scenario and applications. Reduced latency is particularly important to many applications where fast responses to sudden braking or escaping some physical obstacle are required [20].

11.6.1.2 Bandwidth Constraints

V2X communication is impacted by the challenges to bandwidth allocation, cybersecurity, and latency. Due to their dependence on large data volumes and requiring large response times, these systems rely on dedicated short range communication (DSRC) or other cellular networks such as 5G and ensure there are fast and reliable connections. Data security is still a big problem because V2X communication is susceptible to cyber attacks like jamming, data spoofing, and unauthorized data interception. For V2X interactions, secure V2X communications and protectability are crucial and thus require strong encryption protocols and intrusion detection systems [10,14].

11.6.2 Potential for enhanced safety and efficiency

All of this has potential for improved safety and efficiency as V2X communication can improve security and traffic efficiency. Data-agreed AVs will be able to communicate, anticipate, and control traffic flow,

Table 11.3 Overview of the key types of V2X communication

V2X communication Type	*Application*	*Challenges*
V2V (vehicle-to-vehicle)	Collision avoidance, lane changes	Data security, bandwidth limits
V2I (vehicle-to-infra)	Visual perception, object ID	Latency, infrastructure costs
V2P (vehicle-to-pedestrian)	Detects object velocity, distance	Privacy concerns, communication delay

decreasing abrupt stops and car collisions with continuous data exchange. V2X also makes coordinated responses possible for AVs in busy locations such as intersections that can communicate with the infrastructure to avoid formation and increase the efficiency of traffic. As 5G and edge computing technologies are growing, V2X is going to be important for AV networks by being able to seamlessly place into urban scenarios and settings [12]. Edge computing is supported by V2X communication to equip AVs to navigate difficult environments safely and effectively.

In Table 11.3 below, we provide an overview of the key types of V2X communication, their applications, and challenges in learning for the AV setting.

Such progress in connectivity is also essential for AVs where there will be an increased awareness of situations along with secure and efficient operation. However, further work has to be done in data security, regulation support, and control of latency before it is taken more broadly.

11.7 CHALLENGES IN AI-DRIVEN CPS FOR AUTONOMOUS VEHICLES

As great as that sounds, there are yet numerous challenges to be managed and hurdles to be overcome to unlock the full potential of CPS in AVs. Technical issues of the reliability of model and accuracy intertwine with ethical, regulatory, and cybersecurity problems, which must be handled in order to make AVs safer, trustworthy and approved by regulatory bodies and people.

- **Safety and Reliability:** Significant technological challenges include safety and reliability in a wide set of driving conditions. The AI models running inside an AV must be reliable under multiple lighting conditions, differing weather conditions, and other unforeseen scenarios such as road construction or accidents. Adverse weather conditions such as heavy rain or dense fog may cause interference in the accuracy of a sensor, leading to errors that might make the sensor to misread the environment, thereby causing potential safety issues [7,17]. A very important protection for AV operations is maintaining model robustness against adversarial attacks; these are attacks made against AI systems by the manipulation of input data to deceive the systems [18].

- **Cybersecurity Risks:** Another big concern is cybersecurity because of the dependence of AVs on V2X communication, which makes them vulnerable to hacking and unauthorized access to data. Some cyber risks include jamming of signals, data spoofing, and unauthorized entry to control systems, all of which may pose problems for vehicle functions, putting passengers at risk. Studies and researchers are working on creating data encryption protocols and intrusion detection systems especially for AVs in order to reduce risks by monitoring and providing real-time response to potential threats. Nevertheless, strong cybersecurity measures increase the computational workload and necessitate updates to tackle changing threats.
- **Data Privacy and Ethical Challenges:** One of the significant challenges faced by AVs is ensuring data privacy as they collect, analyze, and share large amounts of data, which may include sensitive and personal details about passengers and their environment. Strict protocols for data management are essential for ensuring the safe storage and anonymization of data to protect the privacy of individuals. As an example, an AV might have to decide between two undesirable results, like swerving to avoid hitting a pedestrian but putting the passengers in danger, creating ethical and legal issues. Transparent, reliable, safe, and ethically based frameworks for decision-making and regulatory oversight are needed to guide the behavior of AVs [21].
- **Regulatory and Standardization Barriers:** The deployment of AVs is hindered by regulatory challenges as current traffic laws and safety standards are not fully suited for AVs It is essential to standardize AV testing, cybersecurity protocols, and decision-making guidelines to achieve widespread adoption. Regulators around the world are creating frameworks to meet these needs, but development differs by region, and discrepancies may hinder the cross-border operation and integration of AVs. Guidelines that are standardized and thorough regulations are crucial for the safe and widespread use of AV technology [15].

Table 11.4 shows challenges in the technical and social relations with a rise of AI CPS. This becomes vital on factors that need to develop safe, dependable, and ethical autonomous automobiles. As the technologies progress, so will the requirements for interdisciplinary research and cooperation among developers and regulators, guaranteeing fulfillment of such needs.

11.8 FUTURE DIRECTIONS AND TRENDS

Although the work presented in the paper encourages the use of CPS technologies for the future of AVs, the development of AI for networked control is still essential. Therefore, one can state that the trends such as

Table 11.4 Challenges in the technical and social relations with a rise of AI CPS

Challenge	*Description*	*Implications*
Safety and reliability	Model robustness, sensor limitations	Potential safety risks in adverse conditions
Cybersecurity	Vulnerabilities in V2X communication	Risk of hacking, unauthorized data manipulation
Data privacy	Collection and transmission of sensitive information	Privacy concerns, regulatory compliance issues
Ethical concerns	Decision-making in unavoidable collision scenarios	Public trust, ethical and legal responsibilities
Regulatory barriers	Lack of standardized safety and cybersecurity protocols	Limits on AV deployment and cross-region use

quantum computing, federated learning, and synergy with smart cities are going to definitely meet the present difficulties and open new avenues in AV systems.

11.8.1 Federated learning for data privacy

A special method in AI allows self-driven cars to enhance the developed models but with experience and without accumulating valuable information in one place. The co-operative architecture entailed self-driving cars connecting in training architectures, raw inputs stored locally on the vehicles for privacy concerns, and the potential leakages [12]. There might also be an answer to the question on how much privacy can be afforded to the sharing of AV data, especially in those regions that have closed data protection laws. Here, federated learning would mean that the knowledge base of these AVs would feed into the others for updates without having to invade the privacy of the users—a way of achieving scalable improvement in the technologies, which is definitely going to deal with the current challenges and open fresh possibilities in AV systems [22].

11.8.2 Quantum computing in CPS optimization

The application of quantum computing can be developed to improve complex calculations in the structures of CPS. This has a lot of potential in handling and changing heavy traffic scenarios by using sensor fusion, path planning, and decision making in real time by processing large data at one go. Quantum algorithms can potentially address optimization difficulties when it comes to real-time navigation more quickly than traditional algorithms, which leads to improved responsiveness and adaptability in AVs.

Although quantum computing is still being studied, it is anticipated to change the boundaries of computation in AV systems as it advances [16].

11.8.3 Integration with smart cities

As cities evolve into 'smarter' environments with the implementation of IoT devices and digital infrastructure, AVs are expected to have a crucial part to play. In a smart city setting, AVs could communicate seamlessly with traffic lights, road sensors, and emergency services, supporting smooth traffic movement and improved public safety. V2X communication and edge computing are crucial in this scenario, allowing AVs to utilize real-time information on road conditions, weather, and pedestrian traffic for route optimization and congestion reduction [10]. The incorporation of AVs into intelligent cities is expected to reduce city traffic and emissions, promoting sustainable urban movement.

11.8.4 Advanced safety and ethical frameworks

With the advancement of AV technology, it is very important to create strong ethical frameworks for making decisions in collision. This involves creating AV algorithms that focus on safety and fairness while adhering to regulatory requirements. The expected ethical AI frameworks involve combining complex ML methods with transparency, which will allow AVs to communicate and explain decisions in critical situations [23]. These frameworks are important for building trust and approval of AV technology among the people [6].

Table 11.5 shows emerging trends and highlights their anticipated impacts on the future of AV systems.

The above observed patterns indicate the rising interconnectedness and technological progress in AVs, establishing a base for a future where autonomous self-driving vehicles are extensively woven into the society's infrastructure. Nevertheless, looking into these possibilities necessitates ongoing interdisciplinary studies and coordination of regulations.

Table 11.5 Emerging trends and highlights their anticipated impacts on the future of AV systems.

Emerging technology	*Potential application*	*Anticipated impact*
Federated learning	Privacy-preserving model training	Enhances data privacy and supports scalability
Quantum computing	Real-time processing, optimization	Improved sensor fusion and faster decision-making
Smart city integration	V2X, IoT-enabled urban navigation	Traffic flow optimization and reduced congestion
Ethical frameworks	Decision-making in critical scenarios	Increased public trust and compliance with ethics

11.9 CONCLUSION

Driven by the integration of CPS and AI/ML algorithms, the introduction of self-driving cars signifies a revolutionary change in the transportation and automobile industry. This concept of AI-based CPS has improved AVs by allowing them to collect, interpret, analyze, and react to intricate data and environmental information, leading to advancements in real-time navigation, detection of dangers, and effective route mapping. This chapter analyzes and examines how the CPS architecture of AVs, with the help of edge computing and V2X communication, allows for safe and flexible operation in different varying driving conditions. However, significant obstacles still exist, from technical problems like model stability and cybersecurity to societal issues related to data privacy and regulatory adherence. The anticipated developments in fields such as federated learning, quantum computing, and smart city integration will impact the future of AVs. Regulators, technology developers, and ethicists must collaborate and work together in order to guarantee that AV technology progresses in a responsible manner and advances with safety and public trust as top priorities. CPS and AI have greatly improved AV technology, but further research and regulatory adjustments are necessary and required to fully unlock the potential of AVs. The future of CPS in self-driving cars has a lot of potential, but it needs a comprehensive strategy mixing creativity with responsibility. AVs can be successfully incorporated into the society, finally transforming the transportation industry by creating and promoting a multidisciplinary approach to studying, designing, and implementing policies.

REFERENCES

1. E. A. Lee and S. A. Seshia, *Introduction to Embedded Systems: A Cyber-Physical Systems Approach*, 2nd ed., Cambridge, MA: MIT Press, 2017.
2. S. Shalev-Shwartz, S. Shammah, and A. Shashua, "On a formal model of safe and scalable self-driving cars," *arXiv preprint arXiv:1708.06374*, 2017.
3. Y. LeCun, Y. Bengio, and G. Hinton, "Deep learning," *Nature*, vol. 521, no. 7553, pp. 436–444, 2015.
4. W. Schwarting, J. Alonso-Mora, and D. Rus, "Planning and decision-making for autonomous vehicles," *Annu. Rev. Control Robot. Auton. Syst.*, vol. 1, pp. 187–210, 2018.
5. C. Karolemeas, S. Tsigdinos, E. Moschou, et al. "Shared autonomous vehicles and agent based models: A review of methods and impacts," *Eur. Transp. Res. Rev.*, vol. 16, p. 25, 2024. https://doi.org/10.1186/s12544-024-00644-2
6. J. K. Goodall, "Machine ethics and autonomous vehicles," in *Road Vehicle Automation*, G. Meyer and S. Beiker Eds., Cham: Springer, 2014, pp. 1–6.
7. D. Zhao, "Safety and robustness in autonomous driving: An adversarial perspective," *IEEE Trans. Intell. Transp. Syst.*, vol. 21, no. 9, pp. 3758–3775, 2020.
8. A. R. Mostaani and H. Adeli, "Sensor fusion for intelligent transportation systems," *IEEE Trans. Intell. Transp. Syst.*, vol. 22, no. 2, pp. 1248–1258, 2021.

9. J. Redmon and A. Farhadi, "YOLO9000: Better, faster, stronger," in *Proceedings of the IEEE Conference on Computer Vision and Pattern Recognition (CVPR)*, Honolulu, 2017, pp. 7263–7271.
10. P. W. J. Green and B. T. L. Paul, "Cybersecurity challenges in V2X communication," *IEEE Commun. Mag.*, vol. 59, pp. 126–132, 2021.
11. P. Koopman and M. Wagner, "Challenges in autonomous vehicle testing and validation," *SAE Int. J. Transp. Saf.*, vol. 4, no. 1, pp. 15–24, 2016.
12. A. Dosovitskiy, G. Ros, F. Codevilla, A. Lopez, and V. Koltun, "CARLA: An open urban driving simulator," *Proceedings of the First Annual Conference on Robot Learning, USA*, 2017, pp. 1–16.
13. M. Bojarski et al., "End-to-end learning for self-driving cars," *arXiv preprint arXiv:1604.07316*, 2016.
14. M. Bansal, A. Krizhevsky, and A. Hinton, "ImageNet classification with deep convolutional neural networks," in *Advances in Neural Information Processing Systems*, F. Pereira, C. Burges, L. Bottou, and K. Weinberger, Eds., vol. 25, New York: Curran Associates, Inc., 2012, pp. 1–9.
15. S. Fayaz, S. A. Parah, G. J. Qureshi, J. Lloret, J. D. Ser, and K. Muhammad, "Intelligent underwater object detection and image restoration for autonomous underwater vehicles," *IEEE Trans. on Vehicular Technol.*, vol. 73, no. 2, pp. 1726–1735, Feb. 2024, doi: 10.1109/TVT.2023.3318629
16. K. Bangalore, S. Ibrahim, T. Victor, M. Patrick, S. Ahmad, Y. Senthil, P. Patrick. "Deep reinforcement learning for autonomous driving: A survey," *IEEE Trans. Intell. Transp. Syst.* pp. 1–18, 2021. doi: 10.1109/TITS.2021.3054625.
17. D. Zhao and L. Su, "Trajectory prediction models for autonomous vehicles," *IEEE Trans. Intell. Transp. Syst.*, vol. 20, no. 3, pp. 760–770, 2020.
18. E. Farsimadan, L. Moradi, and F. Palmieri, "A review on security challenges in V2X communications technology for VANETs," *IEEE Access*, vol. 13, pp. 31069–31094, 2025, doi: 10.1109/ACCESS.2025.3541035
19. K. H. Muraro Gularte; J. P. Javidi da Costa; J. A. Ruiz Vargas; A. S. da Silva; G. A. Santos; Y. Wang. "Integrating cybersecurity in V2X: A review of simulation environments," *IEEE Access*, vol. 12, pp. 177946–177985, 2024. doi: 10.1109/ACCESS.2024.3504404
20. Khan, S., and Lee, Y., "5G technology in V2X and AV applications," *IEEE Trans. Commun.*, vol. 68, no. 4, pp. 251–263, 2020.
21. A. Zanella, N. Bui, A. Castellani, L. Vangelista, and M. Zorzi, "Internet of Things for smart cities," *IEEE Internet Things J.*, vol. 1, no. 1, pp. 22–32, 2014.
22. K. Upreti, V. Kumar, D. Pal, M. S. Alam, and A. K. Sharma, "Design and development of tracking system in communication for wireless networking," in *Intelligent Sustainable Systems*, A. K. Nagar, D. S. Jat, G. Marín-Raventós, and D. K. Mishra, Eds., vol. 334, Lecture Notes in Networks and Systems. Singapore: Springer, 2022.
23. U. K. Singh, A. Sharma, S. K. Singh, P. S. Tomar, K. Dixit, and K. Upreti, "Cyber physical systems," in *Cyber Physical Systems*, Boca Raton, FL: Chapman and Hall/CRC, 2023, pp. 141–164.

Chapter 12

Real-time player performance analytics

Leveraging smart cyber-physical systems and AI in sports

Pratham Nijhawan, Faraz Ahmad, Kamal Upreti, Shitiz Upreti, and Ramesh Chandra Poonia

12.1 INTRODUCTION

Recently, it has been demonstrated that integrating smart cyber-physical systems (SCPS) in elite sports can revolutionize player performance analytics by developing advanced data acquisition and analysis frameworks. Through the powerful combination of physical systems, sensors, and data processing technologies, CPS enables continuous monitoring, analysis, and real-time response to dynamic sports environments, representing a significant stride forward in sports technology. CPS frameworks emerge as an arrangement of sensors, Internet of Things (IoT) devices, data processing units, and high-speed networks, with the predominant ability to combine and process large amounts of data during training sessions and competitive games to offer actionable insights on player performance, health, and strategic adjustments. For example, in professional football, teams such as Manchester City use powerful CPS tools within their performance analysis teams to supply real-time coaching decisions based on their heart rates, speed, and exertion levels to avoid performing to maximum levels and thereby prevent injury due to fatigue [1].

CPS is deployed in sports to facilitate real-time tracking and analytics. For instance, it is a multidimensional view of the state of athletes physiologically and biomechanically, which assists in increasing the performance and reducing the chance of injury. Wearables are fitted with sensors (GPS trackers, wearable devices, and smart textiles) in the body to capture data regarding movement, speed, and other physiological metrics [2], with the possibility of transmitting it quickly and efficiently over edge computing, cloud processing, and fast networks like 5G or even 6G [3]. Although CPS benefits are obvious, it is challenging to implement it in high-stakes sports such as security, computational load, and latency [4]. Safeguarding sensitive athlete information in compliance with privacy regulations and optimizing data transmission are ongoing concerns [5]. Additionally, the high-volume, high-velocity data generated by CPS systems can strain processing power

 DOI: 10.1201/9781003542513-12

and network capabilities, making efficient data integration and processing solutions essential for real-time analytics. As CPS continues to reshape elite sports, its role in transforming coaching methodologies and enabling precision-driven performance strategies increases, signaling a new era in sports technology and analytics.

12.2 LITERATURE REVIEW

See Table 12.1.

Table 12.1 A systematic literature overview

No.	Domain	Topic focus	Key contributions
1	Wearable technologies	Wearable technologies in sports	Use of sensors and IoT devices to monitor player movements, biometrics, and workload metrics; foundational research on continuous data acquisition for predictive models related to fatigue, injury risk, and performance peaks [6].
2	Smart wearable CPS	Smart wearable CPS systems	Smart textiles and GPS trackers capture granular data, provide coaches with actionable feedback, and reduce latency in CPS through advanced data handling [2].
3	CPS-enabled sports analytics	Challenges and opportunities in CPS	Discusses the role of edge computing in CPS to minimize latency, facilitating real-time sports analytics [3].
4	Predictive analytics	Predictive modeling in sports	Analysis of MilanLab's predictive modeling system, which uses historical and real-time data to estimate injury risk and fitness levels, highlights algorithms for predicting player readiness and game outcomes
5	AI in sports dynamics	AI models in game dynamics	It focuses on AI models for predicting game dynamics and player responses using ML models trained on historical data for tactical forecasting [7].
6	AI applications	AI-generated heatmaps	Development of AI-driven heatmaps for movement prediction and strategy planning; assists coaches in anticipating opponent strategies and preparing for games [8].
7	Reinforcement learning	RL in NBA analytics	A study of RL models used in the NBA for dynamic game strategy adjustment evaluates defensive and offensive strategies in real-time, providing adaptability during matches [9].
8	Adaptive models	Real-time adaptive frameworks	The framework for continuously recalibrating predictions based on live data allows for on-the-fly adjustments to sports strategies and formations [10].

(Continued)

Table 12.1 (Continued) A systematic literature overview

No.	*Domain*	*Topic focus*	*Key contributions*
9	Machine learning	ML in high-stakes sports	Case studies on RL applications in professional football; real-time ML model updates for strategic agility under fluctuating game conditions [11].
10	Data fusion	Data fusion techniques in sports	Techniques for merging wearable and environmental data enhance real-time decision support and provide a holistic analysis of game dynamics [12].
11	Edge and cloud computing	High-speed networks and cloud solutions	The importance of synchronized data processing through edge cloud frameworks supports rapid data integration to minimize latency and enhance real-time analysis [13].
12	Data security	Data security in sports analytics	Discusses data privacy, encryption, and regulatory compliance in CPS systems; emphasizes secure handling of athlete data [14].
13	Fairness in AI	Fairness in AI algorithms	Exploration of fairness and bias in AI models used in scouting and player evaluation suggests governance frameworks for responsible CPS implementation [4].
14	Ethics in AI	Ethics in AI for sports Analytics	Calls for transparency and fairness in AI-driven sports analytics address ethical concerns and the importance of unbiased algorithmic decisions [15].
15	Network infrastructure	High-speed networks for real-time analysis	Analysis of the role of 5G/6G networks and cloud-based ML models in CPS highlights how reduced latency enables seamless data flow and real-time decision-making [16].

Source: Author's Interpretation.

12.3 RESEARCH METHODOLOGY

This chapter's research methodology adopts a qualitative research design, aiming to comprehensively analyze the integration of CPS and AI within sports performance analytics. Given CPS's dynamic and evolving role in high-performance sports, a systematic approach was necessary, guided by the Preferred Reporting Items for Systematic Reviews and Meta-Analyses (PRISMA) to ensure rigorous data collection, selection, and synthesis. Data collection targeted peer-reviewed journals, conference proceedings, and industry reports published between 2020 and 2023, focusing on studies that reflect current advancements and applications of CPS and AI within sports. Strict inclusion criteria were defined, emphasizing studies discussing CPS architecture, applications of AI and ML in real-time sports analytics, reinforcement learning, predictive modeling, and data-driven insights in high-stakes sports environments. Empirical evidence of AI-driven CPS frameworks in elite sports, real-world examples for elite sports, and technical challenges and solutions for AI-driven CPS frameworks in elite sports were provided. This helped establish a literature base that can be deeply examined with a more detailed focus on CPS as an agent of sports performance transformation. Due to the subjective and qualitative nature of the data, the process of data extraction took place according to a structured coding scheme to extract critical details on the components of the CPS systems, particular AI applications seen, predictive and reinforcement learning techniques applied, and the patterns in player management and game strategy optimization. A structured approach facilitated a detailed capture of the complexity of CPS implementation in sports settings. After data collection, the research utilized a multidimensional methodology based on thematic analysis, meta-analysis, and comparative case study analysis. Literature was thematically analyzed to identify recurring themes, including injury prediction and game strategy optimization, as well as monitoring player fatigue, cost data security concerns, aspects such as latency, and ethical concerns in decisions wherein AI is used. These repeated patterns matched recent findings by Patel and Watson [7] and Richardson and Steven [11] regarding the critical role of CPS in augmenting player management and tactical decision-making.

To validate, a meta-analysis was conducted by synthesizing quantitative findings from studies that measured the value of AI-assisted CPS in in-game outcome optimization, game injury risk prediction, and game decision-making efficiency improvement. Using the study by Yamamoto and Shen [12], this meta-analysis confirmed the observations of those prior studies using CPS-based predictive and reinforcement learning models and how AI-based predictive analytics reduce injury rates and lead to creation of better tactical decisions. This approach confirmed the qualitative findings with empirical support for the thematic trends found.

Further analysis deepens the study by applying a comparative case study methodology; the real-world cases of CPS deployments are examined, for example, MilanLab's application of predictive analytics for player health monitoring, Liverpool FC's work with Stats Perform on predicting players' performance, and the NBA's use of reinforcement learning models for on-the-fly adaptations in the in-game strategy. The inductive learning in these case studies across different sports provided insights into the critical factors required to implement successful CPS in sports, from technical infrastructure and organizational readiness to the specific algorithms and data processing architecture used to support real-time insights and tactical flexibility. The study examines these examples and learns patterns of CPS deployment and their advantages, as well as the resource allocation challenges that teams face, as in the works by Bhat and Desai [15].

The synthesis process involved matching findings from numerous sources to find trends, validate patterns, and examine divergent findings to look at regional or sport-specific differences in how CPS gets applied.

For example, these differences in CPS adoption in the football, basketball, and cricket leagues were found to be a function of technological readiness and infrastructure variations that determine the effectiveness of the CPS in an ongoing real-time environment. The analysis was enriched with this process, enabling a more subtle understanding of the role of CPS and AI in sports based on examining technical and regulatory factors that shape the success and adaptability of these systems.

One key hypothesis of this study is that using AI-informed CPS frameworks, sports analytics significantly improve player management, in-game adaptability, and injury prevention. The hypothesis also posits that CPS integration continues to be only successful if robust data processing architectures and ethical frameworks are developed to respond to new challenges related to data security, latency, and the problem of algorithmic fairness. The literature provides evidence for this hypothesis: studies of CPS benefits in sports analytics and its need for secure data handling and algorithmic transparency. Works by Chu and Park [13] and Turi and Patel [4] further emphasize that while CPS applications offer measurable benefits, they require sophisticated safeguards to manage risks and maintain competitive integrity in sports environments.

The methodology follows a sequential approach, as depicted in the flow diagram below, illustrating the process from defining research objectives to synthesizing findings and formulating the hypothesis.

Figure 12.1 outlines the flow, beginning with objective definition and literature identification, followed by PRISMA-guided selection, structured data extraction, and layered analysis. Key findings are derived through synthesis, supporting the hypothesis formulation that links CPS effectiveness with the need for secure, ethical, and adaptive sports analytics systems.

12.4 THE ARCHITECTURE OF CPS FOR SPORTS ANALYTICS

When the data flow is layered, from acquisition to processing, and proceeds sequentially through an open loop to real-time analytics, it forms a cyclic feedback system. Wearable and environmental sensors capture and process various metrics at the edge with minimal latency first. By enabling this immediate feedback, the data processing is localized, and data can be stored and analyzed in the clouds to form longitudinal datasets. Data fusion techniques then join real-time player data with historical insights for coaches to make tactical adjustments in response to in-game changes. By building upon this architecture, a dynamic, responsive analytics environment is created within elite sports to enable coaches and analysts to immediately benefit from data insights that are a notable departure from traditional post-game video analysis [5].

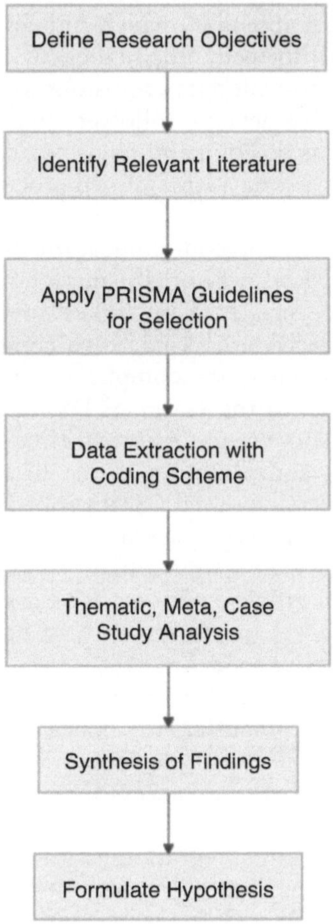

Figure 12.1 Research methodology's sequential flow diagram. (Authors' interpretation.)

However, there are challenges to using these technological advancements for sports analytics. The sensitive nature of athlete data, especially cohorts assembled with a high volume and granularity of wearable data, creates security and privacy concerns in CPS environments. But it's also about staying compliant with regulations like GDPR and having robust encryption and access controls for the data itself. Furthermore, high-stakes sports add particular demands on the network infrastructure since video streaming of these events generates so much data per second, which can consume network bandwidth and computational resources, making network design and edge cloud collaboration particularly important [2]. In addition, processing large datasets in real time at high speed, accuracy, and low latency is required for real-time integration with various data sources. Without overcoming these technical challenges, realizing the full potential of what CPS has to offer in the sports domain is impossible, which implies the ability to optimize performance and make informed, precise tactical decisions.

Figure 12.2 illustrates how the data acquisition system operates, beginning with the setup of wearable devices, followed by continuous metric collection. Once metrics such as position and physiological data are tracked, they are stored temporarily, creating a streamlined process that prepares data for the next transmission phase.

Once data are collected, the communication infrastructure transfers it seamlessly across high-speed networks using a combination of edge computing and cloud platforms. In high-stakes sports environments, with latency important for decision-making, edge computing enables immediate, localized processing where the complex algorithms remain optimized and persist. For example, in the National Football League (NFL), Zebra Technologies RFID sensors are used for real-time tracking of data in spatial coordinates, speed, and direction at the field with edge devices for rapid pre- and post-processing of data. Data transfer speeds for high-speed 5G and forthcoming 6G networks are accelerated so coaches can obtain near-instant intelligence during live games [17]. It allows for real-time decision-making and an efficient process for data management on cloud platforms where historical data can be captured and used for performance trend analysis and predictive modeling [16].

The communication flow diagram (Figure 12.3) shows data being processed through edge computing to reduce latency and sent over an ultra-high-speed network to cloud platforms. The design facilitates rapid data exchange, affording real-time access to information that can lead to tactical adjustment.

Finally, data processing and integration uses advanced fusion to assemble heterogeneous data streams from wearable sensors, field cameras, drones, and other IoT-enabled devices. Data fusion combines such varied input into a combined player and game dynamics analysis. The Hawk Eye system used in tennis and cricket is an example of an exemplary application of data integration, where real-time tracking of ball trajectories, player positioning, and movement patterns leads to meaningful line calls and coaching

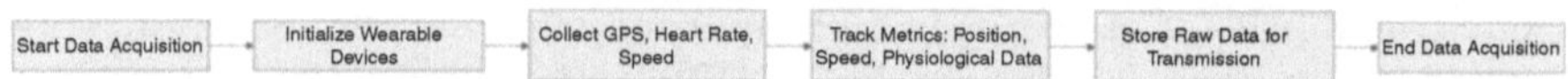

Figure 12.2 Data acquisition systems flow diagram. (Authors' interpretation.)

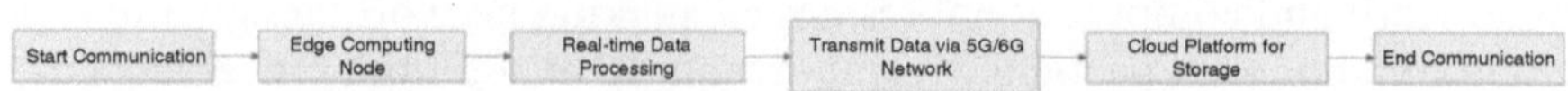

Figure 12.3 Communication infrastructure flow diagram. (Authors' interpretation.)

strategies [13]. By integrating different layers of abstraction, data processing algorithms can exploit data to derive interesting game-specific metrics, such as player positioning, team formation, and fatigue prediction, by optimizing both in-game tactics and post-game analysis. ML models pick up trends and anomalies and help the coaches refine their strategies and individual roles of players based on detailed, data-driven assessments [5].

The data processing and integration phase shown in Figure 12.4 receives raw data from devices and performs data fusion procedures that integrate wearable and environmental data to derive unified analytics. This unified analysis will provide immediate feedback on the field and support performance optimization.

When the data flow is layered, from acquisition to processing, and proceeds sequentially through an open loop to real-time analytics, it forms a cyclic feedback system. Wearable and environmental sensors capture and

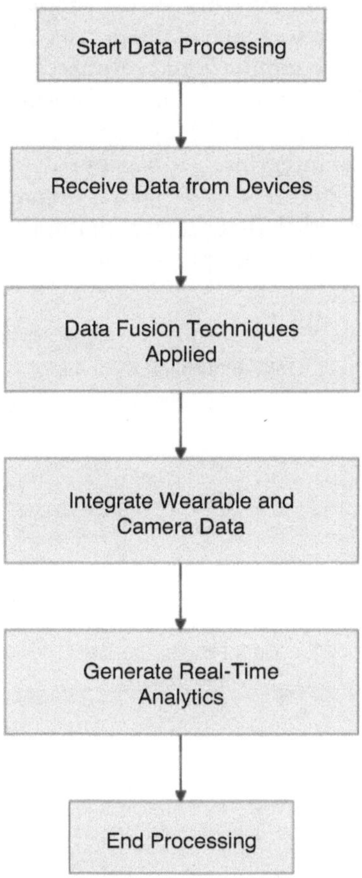

Figure 12.4 Data processing and integration flow diagram. (Authors' interpretation.)

process various metrics at the edge with minimal latency. By enabling this immediate feedback, the data processing is localized and data can be stored and analyzed in the cloud to form longitudinal datasets. Data fusion techniques then are integrated with real-time player data with historical insights for coaches to make tactical adjustments in response to in-game changes. By building upon this architecture, we create a dynamic, responsive analytics environment within elite sports to enable coaches and analysts to immediately benefit from data insights that are a notable departure from traditional post-game video analysis [5].

However, there are challenges to using these technological advancements for sports analytics. The sensitive nature of athlete data, especially cohorts assembled with a high volume and granularity of wearable data, creates security and privacy concerns in CPS environments. However, it is also about staying compliant with regulations like GDPR and having robust encryption and access controls for the data itself. Furthermore, high-stakes sports add particular demands on network infrastructure since video streaming of these events generates so much data per second, which can consume network bandwidth and computational resources, making network design and edge cloud collaboration particularly important [2]. In addition, processing large datasets in real-time at high speed, accuracy, and low latency is required for real-time integration to various data sources. Realizing the full potential of what CPS offers in the sports domain is impossible without overcoming these technical challenges, which implies the ability to optimize performance and make informed, precise tactical decisions (Figure 12.5).

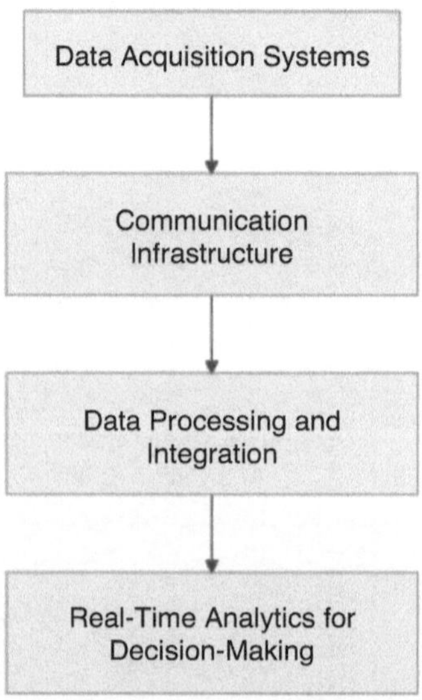

Figure 12.5 A comprehensive flow diagram of CPS architecture for sports analytics. (Authors' interpretation.)

First, data acquisition systems in this architecture take measures directly from the athletes. With edge computing and cloud storage, such data are transmitted over the communication infrastructure with negligible latency. Data processing and integration techniques integrate inputs from various sources into a single source of real-time analytics for coaches and analysts during real-time games and training.

12.5 AI AND ML IN CPS FOR SPORTS PERFORMANCE

The use of AI and ML algorithms will help monitor player movement patterns and predict fatigue along with generating insights, which help in identifying performance peak and injury patterns. CPS systems are able to identify hidden patterns in in-game data using both supervised and unsupervised learning models. They can identify subtle changes in player physiology or behavior, which are key to effective load management at an individual level. Unsupervised learning is particularly useful for clustering and anomaly detection to expose trends such as early fatigue indicators or changes in biomechanics that affect player safety. For example, Liverpool FC is working with Stats Perform, which uses AI models to understand player data regarding fatigue and recovery. This architecture showed data collection and preprocessing through edge devices in a CPS system. Information such as GPS data, heart rate, and speed is collected and filtered out before being sent to the ML modeling or pattern recognition phase for the identification of patterns and insights [7].

Figure 12.6 demonstrates how AI-driven CPS systems gather data from wearables, process it at the edge to reduce latency, and apply ML models to generate actionable insights for player management and in-game decision support.

Predictive analytics further extends AI's utility within CPS, enabling coaches to forecast game outcomes, optimize training loads, and anticipate opponent strategies. These models combine historical and real-time data to forecast player fitness, injury susceptibility, and team performance. Teams like AC Milan's MilanLab use AI to predict fitness and manage injury risks through CPS-based predictive systems, adjusting player workload dynamically to mitigate risks and enhance recovery. Here, data from wearables and other IoT devices are aggregated and analyzed in predictive ML modeling, which merges diverse data types for injury prediction, heatmap generation, and movement forecasting. Coaches access real-time insights on tactical adjustments and workload distribution to ensure optimal player utilization.

The predictive analytics architecture integrates data from multiple sources, which flows through predictive ML models to forecast game dynamics and player conditions. These insights are relayed instantly to coaches, enabling them to make proactive decisions on player rest, positioning, and team tactics (Figure 12.7).

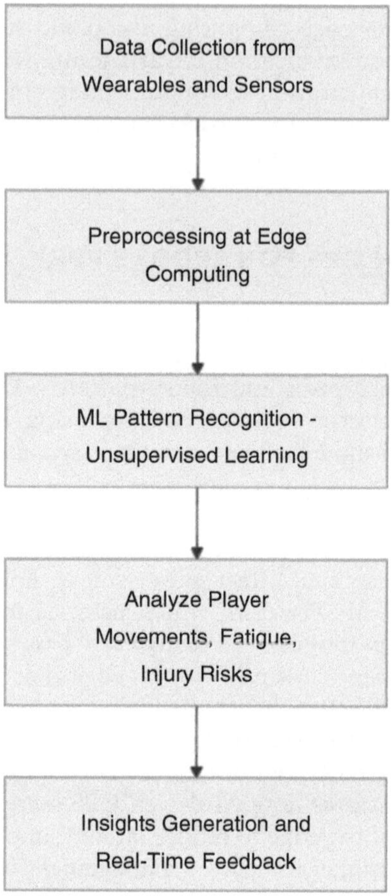

Figure 12.6 Data-driven insights flow diagram. (Authors' interpretation.)

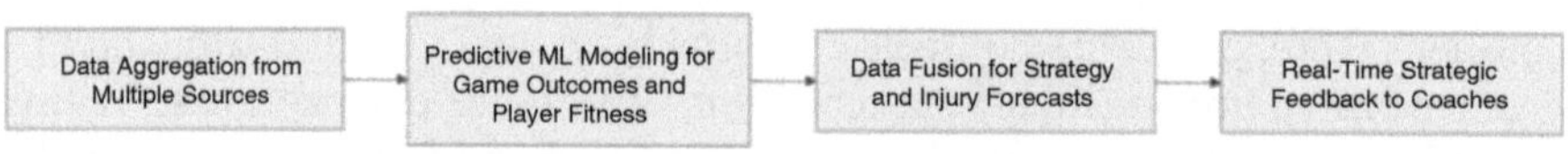

Figure 12.7 Predictive analytics system architecture diagram. (Authors' interpretation.)

Reinforcement learning (RL) is another advanced AI application in CPS for sports, which enables models to learn in real time and enhance predictions during live matches. In the NBA, RL models receive dynamic information about game situations to adapt shot and defensive strategies. The CPS's RL framework involves real-time input processing (RIP), where the game metrics (like player position, shot probability) get updated continuously, followed by the RL model processing layer that updates the predictions. Coaches receive information about refined strategies established during the feedback loop. RL will help teams optimize strategies

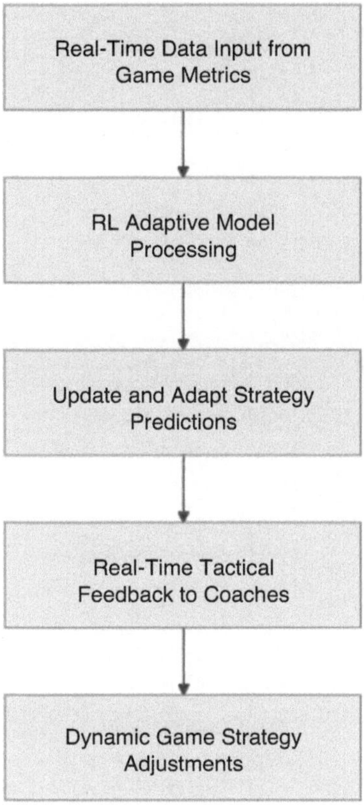

Figure 12.8 Reinforcement learning system architecture diagram. (Authors' interpretation.)

adaptively with a data-driven foundation for in-game decisions based on environmental and game state dynamics.

This RL model uses in-game data to feed algorithms that constantly update predictive information based on the game context and feedback coaches in real time. This supports real-time defensive structure, shot selection, and team tactics decisions (Figure 12.8).

A comprehensive AI-cyber-physical systems (AI-CPS) architecture is integrated to achieve real-time decision-making in high-stakes sports through data-driven insights, predictive analytics, and RL. This system collects data from sensors and wearables, uses edge-based processing, and uses cloud-based AI processing. The processing done at the edge level filters the raw data, whereas AI-based processing creates more advanced analytic techniques like ML and RL. This CPS architecture enables insights to be communicated in real-time to the coaching teams, allowing adaptation of actions during the game in return for a feedback loop, while also turning data into a competitive advantage to enhance the performance and player safety [5].

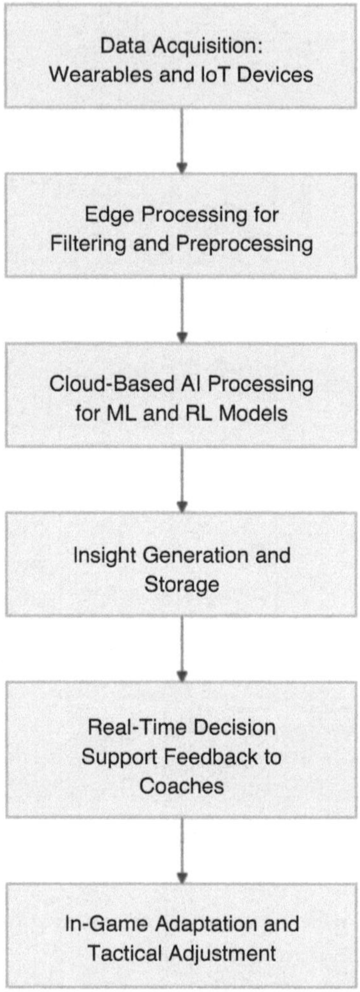

Figure 12.9 Comprehensive AI-CPS system design architecture diagram. (Authors' interpretation.)

As shown in Figure 12.9, the overall flow of an AI-based CPS starts with data acquisition, then edge processing to reduce latency, and finally, cloud AI processing for real-time analysis. The insight generation and feedback loop enable coaches to make adjustments based on immediate, data-driven intelligence.

AI-based systems in sporting environments are a game changer as they enhance the performance of players and coaches by using real-time information to refine training and recovery, injury prevention, and optimization of in-game tactics. Through experts in ML, predictive analytics, and RL integration, continuous monitoring and updating with a common platform for performance optimization with data are performed. The CPS architecture, which is intelligent and adaptive, will support competitive play, promote the health of athletes, and relabel the performance of sports analytics in competition as a future-oriented initiative for elite performance services.

12.6 ADVANCED HUMAN–MACHINE COLLABORATION IN SPORTS ANALYTICS

AI-driven CPS frameworks are designed to provide real-time actionable insights for transforming tactical decisions as coaches can use complex analytics, visualization tools, and real-time performance data. Football clubs like Chelsea and Arsenal are the best examples: they use AI-powered data visualization platforms during matches to modify strategies dynamically in real-time. Such systems can be enabled on tablets and accessed in real-time during live gameplay, allowing a coach to make tactics on the fly based on immediate performance data to enhance real-time strategic agility and interactivity. This capability for real-time, AI-augmented decision-making is situated on a layered architecture for data collection, real-time analysis, insight generation, and visualization, all of which are developed to work together to enable capability on the field for decision boosting [6].

In AI-augmented decision-making, the CPS frameworks collect data from multiple sources, including wearable devices, in-field cameras, and environmental sensors, which continuously capture the key performance indicators on movement patterns, energy expenditure, or degrees of player fatigue. The data can be seen to flow through the edge computing nodes to immediately process much of it locally, thereby eliminating latency and enabling quick feedback loops. This will be followed by a layer of data processing and AI analysis, which will help apply ML and predictive models to generate insights fed into real-time visualization tools, creating an interface on tablets or AR-enabled headsets that visualizes metrics in real-time and allows the coach to adjust accordingly based on insights into player positions or fatigue or tactical effectiveness. For example, the coaches at Chelsea use AR-enhanced data visualization tools to review player performance metrics immediately and make changes in formations based upon that review, an excellent example of how CPS facilitates human–machine collaboration in high-stakes sports environments [7].

This architecture minimizes latency by feeding data straight into edge processing and applying AI model analysis with ML algorithms to generate tactical predictions. They visualize those insights in real time that so the coach can make immediate adjustments, showing how human–machine collaboration is operationalized in sports (Figure 12.10).

Besides tactical change, the CPS has helped shape athlete-specific training programs by providing insights from AI in real-time monitoring of

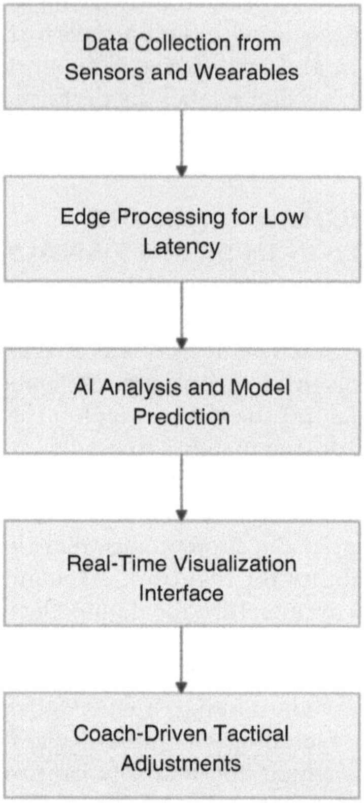

Figure 12.10 AI-augmented decision-making system architecture. (Authors' interpretation.)

athlete load management, recovery optimization, and nutrition planning. This is relevant in high-performance situations wherein the training needs to be adjusted according to the physiological state of each athlete and the volume of work done. One of the most prominent examples is Athos, which Lindsey Vonn uses to monitor muscle activity and receive instant feedback on muscle engagement, fatigue, and strain during workouts. These wearable systems capture detailed physiological metrics, such as muscle activation and exertion, feeding the data into a centralized processing system where ML algorithms assess the optimal workload for each athlete. The insights are provided to the trainers, who can alter training routines in real-time, hence achieving performance targets while minimizing the risk of injury [7].

The architecture backing the personalized athlete training is a layered design. The data acquisition layer gathers metrics through smart clothing and wearable devices and then processes them at edge nodes, filtering and categorizing the data. This layer of personalization algorithms

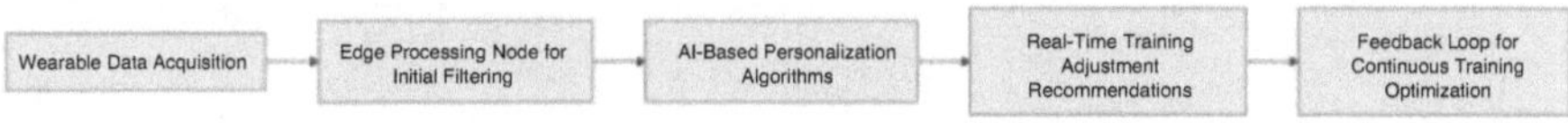

Figure 12.11 Personalized training system architecture diagram. (Authors' interpretation.)

uses AI-based models that analyze an athlete's current workload, muscle activity, and recovery needs to provide personalized recommendations for adjustments in training, rest, or nutritional intake. The result is a dynamic training environment that adapts to the physiological state of each athlete and training that is safe, effective, and aligned with performance.

Figure 12.11 also shows that the data collected from the smart wearables – in our case, Athos – are inspected at the edge to reduce latency before further AI-based personalization algorithms are called to analyze it. From this extracted insight, trainers could receive continuous feedback in making live adjustments on the training loads, rest intervals, and nutritional plans, thereby setting up a highly personalized training ecosystem.

These components are woven together in an integrated framework in the architecture of advanced human–machine collaboration in CPS applications concerning sports, from real-time decision-making to personalized training, in an overarching framework that starts from data collection and transmission via wearables and sensors, and then through real-time processing for immediate feedback and analysis, culminating in AI-augmented insight generation in the service of tactical decisions and athlete-specific changes to training. This closed-loop architecture supports seamless feedback such that real-time insights form strategic gameplay decisions and individual athlete management to create an environment of collaboration that maximizes team effectiveness and individual performance [16].

Data collection, based on wearable devices, is processed for low-latency responses through edge computing. At the same time, the AI and ML processing layer creates real-time information regarding tactical and training insights to be visualized in support of immediate tactical adjustments by coaches and individualized training modifications by trainers. CPS architecture can bridge human–machine collaboration in elite sports in this interconnected system by making AI-augmented insights accessible and actionable for all stakeholders (Figure 12.12).

12.7 EMERGING TRENDS IN CPS AND SPORTS ANALYTICS

12.7.1 Edge AI for real-time processing

Edge AI in CPS, therefore, will highlight and accentuate the shift to decentralized processing wherein data analysis is rather near the point of origin, which fosters lower latency and expedient insights. This capability is

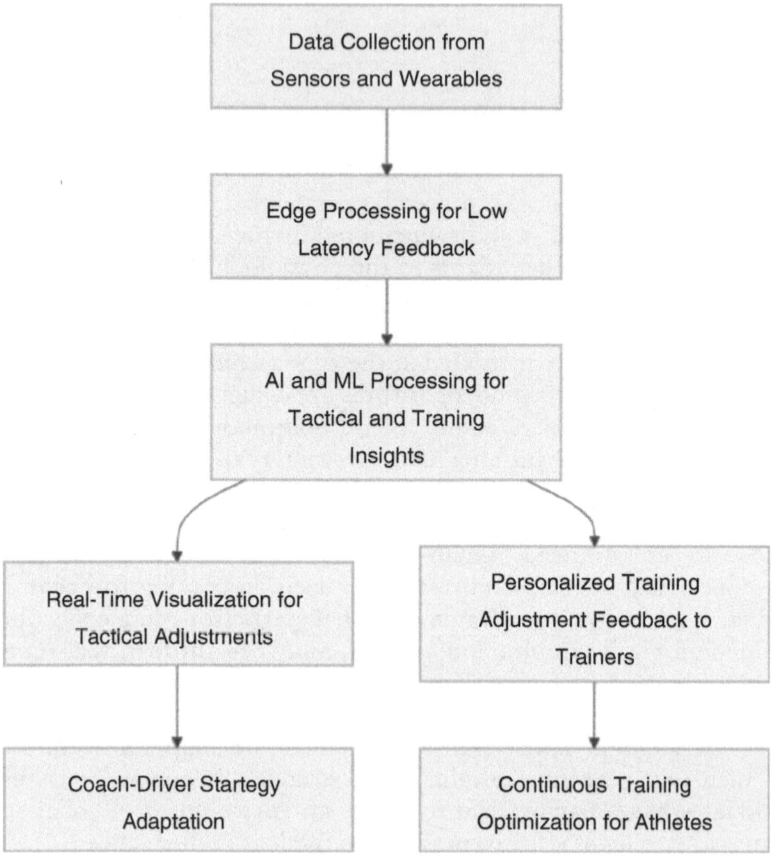

Figure 12.12 Comprehensive human–machine collaboration CPS architecture diagram. (Authors' interpretation.)

critical in high-stakes sports environments because real-time decisions have a bearing on in-game outcomes. For instance, the edge AI technology has been integrated into MLB to process real-time statistics concerning pitcher–batter interactions; this brings immediate feedback, guiding adjustments on coaching based on situational analytics. This will begin with detailed data on pitcher velocity, batter stance, and reaction time captured by wearable devices and in-field sensors in these systems. All the processed data are fed through edge AI nodes within the stadium and analyzed there for patterns and anomalies; consequently, predictions are provided directly to the coaching staff on tablets. This architecture also ensures that data-driven insights are available within seconds, so teams do not have to wait through the lapse of time associated with cloud processing for strategic decisions [2].

This architecture depicts how edge AI nodes can immediately process data collection from on-field devices for latency-free analysis. This is followed by

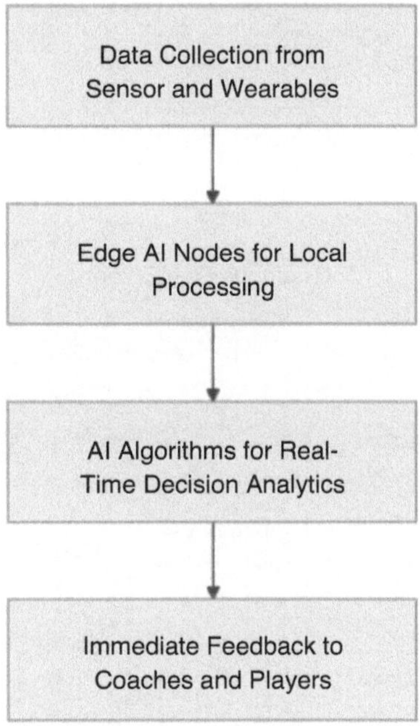

Figure 12.13 Edge AI system architecture diagram. (Authors' interpretation.)

the layer of AI algorithms that input their insights directly to the coaching staff for real-time, data-driven adjustments in the game (Figure 12.13).

One of the reasons edge AI has been essential for sports decisions in the game, particularly where quick reactions are key, is because it allows for real-time analytics without having to be based on data centers in the cloud. Earlier work by Turi and Patel on edge computing in CPS [4] highlights how proximity-based processing substantially reduces delay and provides real-time actionable intelligence for the role of edge AI in enhancing the game strategy with response times.

12.7.2 Autonomous systems in sports

Autonomous systems, including drones, add a new dimension to sports analytics with views and visual data never heard or seen before. In the NFL, real-time expansive views of formations and movements of players are captured through the enhancement of performance analysis and fan engagement through performance analysis and fan engagement through autonomous drones that automatically follow players in aerial footage for review of games and instant replays, enabling coaches and analysts to assess strategies from complete field angles. Such drones with AI-based tracking software are constantly adjusting their positioning through player movement, thus providing live visuals that otherwise an ordinary camera fails to capture.

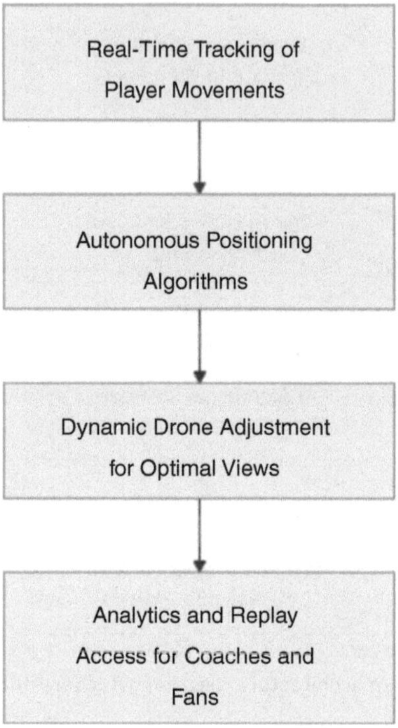

Figure 12.14 Autonomous systems in sports architecture diagram. (Authors' interpretation.)

The real-time tracking layer is the autonomous sports drone system architecture through which sensors identify the positions and movement patterns of the players. It then processes data through autonomous positioning algorithms to adjust drone paths for proper viewing. Finally, the captured data are fed to an analytics and replay layer, which provides the coach and broadcasters with live views of the field or exactly the same video to enable instant replay analysis. Therefore, autonomous systems provide a dynamic and adaptive view for enhanced analytical depth and audience engagement [12].

Figure 12.14 depicts how drones would use real-time tracking for the documentation of players' movement and autonomous positioning algorithms for the dynamic capture of key action moments. That information then becomes available for analytics and replay and accessible to coaches, fans, and broadcasting platforms, making for multifunctionality in both strategic analysis and spectator connection with the game.

Yamamoto and Shen [16] conducted research on autonomous systems, indicating how drones in sports provide broad and flexible views of the visual that can change old ways of thinking about player performance analysis. This system allows a coach or analyst to see the overall formations and strategies that better inform them about team behavior and make time for real-time adjustments.

12.7.3 Next-gen fan engagement

The increasing role of CPS in sports also now changes the behaviors of fans, increasing participation and even making participants have more immersive experiences. AI and AR are now integrated into the broadcasts enabled by CPS, giving viewers real-time analytics and predictive modeling as well as interactive displays. For instance, La Liga will utilize the AR technology with games where real-time statistics, player heat maps, and tactical insights are broadcast into games, and for this experience, fans will be at the center. It enhances the fan experience with the integration of data collection received from in-game sensors in a favorable analysis by AI prediction and visualization models for insights like where the player is positioned or the likelihood of scoring. Appropriate insights are then cast using AR displays to create an interactive layer with which viewers can look in relation to game dynamics and action in real-time [14].

This architecture illustrates how data collection can be instrumental in enabling AI prediction and visualization, which eventually produces a real-time AR display that will heighten the fan's experience of engaging with the game events. The fans will thereby receive real-time statistics and predictive insights into the game. Such passive viewing could then become an engaging experience (Figure 12.15).

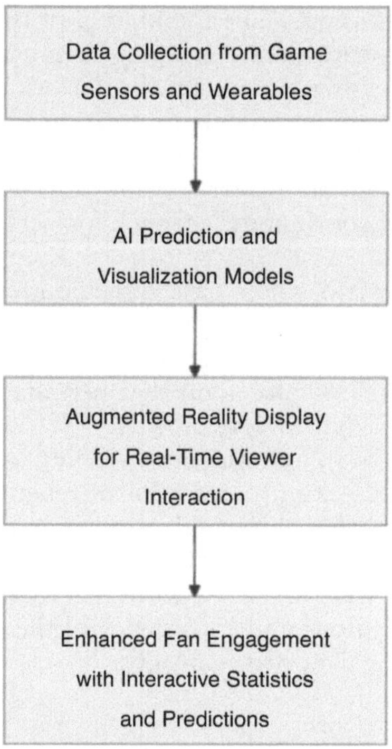

Figure 12.15 Next-gen fan engagement system architecture diagram. (Authors' interpretation.)

Real-time analytics used in fan engagement significantly increases the involvement and satisfaction of the viewer. According to Bhat and Desai [15], the use of AI and AR applications in sports broadcasts enhances the viewing experience and further invests the emotions in the game. This is a progressive application of CPS for engaging fans as it demonstrates how real-time data and predictive analytics can make sports entertainment engaging and invest more energy from fans with the team and its players.

12.8 FUTURE DIRECTIONS: THE EVOLVING ROLE OF CPS IN SPORTS

12.8.1 AI-driven autonomous coaching systems

An important leap in the CPS application is the autonomous coaching system, where AI acts like a "co-coach" who can assist or even independently be an assistant coach in making coaching decisions. Such systems rely on ML algorithms to analyze player data, trace patterns of performance, and give feedback in real-time. Such platforms as HomeCourt AI lead basketball regarding real-time feedback on shooting techniques and footwork that professional teams now consider as additional training tools during practice sessions. The foundation of the CPS architecture of such platforms begins with data collection from cameras and wearables, followed by edge processing for latency minimization. These data are then processed using AI coaching models to make available insights into the mobile devices of coaches and players. Indeed, the system can fine-tune recommendations based on player-specific tendencies, creating a loop that bestows a form of continuous feedback for personal and autonomous coaching guidance [2].

This architecture describes how data collection feeds into edge processing to reduce latency: the AI coaching models are enabled to give immediate feedback on performance. These insights get relayed to players and coaches, thus creating a natural feedback loop that permits real-time adjustments and learning by doing (Figure 12.16).

There is vast potential in autonomous coaching as applied to the sports arena. These AI systems can provide a lot of repetitive feedback, thereby creating time for the coaches at the high-strategy levels rather than detailed efforts at the low level. Studies about CPS and AI in autonomous coaching, as shown by Patel and Watson [7], evidence that these systems can enhance the rates of skill acquisitions and consistency of the practice environments as they deliver real-time data-driven insight.

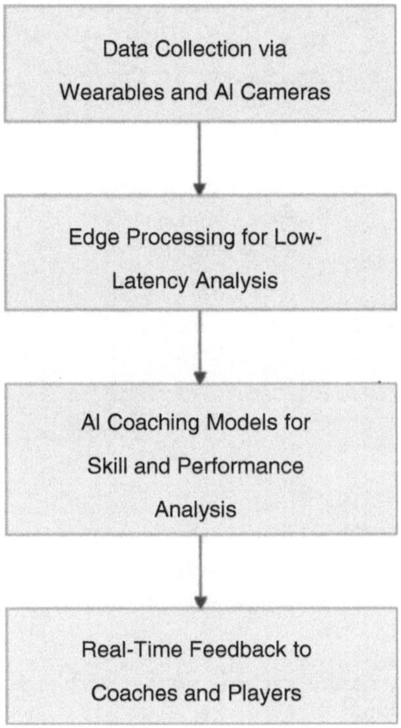

Figure 12.16 AI-driven autonomous coaching system architecture. (Authors' interpretation.)

12.8.2 Augmented reality for on-field tactical adjustments

Augmented reality (AR) is poised to enhance real-time strategic adjustments, potentially allowing players and coaches to access visual feedback directly on the field. In American football, for example, AR-enabled glasses are being developed to assist quarterbacks in reading defensive formations in real-time, providing instant feedback on likely plays or adjustments. This application of CPS architecture integrates data collection from in-game cameras and sensors, which are processed through edge computing nodes for immediate feedback. The processed data flows to AR displays that players or coaches wear, overlaying tactical information directly into their field of view, enabling quick and informed decision-making based on the opponent's formation or field dynamics [6].

This architecture starts with data collection and proceeds to edge computing, ensuring real-time analysis before the data are sent to the AR displays players wear. These displays overlay relevant information in the player's line of sight, offering on-the-fly feedback to enhance in-game decision-making (Figure 12.17).

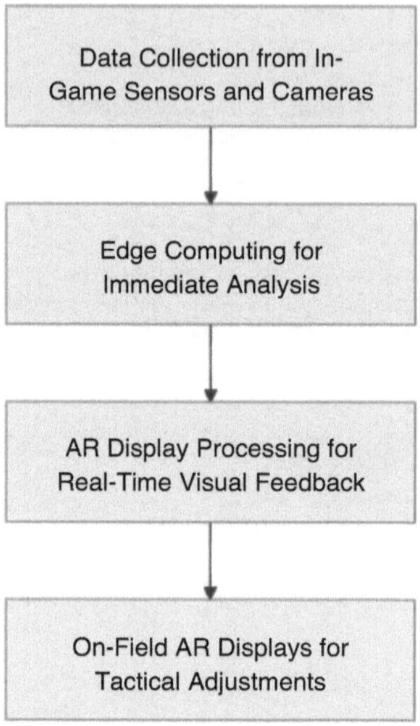

Figure 12.17 Augmented reality tactical system architecture diagram. (Authors' interpretation.)

Augmented reality has the potential to fundamentally change the sports strategy by making complex analytics instantly accessible. Research by Chu and Park [14] supports that AR-enabled CPS can enhance on-field awareness and situational understanding, making the sports strategy more agile and responsive.

12.8.3 Predictive injury management

Predictive injury management is one of the most critical applications of CPS in sports, providing real-time health assessments and dynamically adjusting training loads to prevent injuries before they happen. In the NBA, predictive AI tools analyze player movement patterns, collision data, and historical injury records to forecast potential injuries and recommend preemptive actions. By integrating data collection from wearables that track biomechanics and physiological metrics, CPS architectures analyze these inputs through predictive AI Models that assess stress levels, fatigue, and collision impact. This system's real-time feedback component alerts coaches and medical staff, allowing them to make load adjustments and mitigate risks proactively, reducing the likelihood of injuries [4].

In Figure 12.18, data collection from wearables is processed through edge processing for immediate analysis. Predictive AI models then assess

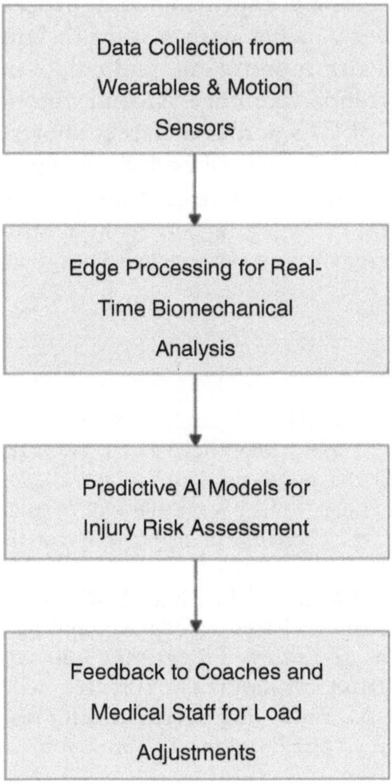

Figure 12.18 Predictive injury management system architecture diagram. (Authors' interpretation.)

injury risks and send alerts to coaches and medical staff, allowing them to adjust training loads dynamically to prevent injuries.

Predictive injury management systems have shown significant promise in injury prevention by tracking biomechanical metrics and highlighting potential risk factors. Studies by A. M. Turi and N. M. Patel [4] confirm that such systems have reduced injury rates by identifying high-risk conditions early and enabling proactive responses, contributing to longer, healthier careers for athletes.

12.9 CONCLUSIONS

Integrating CPS and AI in sports revolutionizes player performance, strategy, and fan engagement. From AI-driven autonomous coaching and augmented reality (AR) for real-time tactical adjustments to predictive injury management, the CPS technology enables new precision, adaptability, and interactivity levels. By capturing and processing real-time data, these systems empower coaches with instant insights, safeguard athletes through predictive health monitoring, and create immersive experiences for fans. Emerging trends like edge AI and autonomous drones showcase the adaptability of CPS across different sports, emphasizing its role in enhancing on-field performance and elevating audience engagement. As CPS in sports continues to evolve, it promises to make sports safer, more strategic, and more engaging, creating a future where data-driven insights are at the heart of competitive advantage and athlete wellbeing.

REFERENCES

1. M. B. Grayson, R. S. Ahmad, and T. K. Ramalingam, "Real-time data processing in CPS for sports analytics," *IEEE Transactions on Computational Social Systems*, vol. 10, no. 4, pp. 1203–1212, 2022.
2. J. C. Liu and P. V. Lee, "Smart wearable CPS systems for player monitoring in professional sports," *Journal of Applied Sport Technology*, vol. 8, no. 3, pp. 174–188, 2021.
3. Xiaowei Tang, Bin Long, and Li Zhou, Real-time monitoring and analysis of track and field athletes based on edge computing and deep reinforcement learning algorithm, *Alexandria Engineering Journal*, vol. 114, pp. 136–146, 2025. ISSN 1110-0168, https://doi.org/10.1016/j.aej.2024.11.024.
4. A. M. Turi and N. M. Patel, "Edge computing for real-time sports analytics: a comprehensive survey," *IEEE Internet of Things Journal*, vol. 11, no. 2, pp. 876–889, 2022.
5. Y. Shen, M. A. Rahman, and S. W. Lee, "AI and machine learning integration in CPS for sports performance enhancement," *Journal of Sports Analytics and AI*, vol. 9, no. 5, pp. 305–319, 2023.
6. M. J. Burke and R. K. Fox, "Wearable technologies in sports analytics: from player monitoring to performance optimization," *IEEE Sensors Journal*, vol. 19, no. 9, pp. 3201–3208, 2021.
7. R. K. Patel and S. A. Watson, "Unsupervised learning in sports CPS: detecting hidden patterns in athlete behavior," *International Journal of Machine Learning in Sports*, vol. 8, no. 4, pp. 209–222, 2022.
8. Q. Zhang, F. Wu, and J. Tan, "AI-generated heatmaps in sports: movement prediction and strategy planning," *IEEE Transactions on Artificial Intelligence*, vol. 11, no. 1, pp. 500–510, 2023.
9. P. Gupta and T. Shen, "Reinforcement learning in NBA analytics: dynamic game strategy adjustment," *IEEE Transactions on Sports Science*, vol. 10, no. 6, pp. 485–497, 2022.
10. L. Yu, A. B. Clarke, and M. H. Davies, "Real-time adaptive models in sports analytics using RL," *Journal of Computational Sports Research*, vol. 15, no. 4, pp. 222–239, 2023.
11. A. H. Richardson and J. C. Stevens, "Machine learning-driven sports analytics: predictive models in professional football," *IEEE Transactions on Sports Science*, vol. 15, no. 2, pp. 92–108, 2022.

12. T. Yamamoto and S. K. Shen, "Data fusion techniques in sports CPS: applications and future directions," *International Journal of Sports Analytics*, vol. 9, no. 4, pp. 176–185, 2022.
13. A. T. Guzman, "High-speed networks and cloud solutions in CPS sports applications," *Computing in Sports Analytics*, vol. 8, no. 5, pp. 497–503, 2023.
14. K. Y. Chu and D. L. Park, "Data security in CPS sports analytics: regulatory and technical perspectives," *IEEE Transactions on Data Privacy*, vol. 11, no. 3, pp. 411–420, 2023.
15. S. Bhat and M. Desai, "Fairness and transparency in AI for sports analytics," *Journal of Ethics in Sports Technology*, vol. 6, no. 1, pp. 55–63, 2023.
16. S. P. Srinivas and D. G. Grant, "The role of edge computing in high-stakes sports: real-time analysis and performance enhancement," *Journal of Sports Technology*, vol. 13, no. 2, pp. 58–66, 2022.
17. R. Lee and J. K. Patel, "RFID and real-time sports analytics: enhancing player tracking and game strategy," *IEEE Internet of Things Journal*, vol. 10, no. 3, pp. 510–519, 2023.

Chapter 13

Navigating the risks

Cybersecurity challenges in cryptocurrency and blockchain solutions

Bhavana Bharati, Kamal Upreti, and Ramesh Chandra Poonia

13.1 INTRODUCTION

With the advent of Industry 4.0, with every passing day and advancement in technology, mechanization, and automation, it is necessary to look at the financial implications arising from all these changes. Large automation and mechanization, cloud computing, and digitization demand huge investments in not only physical systems but also human capital. With this development comes the new age of payments, cryptocurrency. The security and scalability of cryptocurrencies have come into question numerous times; nevertheless, quite a few cryptocurrencies are popularly in use; stablecoins are supposed to reach over a market capitalization of $150 billion by 2024 [1]. Since cryptocurrencies have developed and shown themselves to be a disruptive technology that is immune to the existing protocol and limitations, as well as a potential swap in for the incumbent state's concept of fiat money, this raises concerns about security if they are to become accepted in this digital age [2]. This anonymity has facilitated the use of Bitcoin and other cryptocurrencies by different agencies to aid illicit acts that pose a threat to the society on a whole [2]. The chapter aims at shedding light upon the varied risks and fraud that exist with this mode of virtual currency, which is built on a secured and protected blockchain platform. It tries to clarify on how such risks may be detected and navigated and also examines the pros and cons of implementing this mode, which is now reportedly gaining much validation as a mode against fiat money [2].

13.1.1 Overview of Industry 4.0

Any flourishing economy interested in big production and efficient mass manufacture of mechanized products gives shape to the word "Industry." Since the beginning of industrialization, technological revolutions have set large-scale movements, now popularly known as "industrial revolutions." The first industrial revolution was by machinery, and the second was about the wide use of electricity. The third one has witnessed massive digitalization. Indeed, a high degree of factory digitalization offers a new

 DOI: 10.1201/9781003542513-13

major shift in industrial production by integrating Internet technologies and progressive innovation in "smart" machines and products. Thus, the future of manufacturing presents modular and efficient systems of different scenarios where products manage production processes autonomously. This would allow one to have custom products in batches of one but yet experience an economic benefit of mass production. Inspired by such potentials, the term "Industry 4.0" was even expected long beforehand for describing what became referred to as the "4th industrial revolution," and therefore, being somehow likened to some version number of software [3].

13.1.2 Emergence of cryptocurrency

Cryptocurrency utilizes blockchain technology. Blockchain technology avoids intermediaries but provides safe, fast, and inexpensive transactions [4]. The cryptocurrency market has dramatically expanded since its launch of Bitcoin in 2009. Its value is increasing because of technological improvements and expanding its scope of applications. Some of them have begun to gain significant mainstream users and investors. These include uses like payment, gaming, social, and progress on decentralized identification verification and infrastructure networks. For instance, on 11 January 2024, financial history was recorded when the United States approved the first Bitcoin Exchange Traded Fund. It strongly recognizes cryptocurrencies, especially Bitcoin, as investments [5].

Cryptocurrencies present both opportunities and challenges for state regulation. Approaches range from complete bans to full legalization, with most jurisdictions applying existing financial, anti-money laundering, and consumer protection laws [6]. Others have outright banned them, such as China. Australia and Switzerland, on the other hand, are much more facilitative towards innovating with cryptocurrencies. However, regulatory concerns lie majorly in volatility, criminal abuse, and investor protection [6]. The advantages of these cryptocurrencies are offset by their potential threats against national currencies and financial systems, thus requiring states to weigh between the two opposites [7,8]. However, some sovereign laws failed to completely control the use of cryptocurrencies and, thus, require a more nuanced, global regulatory standard [2]. Cryptocurrency adoption, regulation, and prohibition are all options available to the state; each has an impact on both domestic and global financial relations [7,2].

13.1.3 Blockchain's function in the creation of cryptocurrencies

Constituting the framework of distributed cryptocurrencies, blockchain is an immutable ledger that enhances transparency, data protection, and security in online virtual financial dealings [9,10]. Cryptocurrencies are blockchain-based virtual currencies such as Ethereum and Bitcoin [11].

Blockchain was adopted as the foundation of cryptocurrencies due to its numerous salient features. Immutability in the blockchain means that once a block has been generated, customization is prohibited. Since a block contains a hash value derived from its contents, every change will look conspicuous. Another very important aspect is anonymity; people are addressed through public keys and never reveal their details. A decentralized network system helps one make direct, third-party-free transactions; therefore, it is financial efficiency in its best sense. In a distributed ledger, records are identical and accessible in the network; this means total transparency. Persistence ensures that the time valid transactions are accepted into the blockchain, then they cannot be deleted or overwritten. Transactions are recorded verifiably and securely, thus simplifying auditability, doing away with the conventional auditing methods, similar to those applied in the bank statements [12].

13.2 ARCHITECTURE OF CRYPTOCURRENCY

13.2.1 Definition and technology of cryptocurrency

A system can be considered a cryptocurrency when it is a distributed consensus nexus that manages a record of all cryptocurrency units and respective owners, negating the need for a central authority. The system must approve all transactions after their ownership is cryptographically verified. The system is framed to tailor to one of two instructions entered concurrently with the purpose of changing their ownership.

Blockchain inspires the technological innovation that underlies many cryptocurrencies to allow fund transfers from peer to peer (P2P) in a trustless computing environment without a central authority [13]. For the case of cryptocurrency, new units cannot be produced, and so far, companies or governments have not provided support to other firms, banks, or corporate entities holding asset value measured in it [14].

To enable advanced functionalities within the financial network, every cryptocurrency must be comprehended by its system architecture. Advanced technologies such as blockchain systems that allow virtual monetary transactions without any overhead central authority are the primary proponents of cryptocurrency development. It can pose a threat to sovereign institutions due to the presence of drawbacks, such as the lack of any predetermined authority or protocol to steer it. The volatile nature of cryptocurrencies can raise virtual security and personal concerns that result in inconsistent use and overexploitation [15]. However, cryptocurrencies offer a new paradigm for safe finances, testified through their enterprise valuation with drawbacks that may unveil their fallacies and expose them to exploitation [16].

One of the most proficient ways to explain the workings of cryptocurrencies is to examine the functioning of Bitcoin, being the most popular and efficient example. Publicly accessible and with no central authority commanding it, the digital ledger of Bitcoin records transactions with an average total transfer value of approximately $75 million. Since the inception of Bitcoin, more than half a million blocks have been produced, with over 500,000 blocks already created. With a market value of over $500 billion, there are more than 1,500 different kinds of cryptocurrencies. The ability of blockchain technology to create and manage digital currencies is demonstrated by Bitcoin's success in shedding light on how cryptocurrencies operate and its substantial financial presence in markets driven by blockchain technology [17,18].

Figure 13.1 describes how Bitcoin works, the issuance of which relies on proof-of-work mining, with network nodes competing to solve particularly complex puzzles. Winners in this competition then validate and add to blocks these transactions. The blockchain accepts freshly generated blocks and secures the circulation of Bitcoins through cryptography. Each valid transaction is hashed and encrypted before being added to the ledger; Bitcoins can be programmed to have intelligent contracts for automated flows in circulation. The Bitcoin blockchain includes the public ledger, P2P networking, decentralized consensus, economic incentives for miners, and

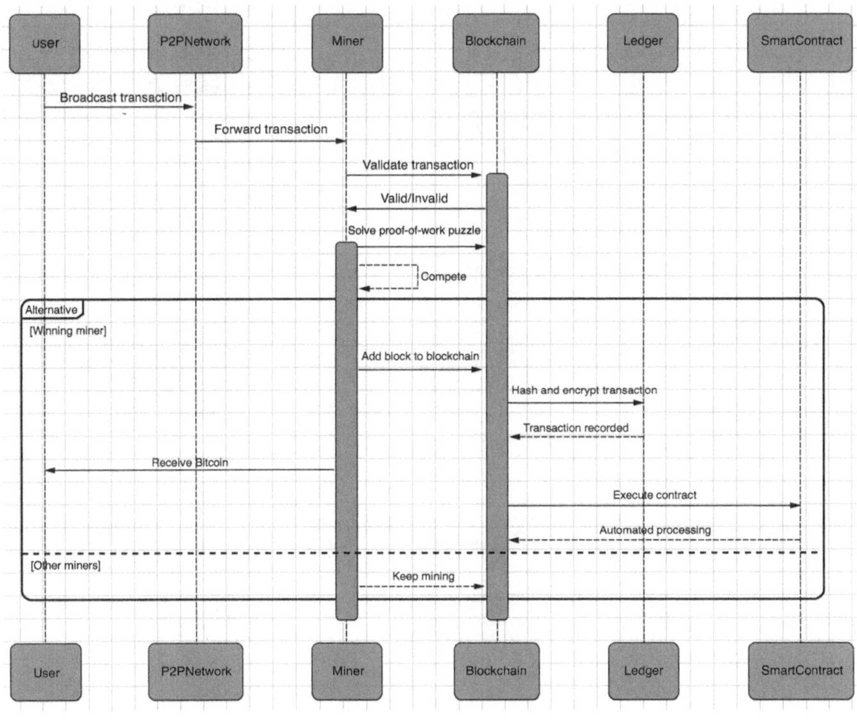

Figure 13.1 The blockchain system of cryptocurrency transaction.

programmable smart contracts. The Bitcoin system incentivizes miners to contribute computing power and ensures secure and automated transaction processing [19].

13.2.2 Primary components

A blockchain (Figure 13.2) system can be deconstructed into the following layers [20,21].

13.2.2.1 Data layer

It manages and processes data from various sources. It includes foundational techniques and tools for handling data. It handles data that originate from three primary areas: cyber space, which refers to digital environments; physical space, which contains real-world data; and social spaces, which refer to human interactions, such as social media or communication platforms [22–24]. The data are kept in the chained blocks in a full node of a blockchain network, which uses encrypted, hashed, and time-stamped Merkle trees. A winning node can create a new block containing time-stamped data to solve conflicts like double spending by choosing one agreed version [19]. Figure 13.3 demonstrates a typical Bitcoin block comprising a meta information header and a Merkle tree body holding verified and hashed data [25]. Blocks are chained chronologically from the genesis block onwards to new blocks. Adding to the dimension of blockchain history per timestream, a Merkle tree along with a time stamp has been an important factor in fast verification as well as traceability of the Blockchain data [19].

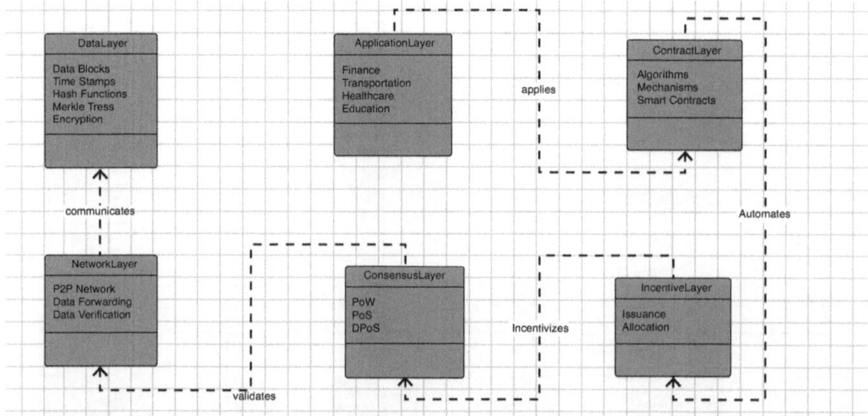

Figure 13.2 Structure of blockchain.

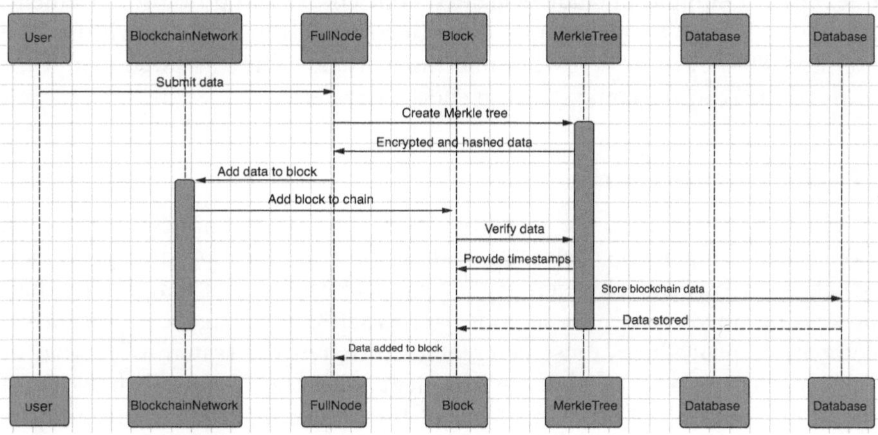

Figure 13.3 Working diagram of a blockchain's data layer.

13.2.2.2 Network layer

An essential part of its system, blockchain systems allow users to communicate directly without third-party interventions and ensure security. Data and resources are transmitted directly in an inter-node nexus. Blockchain finds applications in open, dynamic environments with various connected devices. The blockchain system uses P2P networks in which all nodes have equal privileges and operate without central control. Nodes verify data and blocks, discarding the invalid ones and forwarding the valid blocks to succeeding nodes. The decentralized architecture of the P2P net makes blockchain suitable for the next generation of cloud computing, even ensuring data synchronization and restoration in case of failure at multiple nodes [19,20].

13.2.2.3 Consensus layer

Blockchain employs consensus methods to guarantee distributed nodes' fault tolerance and data consistency. PAXOS is used in primitive applications, such as closed ecosystems, for effective consensus. More complex algorithms are needed for blockchain models, which concentrate on dynamic, open environments with several untrusted participants. For semi-open environments, practical Byzantine fault tolerance is employed, whereas proof-of-X (POX) consensus is employed for relatively unbarred environments, such as cryptocurrency exchanges [26]. Examples of POX algorithms include the most widely used PoW-asking nodes to compete several times to do a difficult mathematical computation to validate data [27]. Nodes that succeed can append a new block on the ledger, and proof-of-stake (PoS) employs those with the most significant stake by predefined stakes to sign the latest one. Other examples include delegated PoS, DPoS, and proof of movement. Among all the POX algorithms, the noncompete-intensive algorithms like PoS and DPoS are more favourable than others. They are well suited for most of the transparent systems in such ecosystems.

13.2.2.4 Incentive layer

In blockchain systems, the incentive layer consists of financial incentives. Consensus competitions drive data validation and block building processes, which is a crowdsourcing activity for participating nodes that donate their processing power. Since these nodes act as self-centred agencies, incentive-compatible procedures must be created to match the system-wide goal of ensuring a safe and reliable environment with the revenue behaviour of individual nodes [28–30]. This layer of the blockchain includes techniques for distributed networking, data transfer, and validation as well as decentralized communication formats. Blockchain applications often function in dynamic, open environments with several linked devices. The P2P-based blockchain network offers all nodes equal privileges and operates without central control. Node verification checks for valid data and blocks by discarding invalid ones and sending valid blocks to the other nodes. This decentralized, peer-to-peer network architecture makes blockchain suitable for next-generation cloud computing, capable of synchronizing and restoring data even in the failure of several nodes [19].

13.2.2.5 Contract layer

Numerous intelligent contracts, processes, and algorithms are packaged by the contract layer. They may be used as a higher level of business logic to activate the blockchain's assets, money, or steady data. Narrowly speaking, a smart contract is a collection of state response rules that are maintained and protected by the blockchain and that are self-verifying, self-executing, and self-enforcing. The parties can create a smart contract, digitally sign it, and broadcast it to the peer-to-peer network for verification after agreeing on some predetermined parameters or guidelines. After then, a block in the ledger will contain the verified contract. The stipulation and associated actions will activate and execute without manual intervention once one or more existing conditions are met.

This feature takes the physical or digital features of assets and transforms them into innovative assets that can then be controlled and managed in a programmable and automated manner, thus substantially reducing the social complexity. More broadly speaking, scripts and algorithms also can be taken to be dynamic decision rules instead of the static nature of the rules of the contract, and therefore, smart contracts on the ledger; this allows improved autonomy as well as programmability [18,20,31,32].

13.2.2.6 Application layer

This layer encapsulates all possible application scenarios and uses cases of blockchain. Blockchain is still in its infancy, although it has rapidly grown and developed through research and industry applications over the last several years. New business models and new use cases arise trying to create a completely decentralized system, which becomes disintermediated by data security [19].

13.3 SECURITY RISKS IN CRYPTOCURRENCY TRANSACTIONS

Another major factor in the widespread popularity of cryptocurrencies like Bitcoin was that it had found acceptance as a currency in the Darknet [33]. For instance, the most infamous marketplace for Darknet academicians, Silk Road, accepted only Bitcoins for its illicit goods and services. Even though the website had reached governments, the Silk Road existed for more than 2 years without any stop until October 2013, when the FBI managed to track down and arrest the developer of the dark website [34]. Cryptocurrency transactions have been linked with many Darknet cases, including assassination for hire, corporate attack, child exploitation and pornography, corporate espionage, drugs, fake passports and IDs, investment schemes, sexual exploitation, and credit card theft [35].

Cryptocurrency trading involves severe risk for investors and the whole financial system. It carries high volatility and downside risk, which demands sophisticated predictive models to predict the value at risk [36]. The major risks are related to technological failure, fraud, and legal uncertainty; the volatile nature of the market is also a threat. Then, there are issues concerning the liquidity and impact caused by COVID-19 [37]. The anonymity of the market, coupled with the absence of reasonable regulations, makes this a weak place for fraud cases, manipulation in the market, and even insider trading [38]. Hacking attacks, malware, loss of access credentials, and erroneous transactions are considered against cryptocurrency businesses and users [39]. These may have profound effects on disrupting monetary systems, economic stability, and transparency in financial markets [40]. These challenges require seriousness from investors and regulators before engaging with or overseeing the cryptocurrency markets.

Peer-to-peer architectures that are inherently unsafe, C and C++ style [41] programming languages, allow easier worm propagation into any infected system, and with every step of the way, system risks were amplified through transactions over the exchanges handled. Almost every trading platform tries to stay unregulated; almost everyone fails with near-perfect predictability, first and foremost because of coincidental theft and fraud. The exchanges can also operate aggressively to manipulate market prices [41].

13.4 PRIVACY MEASURES IN BLOCKCHAIN TECHNOLOGY

Cryptocurrency is subject to a range of risks and threats, such as mining balanced privacy leaks, dual spending, and private key safety [42].

13.4.1 Identification of potential risks

In cryptographic primitives, privacy and anonymity in the blockchain will be ensured, especially while hashing transactions. Regarding the Bitcoin consensus algorithm and Ethereum consensus algorithms, various cryptographic algorithms are used in the mining procedure [43]. The analysis of open data lakes (ODLs) is done to maintain trust management in authentication. Blockchain authentication of property maintains data integrity such that threats are reduced via upholding data integrity. The model is designed with the help of elliptic curve cryptography and is suitable for a Bitcoin exchange. A PKI-based framework that is integrated into a Cecoin-distributed PKI architecture is provided there to enhance the level of security [43]. Protection related to security and privacy becomes essential while implementing blockchain; novel trust-based methods have been proposed for cooperative content uploading. Various identified attacks follow a double-spending attack and an ordering service that is Byzantine fault-tolerant. Blockchain technology does not tamper with odometers, ransomware attacks, and smart contract manipulation. The authors also discuss the secure deployment of drones to improve inter-service interoperability and propose mechanisms to ensure privacy protection with principles based on blockchain technology. Issues discussed here form part of the need to address the challenges that the society faces concerning personal privacy protection [43].

13.4.2 Consensus algorithms

A consensus has to be reached to build trust amongst the world's cryptocurrency traders. Like the decentralized cryptocurrency industry, complete transparency is crucial to controlling a specific coin. It minimizes the possibility that the customer may be a victim of fraudulence [44].

Because consensus algorithms enable all network users to agree on the blockchain state despite the potential for malicious or insecure system members, they are essential in the context of blockchain. Below is a brief explanation of some of the consensus algorithms.

13.4.2.1 Proof of work (PoW)

In the PoW concept, computers compete with each other to solve complex puzzles. The winner is the one who first finds the solution, gets to add the following block of transactions to the blockchain, and is rewarded in the form of cryptocurrency. This process makes blocks challenging to add and requires effort so nobody can quickly take over the network.

13.4.2.2 Proof of stake (PoS)

Instead of wasting some computation cycles, PoS chooses a participant, also known as a validator, who will create the next block based on how much cryptocurrency he has, or "stake." The greater your currencies in your treasury, the higher your chance will be picked. It is much more energy-efficient than the former as solving the puzzle is unnecessary.

13.4.2.3 Delegated proof of stake (DPoS)

The system is similar to PoS but a variant: owners vote for just a few delegates. A few delegates then validate all transactions and maintain the blockchain. That makes the system faster because only few trusted participants do the work, yet on the other hand, voting is what makes it fair.

13.4.2.4 Practical byzantine fault tolerance (PBFT)

It assumes construction in designing a PBFT within a cheating system comprising all participant nodes on its network and operating through a negotiated basis for passing on block selection. Although most cheat, a block will go through consensus when more than half of participants are trustworthy.

13.4.2.5 Raft

Raft is another consensus algorithm in which one leader election is carried out to propose the next block, and others agree. If the elected leader fails again, then once more, an election is carried out for the same. It's very efficient for a small network and is easy to understand and implement.

Each consensus algorithm makes a blockchain secure, trustworthy, and decentralized. They vary according to how the network uses them, specifically in guaranteeing the entire operation's fairness, efficiency, and security.

13.5 UNDERSTANDING THE IMPACT OF CRYPTOCURRENCY

Businesses and people that trade cryptocurrency run a number of dangers. Cryptocurrencies are an appealing online virtual payment method for many because they provide decentralization, security, transparency, reduced transaction costs, and the possibility of large returns that encourage further investment when compared to existing financial systems.

13.5.1 Advantages of cryptocurrency usage

- There will be a reduced reliance on centralized financial institutions since cryptocurrencies will operate on decentralized networks that can boost financial inclusion, especially in areas without access to the banking sector [42,45].
- The basis for cryptocurrencies is blockchain technology, which enables the safe and transparent execution of transactions. Every transaction executed will be written on a publicly visible ledger, making record-tampering challenging to implement without network agreement [46,47].
- Cryptocurrencies can reduce transaction fees compared to traditional banking systems and remittance services, especially for cross-border transactions. This cost-effectiveness is particularly beneficial for individuals and businesses engaged in international trade [45,48].
- The volatility of cryptocurrencies can lead to significant investment opportunities. Many investors are attracted by the potential for high returns, as evidenced by the rapid price increases in various cryptocurrencies [46,48].
- Smart contracts on platforms like Ethereum enable automated transactions and complex financial agreements without intermediaries, enhancing efficiency and minimizing the risk of human error [42,47].

13.5.2 Disadvantages of cryptocurrency usage

- Prices of cryptocurrencies change very dramatically, so there is a possibility of huge losses for investors, and such currencies may become less valuable as stable money for daily operations. Such volatility also leads to huge monetary losses [45,48].
- The lack of specific cryptocurrency regulations creates confusion for both users and investors. Changes in regulatory structures can, indeed, affect the market, consequently influencing the legal status and recognition of cryptocurrencies in most jurisdictions [45,47].
- Although blockchain is secure, its cryptocurrency exchange and wallet are vulnerable to hacking and fraudulent use. The highest profile theft has occurred, losing millions in the hands of users to such systems [42,46].
- Although gaining high acceptance, cryptocurrencies are not accepted as a payment mode. Acceptance in this manner restricts their use in the practice within a routine transaction [42,48].
- For example, consensus protocols like Proof of Work used by Bitcoin consume a lot of energy; hence, it is susceptible to environmental issues since it depends on high electrical power as required in mining operations [45,47].

13.6 CONCLUSION

This chapter discusses the design and security features of the operation of cryptocurrencies and also aims to highlight the various forms of risk and uncertainty during its use, while also imbibing a query into the deeper aspects of the contemporary world, the existing system of technologies, and countries and their national sovereigns. It also aims to educate readers on how to be informed before delving into this realm and weighing their pros and cons. It empowers readers to keep up with contemporary technology and protect their interests.

REFERENCES

1. S. Chauriha, "Stablecoins and their vital roles within the cryptocurrency ecosystem," *The Quint*, 2024. [Online]. Available: https://www.thequint.com/tech-and-auto/stablecoins-and-their-vital-roles-within-the-cryptocurrency-ecosystem#read-more. Accessed: Oct. 20, 2024.
2. R. L. Frebowitz, "Cryptocurrency and state sovereignty," M.S. thesis, Naval Postgraduate School, Monterey, CA, USA, 2018.
3. H. Lasi, P. Fettke, H.-G. Kemper, T. Feld, and M. Hoffmann, "Industry 4.0," *Business & Information Systems Engineering*, vol. 6, no. 4, pp. 239–242, 2014, doi: 10.1007/s12599-014-0334-4.
4. Angelo Aspris, Sean Foley, Jiri Svec, and Leqi Wang, Decentralized exchanges: The "wild west" of cryptocurrency trading, International Review of Financial Analysis,Volume 77, 2021, 101845, ISSN 1057–5219, https://doi.org/10.1016/j.irfa.2021.101845.
5. F. Zhou, "Cryptocurrency: A new player or a new crisis in financial markets?—Evolutionary analysis of association and risk spillover based on network science," *Physica A: Statistical Mechanics and Its Applications*, vol. 648, p. 129955, 2024.
6. B. Custers and L. J. Overwater, "Regulating initial coin offerings and cryptocurrencies: A comparison of different approaches in nine jurisdictions worldwide," *European Journal of Law and Technology*, vol. 10, no. 3, pp. 29, 2019.
7. J. Lansky, "Possible state approaches to cryptocurrencies," *Journal of Systems Integration*, vol. 9, no. 1, pp. 19–31, 2018, doi: 10.20470/jsi.v9i1.335.
8. I. N. Efremenko, V. A. Bondarenko, A. Y. Palant, and G. V. Nazarenko, "Specific features of state regulation of operations with cryptocurrencies in the conditions of digitalization," 2019. [Online]. Available: https://www.um.edu.mt/library/oar/handle/123456789/46244. Accessed: Jan. 1, 2019.
9. V. M. Cvijić and N. Gospić, "Kreiranje kriptovalute korišćenjem blockchain tehnologije," *Tehnika*, vol. 76, no. 5, pp. 665–670, 2021.
10. A. Singh, R. Jalota, and V. Gupta, "Blockchain technology in cryptocurrency: A review," in *2022 International Conference on Machine Learning, Big Data, Cloud and Parallel Computing (COM-IT-CON)*, Faridabad, India, 2022, pp. 715–719.

11. F. A. Cahyadi, A. I. Owen, F. Ricardo, and A. A. S. Gunawan, "Blockchain technology behind cryptocurrency and bitcoin for commercial transactions," in *2021 First International Conference on Computer Science and Artificial Intelligence (ICCSAI)*, Jakarta, Indonesia, 2021, pp. 115–119.
12. A. Gaikwad, "Overview of blockchain," *International Journal for Research in Applied Science and Engineering Technology*, vol. 8, no. 6, pp. 2268–2270, 2020.
13. R. Zhang and B. Preneel, "Lay down the common metrics: Evaluating proof-of-work consensus protocols' security," in *2019 IEEE Symposium on Security and Privacy (SP)*, San Fancisco, 2019, pp. 175–192.
14. B. B. Gupta and M. Quamara, *Smart Card Security: Applications, Attacks, and Countermeasures*, CRC Press, 2019, doi: 10.1201/9780429345593.
15. A. Lohachab, S. Garg, B. Kang, M. B. Amin, J. Lee, S. Chen, and X. Xu, "Towards interconnected blockchains: A comprehensive review of the role of interoperability among disparate blockchains," *ACM Computing Surveys (CSUR)*, vol. 54, no. 7, pp. 1–39, 2021.
16. Blockchain monitoring website, 2016. [Online]. Available: https://blockchain.info/. Accessed: Jan. 8, 2016.
17. Cryptocurrency monitoring website, 2015. [Online]. Available: https://coinmarketcap.com. Accessed: Nov. 24, 2015.
18. Y. Yuan and F.-Y. Wang, "Blockchain: The state of the art and future trends," *Acta* Automatica Sinica, vol. 42, no. 4, pp. 481–494, 2016.
19. Y. Yuan and F.-Y. Wang, "Blockchain and cryptocurrencies: Model, techniques, and applications," *IEEE Transactions on Systems, Man & Cybernetics: Systems*, vol. 48, pp. 1421–1428.
20. Y. Yuan and F.-E. Wang, "Towards blockchain-based intelligent transportation systems," in *2016 IEEE 19th International Conference on Intelligent Transportation Systems (ITSC)*, Rio de Janeiro, Brazil, 2016, pp. 2663–2668.
21. X. Wang, L. Li, Y. Yuan, P. Ye, and F.-E. Wang, "ACP-based social computing and parallel intelligence: Societies 5.0 and beyond," *CAAI Transactions on Intelligence Technology*, vol. 1, no. 4, pp. 377–393, 2016.
22. F.-Y. Wang, Y. Yuan, X. Wang, and R. Qin, "Societies 5.0: A new paradigm for computational social systems research," *IEEE Transactions on Computational Social Systems*, vol. 5, no. 1, pp. 2–8, 2018.
23. T. V. Lakshman and A. K. Agrawala, "Efficient decentralized consensus protocols," *IEEE Transactions on Software Engineering*, vol. SE-12, no. 5, pp. 600–607, 1986.
24. C. L. P. Chen, G.-X. Wen, Y.-J. Liu, and F.-Y. Wang, "Adaptive consensus control for a class of nonlinear multiagent time-delay systems using neural networks," *IEEE Transactions on Neural Networks and Learning Systems*, vol. 25, no. 6, pp. 1217–1226, 2014.
25. X. Wang, X. Zheng, X. Zhang, K. Zeng, F.-E. Wang, "Analysis of cyber interactive behaviours using artificial community and computational experiments," *IEEE Transactions on Systems, Man, and Cybernetics: Systems*, vol. 47, no. 6, pp. 995–1006, 2017.
26. J. Fan, L.-T. Yi, and J. Shu, "Research on the technologies of Byzantine system," *Journal of Software*, vol. 24, no. 6, pp. 1346–1360, 2013.
27. D. Kraft, "Difficulty control for blockchain-based consensus systems," *Peer-to-Peer Networking and Applications*, vol. 9, no. 2, pp. 397–413, 2016.

28. Y. Yuan, F.-E. Wang, and D. Zeng, "Competitive analysis of bidding behaviour on sponsored search advertising markets," *IEEE Transactions on Computational Social Systems*, vol. 4, no. 3, pp. 179–190, 2017.
29. Y. Yuan, D. Zeng, H. Zhao, and L. Li, "Analyzing positioning strategies in sponsored search auctions under CTR-based quality scoring," *IEEE Transactions on Systems, Man, and Cybernetics: Systems*, vol. 45, no. 4, pp. 688–701, 2015.
30. J. J. Li, X. Ni, and Y. Yuan, "The reserve price of ad impression in multichannel real-time bidding markets," *IEEE Transactions on Computational Social Systems*, vol. 5, no. 2, pp. 583–592, 2018.
31. F.-Y. Wang, Y. Yuan, C. Rong, and J. J. Zhang, "Parallel blockchain: An architecture for CPSS-based smart societies," *IEEE Transactions on Computational Social Systems*, vol. 5, no. 2, pp. 303–310, 2018.
32. Y. Yuan, T. Zhou, A.-Y. Zhou, and Y.-C. Duan, "Blockchain technology: From data intelligence to knowledge automation," *Acta Automatica Sinica*, vol. 43, no. 9, pp. 1485–1490, 2017.
33. N. Chistin, "Traveling the silk road: A measurement analysis of a large anonymous online marketplace," in *Proceedings of the 22nd International Conference on World Wide Web* (WWW'13), Rio de Janeiro, Brazil: ACM, p. 1, 2013.
34. P. Vigna and M. Casey, *The Age of Cryptocurrency: How Bitcoin and Digital Money Are Challenging The Global Economic Order*, St. Martin's Press, 2015.
35. L. Trautman, "Virtual currencies bitcoin and what now after Liberty Reserve Silk Road, and Mt. Gox?" *Richmond Journal of Law & Technology*, vol. 20, no. 4, p. 2, 2014.
36. F. Iqbal, M. Zahid, and D. Koutmos, "Cryptocurrency trading downside risk," Risks, vol. 11, no. 7, p. 122, 2023, doi: 10.3390/risks11070122.
37. S. Arsii, S. Ben Khelifa, Y. Ghabri, and H. Mzoughi, "Cryptocurrencies: Key risks and challenges," in Cryptofinance: A New Currency for a New Economy, S. Goutte, K. Guesmi and S. Saadi, eds., World Scientific Publishing Company, pp. 121–145, 2021.
38. J. Veličković, M. Veličković, M. Jovović, and J. Đuričanin, "Risks related to cryptocurrencies," *First International Scientific Conference on Economy, Management and Information Technologies (ICEMIT) 2023*, Blace, Serbia,, 2023.
39. I. Honak, "The risks of operating a cryptocurrency business," Scientific Bulletin of Kherson State University, Series Economic Sciences, 2021.
40. [Online]. Available: https://www.bis.org/publ/othp72.pdf.
41. N. Weaver, "Risks of cryptocurrencies," *Communications of the ACM*, vol. 61, no. 6, p. 20, 2018, doi: 10.1145/3208095.
42. B. K. Mohanta, D. Jena, S. S. Panda, and S. Sobhanayak, "Blockchain technology: A survey on applications and security privacy challenges," *Internet of Things*, vol. 8, p. 100107, 2019. doi: 10.1016/j.iot.2019.100107.
43. H. Wang, D. He, and Y. Ji, "Designated-verifier proof of assets for bitcoin exchange using elliptic curve cryptography," *Future Generation Computer Systems*, vol. 107, p. 854, 2020.
44. S. Biswas, "What is consensus in blockchain?" *ClearTax*, 2023. Available: https://cleartax.in/s/consensus-in-blockchain. Accessed: Feb. 21, 2023.

45. A. Urquhart and L. Yarovaya, "Cryptocurrency research: Future directions," *The European Journal of Finance*, vol. 30, no. 16, pp. 1849–1854, 2023, doi: 10.1080/1351847X.2023.2284186.
46. F. Fang, C. Ventre, M. Basios, L. Kanthan, D. Martinez-Rego, F. Wu, and L. Li, "Cryptocurrency trading: A comprehensive survey," *Financical Innovation*, vol. 8, p. 13, 2022, doi: 10. 1186/s40854-021-00321-6.
47. W. H. Bommer, E. Milevoj, and S. Rana, "The intention to use cryptocurrency: A meta-analysis of what we know," *Emerging Markets Review*, vol. 55, p. 100962, 2023, doi: 10.1016/j.ememar.2022.100962.
48. M. Anas, S. J. Shahzad, and L. Yarovaya, "The use of high-frequency data in cryptocurrency research: A meta-review of literature with bibliometric analysis," *Financial Innovation*, vol. 10, no. 1, p. 1, 2024, doi: 10.1186/s40854-023-00595-y.

Chapter 14

Smart grid and energy management in smart cyber-physical systems (SCPS)

Asmita Sarkar, Bhavya Mathur, Virendra Singh Kushwah, Kamal Upreti, and Shitiz Upreti

14.1 INTRODUCTION TO SMART GRIDS AND ENERGY MANAGEMENT IN AN SCPS

Smart grids can be considered a paradigm transition from traditional power systems, considering the incorporation of advanced CPS features in enhancing energy efficiency, reliability, and interactivity between power producers and consumers [1]. The integration of physical and digital infrastructure in smart grids uses CPS to manage, monitor, and optimize real-time power flows so as to achieve sustainable energetically relevant solutions [2]. The goal is to integrate computational intelligence with physical systems to provide a flexible, distributed control system that will meet the challenges created by fluctuating energy demands and renewable sources of energy [3].

The enabling of energy efficiency, security, and resilience for smart grid evolution is done through advanced sensing, control, and computation technologies enabled by CPS [4]. A smart grid in general is the integration of power infrastructure with information and communication technology, forming a connected network to support distributed generation, real-time monitoring, and load balancing based on forecasted load demand conditions [5]. Smart grids use CPS to meet the demand of energy in real-time and predict the future demands by integrating renewable energy supplies originating from wind and solar sources, both having intermittent supply characteristics that require advanced energy management solutions [6].

CPS integrated into smart grids further emphasizes that energy management must be carried out to control the dynamics of supply and demand. CPS systems are optimized in real time due to inherent capabilities for distributed control and edge computing, which is extremely desirable for sectors with very high demands, such as in an urban power grid and electric vehicle infrastructure [7]. For instance, CPS equipped with advanced sensors monitors the fluctuating demands of energy and acts instantly to stabilize and manage power distribution [8]. Therefore, the rise in the number of prosumers, or consumers who produce energy themselves, constructs an energy ecosystem where the complexity is continuously growing.

DOI: 10.1201/9781003542513-14

This depends on a reliable CPS solution especially in the area of decentralized systems' energy management process (Figure 14.1) [9].

All these factors have accelerated the implementation of CPS in energy management settings due to the global push coming from renewable energy and reduced carbon emissions. Smart grids form the backbone of such green initiatives as they ensure the smooth integration of renewable resources while addressing various challenges in integration such as intermittency and transmission losses [10]. Advanced CPS in smart grids will also enable demand response strategies whereby energy consumption can be adjusted to real-time grid requirements, hence enhancing grid stability in addition to reducing operational costs [11]. Recent studies have indicated that applying CPS to energy grids could potentially reduce peak loads by as much as 15%, thereby greatly reducing excess infrastructure investment and enabling more sustainable energy solutions [12].

Condition-based maintenance and fault detection are a reality in smart grids, thanks to the CPS capabilities in data gathering and analyses that yield increasing grid reliability and resiliency against disruptions at both the

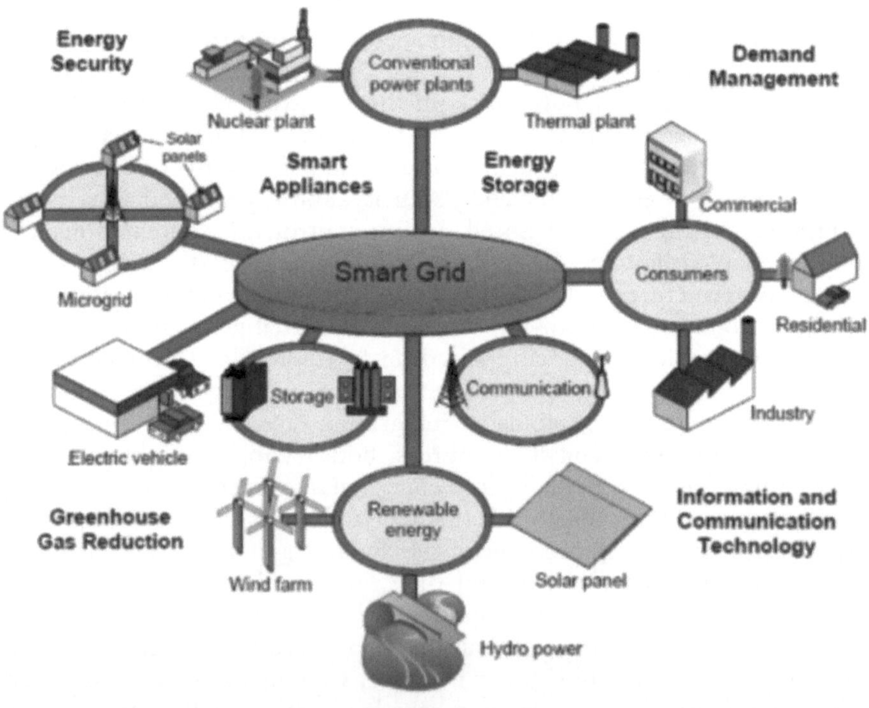

Figure 14.1 (Smart Grids: A cyber–physical systems perspective, This paper presents an overview of challenges for smart grids in the context of CPSs, outlines potential contributions that CPSs can make to smart grids, and points out the implications of current technological advances to smart grids. By Xinghuo Yu, Fellow IEEE, and Yusheng Xue, Member IEEE.)

cyber and physical levels [13]. Data-driven methodologies ensure utilities are well-predicted in energy trends and can trace impending failures of systems, thereby reducing risks in pursuit of self-sustaining, reliable energy management infrastructures [14]. In other words, due to the readiness of smart grids with CPS technologies, a proactive, adaptable, and resilient energy network is realized—so far one of the most critical evolvements toward sustainable energy goals worldwide [15].

14.2 ARCHITECTURE OF CYBER-PHYSICAL SYSTEMS FOR SMART GRIDS

Adaptability, resilience, and security in the smart grid are all built on the architecture of CPS. In the structure of the CPS within smart grids, the embedding of physical infrastructure with computational technologies and communication networks is smooth. This would enable the intelligent grid to respond to real-time events constantly [1]. Sensors, actuators, communication networks, and embedded systems are the usual elements synchronized together for holistic monitoring and control of energy flow throughout the grid [2].

Sensors and actuators are central to CPS architecture in continuously monitoring and controlling several parameters that include voltage, frequency, and energy load. These feed data into DCS and EMS for timely adjustment with an objective of real-time stability of the main grid [3]. Advanced communication protocols like IoT and M2M communication allow grid components to share information effectively, thereby making the system more responsive and adaptive [4].

In smart grids, the architecture of CPS is based on distributed computational models that enhance grid resilience by facilitating local decision-making capabilities in substations and control centers [5]. For example, the control of power distribution over the grid is facilitated by networked control systems through sensor data analysis, best computation of load distribution, and sending the real-time control signals to actuators [6]. These distributed systems work in concert with centralized controls to ensure stability while reducing latency issues arising from centralized processing (Figures 14.2 and 14.3) [7].

There are three major layers in the general architecture of CPS in smart grids: physical layer, cyber layer, and application layer [8]. The physical layer consists of a physical part of the system, such as transformers and power lines to storage systems. The cyber layer consists of cloud and edge computing resources that process large volumes of data generated on the grid. Software applications for energy management, security, and monitoring are finally involved in the application layer [9]. A layered design allows each of these constituents to contribute independently to an integrated robust system capable of responding effectively to challenges from complex power flows and fast-changing demand patterns [10].

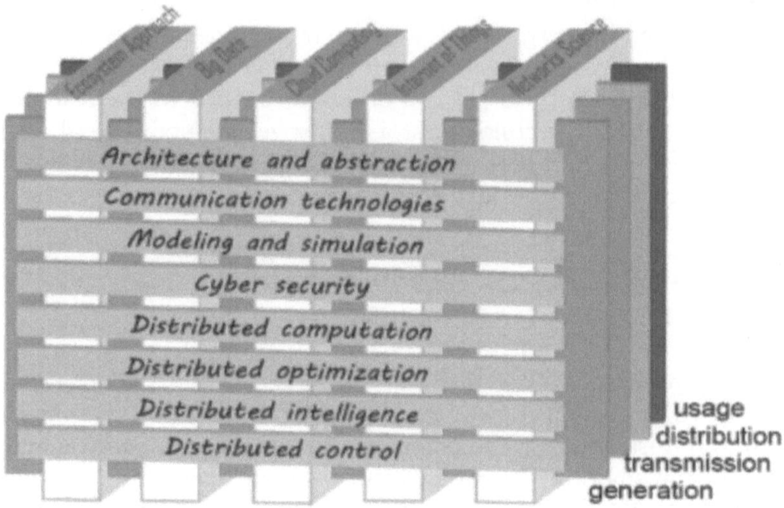

Figure 14.2 SG from a CPS viewpoint. (Smart Grids: A cyber–physical systems perspective, This paper presents an overview of challenges for smart grids in the context of CPSs, outlines potential contributions that CPSs can make to smart grids, and points out the implications of current technological advances to smart grids. By Xinghuo Yu, Fellow IEEE, and Yusheng Xue, Member IEEE.)

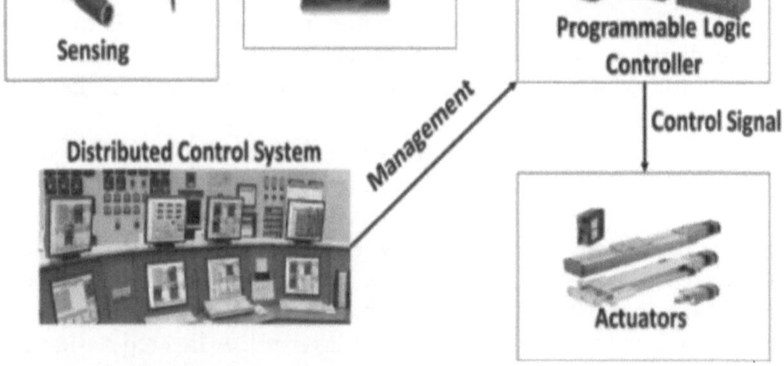

Figure 14.3 Infrastructure of CPS. (cyber-physical systems security: Limitations, issues and future trends, Jean-Paul A. Yaacoub[a], Ola Salman[b,*], Hassan N. Noura[a], Nesrine Kaaniche[c], Ali Chehab[b], Mohamad Malli[a].)

CPS architecture for smart grids have a major role in interoperability and standardization. Interoperability and standardization can allow integration upon diverse technologies and systems to achieve smart grids. The IEEE and the National Institute of Standards and Technology have approved standards for smart grid CPS components, such as IEEE 2030, which enable interoperability[11]. This standard guides a designer to create pieces that allow utilities to incorporate new technologies onto the grid without having to reconfigure it, as was experienced earlier with the grid [12].

Therefore, security issues inherently form one of the most critical aspects of CPS architecture as the networked smart grids are automatically prone to increased vulnerabilities for cyberattacks. For protection against unauthorized access or tampering of data and control signals, security protocols such as encryption and IDS are embedded within each layer [13]. Blockchain technology and several decentralized solutions are being studied for providing more security and transparency in data exchange across the grid (Figure 14.4) [14].

In conclusion, the CPS architecture for smart grids forms a complicated integration of hardware, software, and communication technologies. By unleashing advanced computation, real-time data analytics, and secure communication protocols, CPS discerns responsive, adaptable, and resilient energy systems in smart grids that meet modern energy demands with efficiency and sustainability [15].

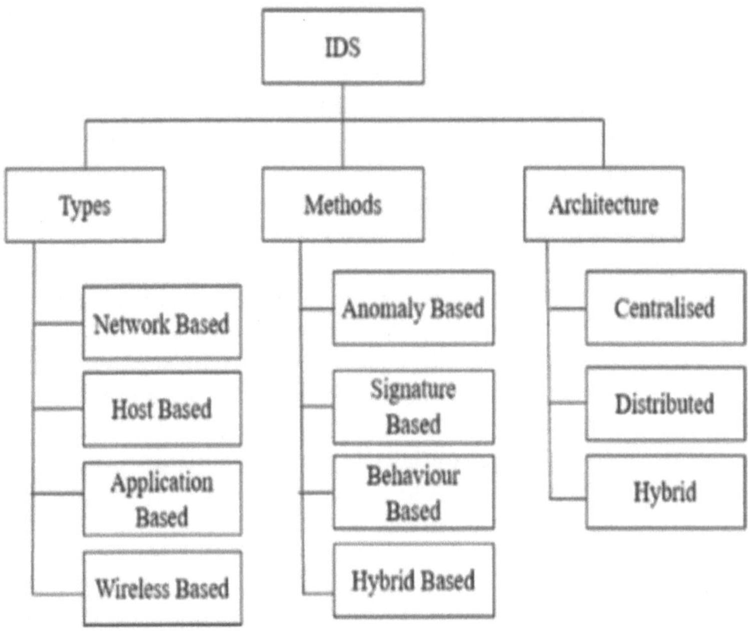

Figure 14.4 IDS structure. (cyber-physical systems security: Limitations, issues, and future trends, Jean-Paul A. Yaacoub[a], Ola Salman[b,*], Hassan N. Noura[a], Nesrine Kaaniche[c], Ali Chehab[b], Mohamad Malli[a].)

14.3 INTEGRATION OF SMART GRIDS WITH SCPS: CHALLENGES AND SOLUTIONS

14.3.1 Intermittency of renewable energy and fluctuations in demand

The most critical challenge one has to face while integrating renewable energy sources, such as wind and solar power, into the SCPS smart grid is the intermittency of sources. The ultimate dependence on weather conditions in renewable energy sources easily leads to fluctuating power generation, further complicating grid management and power quality. These fluctuations affect the stability of the grid since it is required to maintain a supply/demand balance at all times for preventing outages or failures [1].

Real-time monitoring and prediction analytics are critical methods in damping intermittency. Smart grids enabled with SCPS capabilities may further apply the machine learning algorithms to forecast renewable energy generation based on the available outputs of weather forecasting for better planning and balancing of the loads [2]. These types of systems use sensors and data analytics to track the output of renewable sources, updating predictive models in real time and correlated adaptability to changing environmental conditions (Figure 14.5) [3].

It has to be the case that this variation can be facilitated by predictive models and AI by predicting energy supply and shifting consumption using demand response programs. Besides that, there is the variability factor with which renewable energy complicates the issue. Peak demand times generally fall at other times than those when renewable sources are at the maximum output. In order to respond to these variable demands of peak times, smart grids make use of ESS such as batteries, which store excess energy generated at low-demand periods for release whenever there is a spike in the demand. This way, it not only enhances grid reliability but also heightens energy utilization efficiency by reducing wastage [4].

In addition, SCPS will be able to support advanced energy storage systems that can facilitate distributed storage solutions across various nodes within the grid. This, in turn, will enable a more localized response to any fluctuations in energy supply without necessarily relying on a centralized infrastructure system, another crucial remedy to maintain fluctuations includes demand response. With SCPS, utilities will be in a position to adjust consumer loads according to the conditions of the grid by incentivizing users to reduce or shift energy usage during peak usage periods [5].

Demand response technologies rely on real-time communication between consumers and the grid, using smart meters and IoT devices to relay data on consumption and enable instantaneous adjustments (Figure 14.6). This flexibility ensures that there is a balance between demand and supply whenever renewable sources overproduce or fail to produce adequate power. It is also a valuable tool to the extent that it reduces infrastructure that

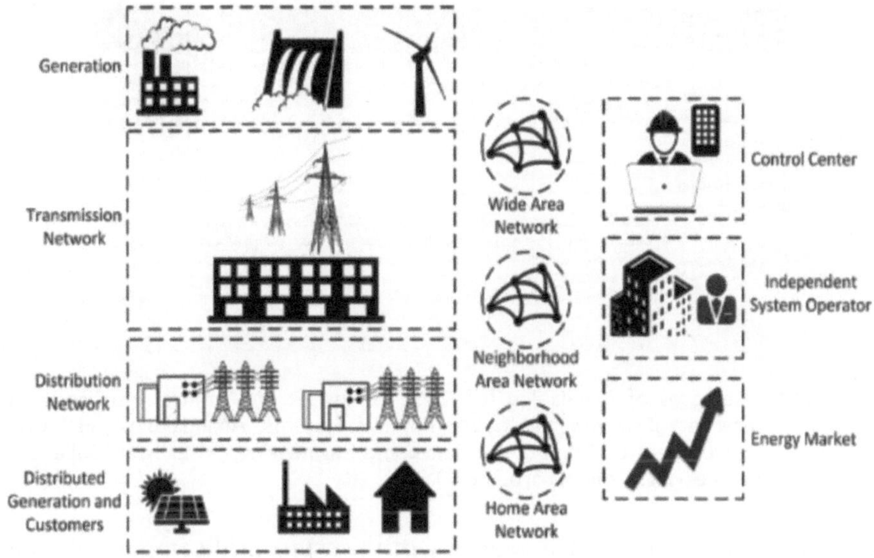

Figure 14.5 Cyber-physical infrastructure of the power grid. (cyber-physical systems: A security perspective, Charalambos Konstantinou*, Michail Maniatakos*, Fareena Saqib†, Shiyan Hu‡, Jim Plusquellic§ and Yier Jin¶, *Department of Electrical and Computer Engineering, New York University - Abu Dhabi, †Department of Electrical and Computer Engineering, Florida Institute of Technology, USA, ‡Department of Electrical and Computer Engineering, Michigan Technological University, USA, §Department of Electrical and Computer Engineering, University of New Mexico, USA, ¶Department of Electrical Engineering and Computer Science, University of Central Florida, USA {ckonstantinou@nyu.edu, michail.maniatakos@nyu.edu, fsaqib@fit.edu, shiyan@mtu.edu, jimp@ece.unm.edu, yier.jin@eecs.ucf.edu}.)

otherwise would be needed to handle peak loads and hence operational costs. Moreover, shifting energy demand to match periods of renewable generation can reduce some of the reliability problems related to integrating renewables. The European integration of demand response, for instance, is said to stabilize grid networks at times when wind and solar output is high, and energy supply could be over the demand [6].

14.3.2 Scalability and interoperability issues of smart grid networks

Consequently, scalability and interoperability are key preconditions for successful deployment of SCPS into smart grids, considering the increased number of renewable energy installations and the need for flexible and adaptive grid structures. Seamless scalability can be achieved in a rather

AI enabled IoT-CPS cases	Brief description
Energy Utilization	Small-scale algorithms have been created for the purpose of decreasing the coffee machine's energy usage (ARIIMA). It may be adapted and used in different contexts to lessen energy use; for instance,in residential temperature control systems to improve efficiency and cut down on waste. The system will learn how to effectively change the temperature in each home to the preferences of its inhabitants.
Traffic/Routing	One area of application for ML is traffic management and routing. You'll be given suggested routes based on a variety of factors such as traffic, road conditions, weather, and more.
Cost Savings	The ability to predict outcomes is very valuable in a manufacturing scenario. Machine learning computations may learn typical operating conditions by accumulating data from various sensors in or on machines. As a result, it can pinpoint the device if anything goes wrong and sound an alert. Doing so would prevent mishaps and save money. One company, Augury, does this very thing by installing ultrasonic sensors and vibrations in their machinery, which allows them to save money by anticipating potential breakdowns.

Figure 14.6 Use cases of AI-enabled IoT-CPS. (A Review on Artificial Intelligence in Internet of Things and Cyber Physical Systems, Anandakumar Haldorai, Department of Computer Science and Engineering, Sri Eshwar College of Engineering, Coimbatore, India.)

complex way due to the heterogeneity of the smart grid components, sensor types, communication protocols, and different control systems. In such cases, interoperability issues arise when these components from different manufacturers need to work cohesively within one system. For this, standardized communication protocols become highly essential [7].

The interoperability framework defined by the Smart Grid Interoperability Panel for smart grids allows implementation of guidelines and standards that will allow better compatibility between different systems. Standards such as IEEE 2030 and IEC 61850 in this regard allow different components to seamlessly work together with each other irrespective of whom they have been provided by. Seamless interaction, in other words, means plug-and-play compatibility and lesser integration costs for the end consumers, as noted. In addition, CPS modular architectures allow smart grids to scale their capacity without having completely to replace the existing infrastructure—in other words, there is an affordable path to scaling [8].

That also addresses the scalability related to data management issues in smart grids, which generate voluminous information from distributed sensors and IoT devices. Scalable data handling solutions such as cloud computing and edge computing must be developed, which can process and analyze the information with the barest minimum of time. For instance, edge computing reduces latency since data may be processed at the edge of the network rather than in a centralized server, thus enabling quicker responses and reducing communication bottlenecks [10]. Decentralized, therefore, edge computing fits the bill of goals laid out by SCPS: enhanced resiliency and adaptability through fast decision-making at different levels of the grid.

Concomitantly, scalable smart grid systems must respond to challenges of expanded networks: with increasing number of renewable energy installations, the topology of the grid becomes increasingly complex and inherently requires real-time communication and coordination at many nodes. This SCPS can accomplish by using distributed computing and multi-agent systems for autonomous control of a segment of the grid to enhance the resilience and operational efficiency. Each node or agent can make local decisions in their immediate vicinity and yet contribute significantly to the overall stability of the grid. This feature is helpful in managing distributed renewable sources, such as residential solar installations [11].

14.3.3 Cybersecurity challenges and privacy concerns

Due to this integration of CPS in smart grids, cybersecurity is now of primary importance because the vulnerability to different cyber threats has grown significantly. In a smart grid, there is immense electronic communication and exchange of data in between its various segments; hence, it is prone to data breach, DoS, and malware infection [12].

Energy should be critical; it has to be very secure, both internally and against external threats, with a view to ensuring continuity in power supply and safeguarding sensitive consumer information. First, encryption provides a first layer of cybersecurity in smart grids enabled with SCPS, which ensures the integrity and confidentiality of data during transmission. The attacks against CPS are getting much more sophisticated, and traditional encryption schemes might be not enough to resist such attacks. Therefore, advanced cryptographic techniques are being developed to provide more robust security frameworks for smart grids [13].

Besides that, MFA and PKI systems verify users and devices accessing the grid to prevent unauthorized access and hence let the critical infrastructure interact with trusted entities only. IDS forms the backbone for identifying and mitigating cybersecurity threats on both a real-time and post-incident basis, along with other anomaly detection mechanisms. The intrusion detection system within an SCPS architecture monitors network traffic for patterns that are out of the ordinary and signals an ongoing cyberattack. If these methods were considered in concert with machine learning algorithms, it would heighten the detection accuracy and reduce response time, enabling utilities to take action before threats impact grid stability [15].

Anomaly detection works especially well for insider threats, which have now become a huge concern with increased digitization of the grids. Consider that in most networks, behavioral patterns are set; it is the deviation from these normal operating states that anomaly detection systems flag and can thus be intervened early using intervention methods [16]. Similarly, the SCPS deployed within smart grids give rise to a number of privacy concerns associated with collecting detailed consumer usage data via smart meters. This is very valuable for optimizing grid operations and personalized energy services; however, such information also raises risks regarding user privacy if not treated responsibly (Figure 14.7).

All privacy-preserving techniques, including data anonymization and differential privacy, are hence highly required to keep personal private information confidential and yet enable data analytics for grid optimization (Figure 14.8). For example, differential privacy adds statistical noise to data queries; utilities may extract useful information without compromising individual privacy [17]. In addition, blockchain technology provides promising applications that lift both privacy and security in SCPS by building a decentralized ledger. Blockchain can make data transactions in the grid secure with transparent and tamper-resistant records. Blockchain ensures efficiency in enhancing data integrity and preventing unauthorized modification, due to its cryptographic properties. Each transaction is verified and encrypted through a set of nodes within the network. Smart

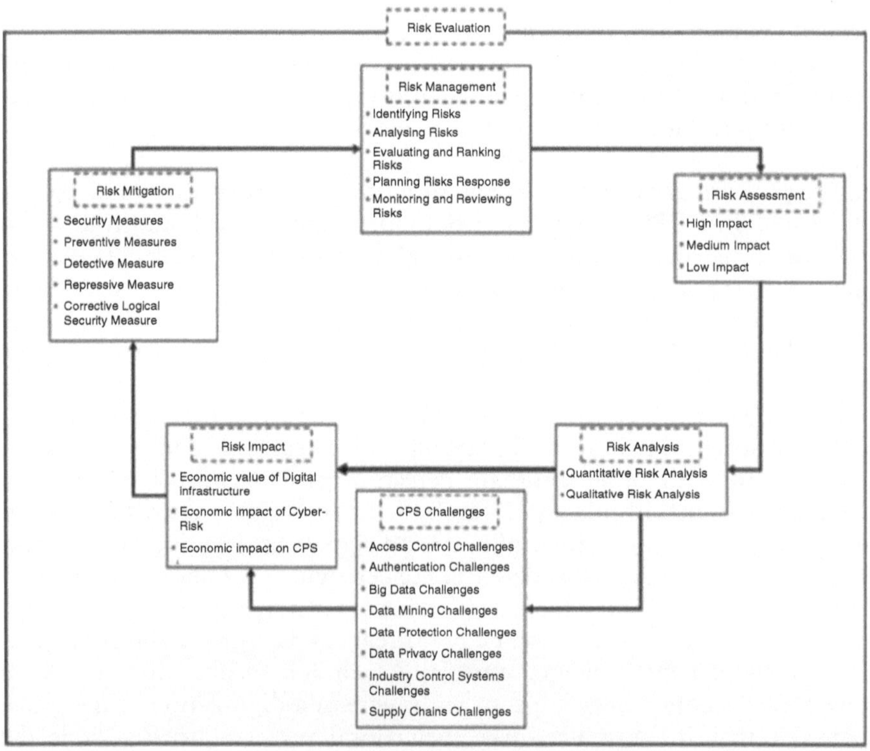

Figure 14.7 CPS risk evaluation. (Cyber-physical systems security: Limitations, issues and future trends, Jean-Paul A. Yaacoub[a], Ola Salman[b,*], Hassan N. Noura[a], Nesrine Kaaniche[c], Ali Chehab[b], Mohamad Malli[a].)

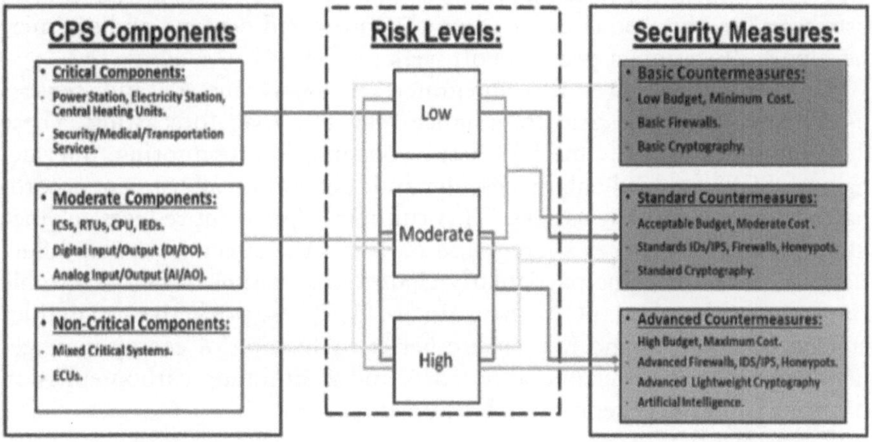

Figure 14.8 CPS component classification and protection. (Cyber-physical systems security: Limitations, issues, and future trends, Jean-Paul A. Yaacoub[a], Ola Salman[b,*], Hassan N. Noura[a], Nesrine Kaaniche[c], Ali Chehab[b], Mohamad Malli[a].)

contracts, running on blockchain, further enhance security by automating energy transactions based on preset conditions, reducing fraudulent activity, and ensuring regulatory standards are met [18].

14.4 COMPUTATIONAL TECHNIQUES IN SCPS FOR SMART GRID MANAGEMENT

14.4.1 Application of AI and machine learning in predicting energy demand

Artificial intelligence (AI) along with machine learning (ML) plays an important role in the development of sophisticated prediction capabilities of SCPS-based smart grids, especially the prediction of energy demand. AI and ML algorithms analyze complex consumption patterns from large data emanating from historical consumption records, weather data, and real-time sensor readings to predict energy needs with a high degree of accuracy. For instance, deep learning, reinforcement learning, and neural networks are useful techniques for predictive modeling for time-series data of nonlinear relationships that characterize energy demand variability. Indeed, deep learning approaches have been applied to energy demand forecasting owing to their capability of discovering patterns in data across long time horizons; these include methods such as convolutional neural networks and LSTMs [2]. To that end, the LSTM can fetch temporal dependencies of energy consumption data much better for grid operators both on short-term and long-term forecasts, therefore aiding in the improvement and optimization of grid operations, increasing the reliability thereof [3]. Another variant of

AI, reinforcement learning algorithms, enables SCPS to adaptively control distributed energy resources based on the forecasted demand and optimize the supply allocation at peak and off-peak periods [4].

ML techniques have also integrated at smart meters and sensors distributed across the grid to enhance real-time load forecasting. These algorithms run all the time for energy consumption monitoring, interfacing with SCPS for load balancing—dynamic generation adjustment according to the consumption pattern. This runtime adjustment reduces reliance on non-renewable energy sources and mitigates the effects of demand fluctuations, which become particularly challenging in times when renewable sources—wind and solar—are much in use [5]. These predictive capabilities enabled through AI and ML ensure better realization of energy management functions at lower operation costs and a minimum carbon emission footprint from the generation of backup energy [6].

14.4.2 Big data analytics and decision-making for real-time grid management

A state-of-the-art SCPS-based smart grid management heavily depends on big data analytics as the backbone of the decision-making process after processing and analyzing large-volume data streams, as generated by the infrastructure of the grid. Yet, increased deployment of IoT-enabled devices, smart meters, and sensors always provides real-time data on consumption, generation, grid health, and environmental information. In these data-rich environments, high-frequency decisions related to maintaining grid stability, energy quality, and fault detection become critical [7].

Real-time management of the grid involves several computational techniques: data mining, statistical analysis, and real-time anomaly detection. Data mining algorithms analyze large volumes of data in search of consumption trends and abnormal patterns that might reveal inefficiencies or faults within the system. For example, clustering algorithms group grid nodes by their consumption characteristics and can thereby effectively place more resources where the local demand is high [8]. Besides, statistical models like regression analysis provide insights into the distribution of loads and hence help in grid balancing by allowing for precise estimation of peak and low-demand periods. Detection of anomalies is one of the major components of big data analytics in SCPS and is quite important in early diagnosis and prevention of cascading failures. These models leverage unsupervised learning methods to detect anomalies in the grid data, which might indicate malfunctioning equipment, security breaches, or grid instability [9]. The anomaly detection system locates deviations from normal parameters of operation, thus assisting the grid operator to intervene before minor issues escalate to major failures [10]. Data visualization, in turn, interprets all of the complex data into operatically understandable formats, therefore presenting situational awareness for decision-making in SCPS-enabled smart grids. Many times, such tools interface with geographic information systems to provide spatial insights into the conditions of the grid, hence allowing the operator to observe health across different locations and make timely, informed decisions about energy dispatch and fault isolation. The edge in cloud computing and computing, therefore, has enhanced this data-driven approach to decision-making in its ability to process and disseminate data quickly across the infrastructure of the grid [11].

14.4.3 Multi-agent systems and distributed optimization technique

It is important to note that multi-agent systems form one of the major approaches toward SCPS, integrated with techniques for distributed optimization, which allow operation of decentralized energy resources and enhance resilience within the grid. Multi-agent systems in the context of the smart grid involve a group of autonomous agents, each normally representing various entities within the grid, including power plants, substations, or prosumers, which may cooperate to achieve the global grid-wide optimization goals. These agents communicate with each other and make local decisions based on their conditions and objectives; thus, they provide a distributed, resilient, and scalable management framework of smart grids [12].

In the case of distributed MPC, the agents independently predict several of the future states of the grid and make a decision on the way to operate optimally, while considering the actions taken by other neighboring agents. It minimizes dependency on centralized control and makes the grid resilient to abrupt changes in demand or generation. Moreover, distributed MPC effectively manages microgrids in the wider smart grid due to the fact that each microgrid is able to operate semi-independently and yet still contribute to the stability of the wider grid [14].

The employment of MAS and distributed optimization in SCPS for smart grids handles the computational issues related to large-scale systems. Such systems reduce computational burdens and allow much higher processing speeds by decomposing complex optimization problems into smaller, manageable tasks designated for different agents. Thus, a decentralized computation methodology has much in common with the basic ideas of the SCPS architecture to enhance such features as flexibility, fault tolerance, and adaptability in interconnected grid components [15].

In the case of distributed optimization, MAS coordinates energy resources in a distributed way thanks to the employment of optimization algorithms, among which are consensus-based algorithms and distributed model predictive control (Figure 14.9). These consensus-based algorithms can achieve agreement among the agents on operational parameters such as voltage levels or load distribution so that locally made decisions by an agent are consistent with global grid objectives. This is quite useful in the current scenarios when the renewable energy sources have a very high penetration and the task of maintaining the balance and stability of the system becomes very complex [13].

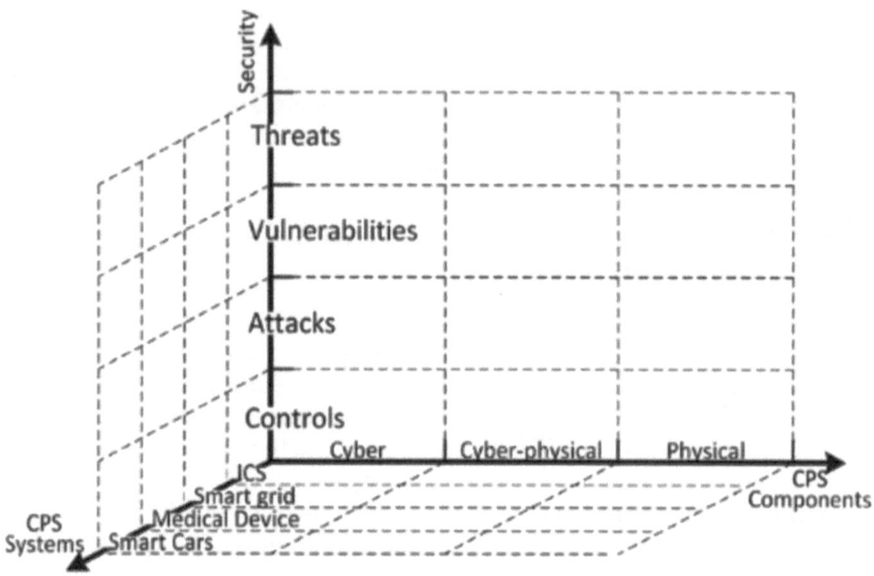

Figure 14.9 CPS security framework with three orthogonal coordinates: security, CPS components, and representative CPS systems. (Cyber-physical systems security—A Survey, Abdulmalik Humayed, Jingqiang Lin, Fengjun Li, and Bo Luo.)

The MAS approach is also of great help in integrating electric vehicles and other flexible loads into the grid. With MAS letting each EV be an independent agent, SCPS can achieve coordination of the EV charging schedule according to grid demand, renewable energy availability, and user preference. Such coordination will reduce peak loads, optimize energy storage use, and increase renewable energy consumption, which in turn will help meet the goals of sustainability for smart grids, as deliberated in [16].

14.5 ADVANCED TECHNOLOGIES IN SMART GRID SCPS

14.5.1 Role of IoT and edge computing in decentralized energy systems

The integration of IoT and edge computing into the field of SCPS for decentralized energy systems is indicative of a paradigm shift toward an integrative, transformative approach in energy management, moving the center of data processing nearer to the point of data generation and thereby enhancing the efficiency and responsiveness of the smart grid. IoT devices, like smart meters, sensors, and actuators, form an important integral part of decentralized energy systems for monitoring and real-time controlling of DERs. Such decentralized network architecture encourages energy management with a distributed model, whereby an SCPS can respond dynamically to changes in energy demand and generation at generation and demand sides [1,2].

Coupled IoT and edge computing perform computations locally with data, and they have faster response times to critical functions needed in the grid, such as load balancing and demand response. Since data are processed at the edge rather than in a point-to-point, centralized data center, latency is reduced. This can significantly improve applications wherein access to real-time data is crucial for grid stability. The edge nodes can facilitate energy-efficient communication protocols, thereby easing the core network by reduced the volume of data transmitted to the cloud or central systems. The above feature is quite relevant to the decentralized energy systems relying on DERs like solar and wind energy, where the variability of energy generation requires real-time adaptability [3].

Also, IoT and edge computing are enabling the integration of renewable energy into the existing infrastructure of the grid to be seamless, with real-time information on the generation of energy, weather conditions, and grid load. This information is processed locally at the edge to allow the smart grid to make autonomous decisions on how to balance energy supply to demand without human intervention. Therefore, edge computing is critical in solving volume and latency problems of data in large-scale IoT implementation in SCPS, enabling the way to decentralization and hence to support the evolution of energy systems [4,5].

14.5.2 Adoption of blockchain and secure data sharing protocols

The adoption of blockchain technology and the adoption of protocols for secure data sharing thus become critical in ensuring security, transparency, and efficiency in all transactions occurring in smart grids. Indeed, the blockchain ledger technology provides a tamper-resistant record of transactions, which makes it very suitable for decentralized energy markets where multiple stakeholders, such as producers and consumers along with regulators, have to track energy exchanges while transiently proving ownership transparently [6,7].

It also empowers blockchain-based systems for P2P energy trading in order to allow a prosumer to sell his surplus energy to other participants of the grid in some decentralized market model. It can be very useful for increasing the penetration of renewable energy and decreasing dependence on centralized energy suppliers [8].

Furthermore, blockchain essentially implements the notion of spreading freedom by the Cypherpunk principle through allowing unauthorized tampering with data, hence increasing data security and integrity throughout the smart grid. A smart contract is rightly a programmed agreement in some blockchain form, which allows for automated energy dealing to be executed when predetermined criteria—including levels of the energy price threshold or grid demand—are met. These self-executable contracts become immutable and carry out energy dealings with ensured reliability and simultaneously reduce administrative overheads and fraud. Further, blockchain applications in SCPS require secure data-sharing protocols supplying methods for secure access control and encryption against cyber threats to keep sensitive grid data safe [9].

Beyond P2P trading, blockchain can support appropriate demand response strategies by providing transparency and accountability in energy consumption data, which is very important to grid operators for load management and preventing grid overload. It therefore enables better forecasting of energy load and efficient utilization of grid resources, hence making blockchain a potent tool for grid stability in decentralized energy systems [10,11].

14.5.2 Emerging trends: real-time energy trading and demand response strategies

The smart grid landscape continues to evolve with technologies that leverage SCPS advances into real-time energy trading and demand response strategies. The consumer and prosumer in real-time energy trading can sell and buy energy openly by responding to the real-time grid condition and energy price. This flexibility contributes to the stability of the grid, in which through opened access into energy redistribution, the upset of the mobile demand in peak load periods will be balanced out. While blockchain and IoT technologies complement each other, they have now made it possible by providing some certainty and up-to-date information on energy production and consumption, hence allowing for transparent and immediate transactions within the energy market itself [13].

On the other hand, demand response strategies enable a dynamic mechanism of managing grid load by enabling consumers to adjust energy usage in response to grid signals, often due to peak demand pricing or some other incentive schemes in place. SCPS enable real-time monitoring, and predictive analytics enable a better implementation of the DR strategy. The system may predict demand trends, and automatic DR signals are delivered to the distributed consumers in order to reduce the load at critical times [14]. These methods are further complemented by the use of machine learning algorithms, which study historical consumption patterns and achieve real-time optimization of DR schedules for supply–demand balance, ensuring grid stability with reduced energy costs [15,16].

14.6 CASE STUDIES: SCPS IMPLEMENTATIONS IN SMART GRIDS

Case study 1: Smart grid implementation in urban environments

The urban environment presents a specific challenge in smart grid implementation due to the high density of population, diversified energy demands, and ecologically complex infrastructures. The famous case study is Amsterdam Smart City, connecting IoT devices to AI and SCPS for recognizing the best manners of energy distribution and facilitating sustainable energy behavior. The project, tapping into a network of smart meters and sensors, is mapping energy usage through residential, commercial, and industrial zones, while making real-time adjustments in grid operations to match up with the oscillating demand patterns. Big data analytics facilitate predictive maintenance of the grid infrastructure, reducing operational costs while enhancing the reliability of service delivery [17].

Again, the Amsterdam smart grid based on SCPS lends further support to edge computing, where local processing minimizes latency and thus offers fast response time to changes in demands. Rooftop solar installations in the city generate renewable energy integrated with and fed into the city's main grid through real-time monitoring and automatic control mechanisms. The grid works by using renewable energy in this setup; that is, the reliable power for urban residents is maintained. SCPS application has enabled this project to achieve goals of energy efficiency and sustainability in highly dense urban areas (Figure 14.10) [18].

Case study 2: Integration of renewable energy and SCPS in rural settings

Rural settings offer particular opportunities and challenges for the implementation of a smart grid based on the SCPS. Indeed, rural settings related to the integration of renewable energies will involve unique areas of opportunity and challenge. The Smart Microgrid Project in rural Rajasthan, India, demonstrates how SCPS enables the deployment of decentralized energy systems that can deliver reliable electricity to regions with limited grid connectivity. This project utilizes SCPS to integrate local renewable resources—solar and wind—with the microgrid infrastructure for energy self-sufficiency in rural communities (Figure 14.11).

Additionally, the microgrid uses IoT-enabled sensors for monitoring energy production and consumption with edge computing, balancing the local supply with demand. This microgrid uses ML algorithms to predict energy production by renewable sources using data about the weather,

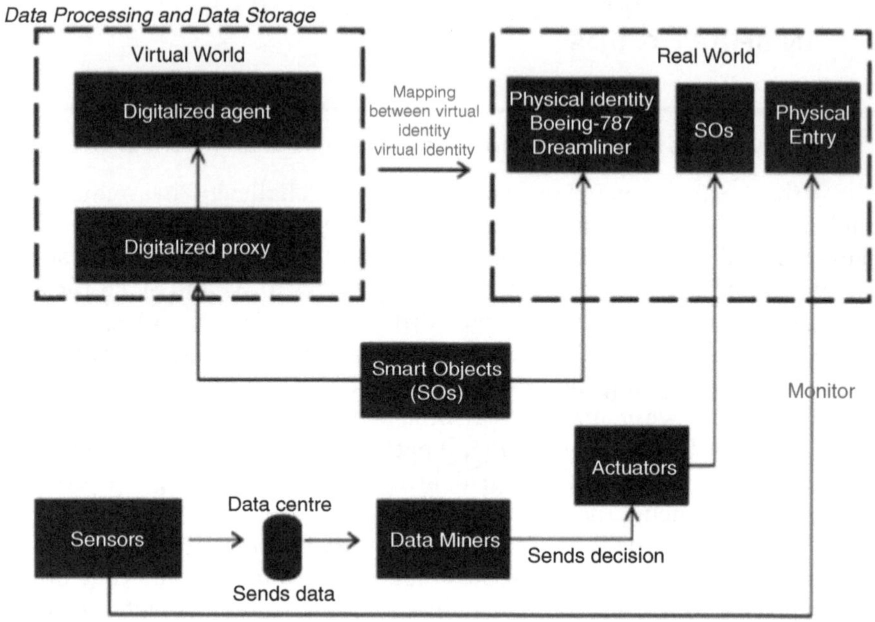

Figure 14.10 The mapping of the virtual world to the real world. (A Review on Artificial Intelligence in Internet of Things and Cyber Physical Systems, Anandakumar Haldorai, Department of Computer Science and Engineering, Sri Eshwar College of Engineering, Coimbatore, India.)

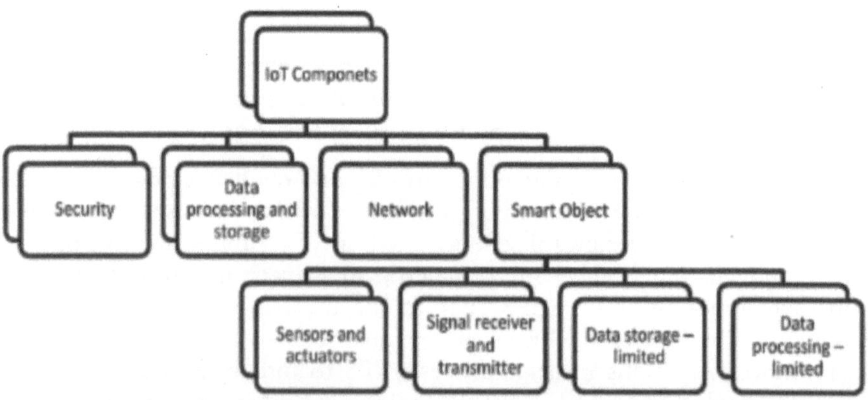

Figure 14.11 Structure of the IoT architecture. (A Review on Artificial Intelligence in Internet of Things and Cyber Physical Systems, Anandakumar Haldorai, Department of Computer Science and Engineering, Sri Eshwar College of Engineering, Coimbatore, India.)

hence enabling changes in the system's strategy for load management. It implements blockchain technology, allowing P2P energy trading in the community; each household can sell excess energy to their neighbors. Therefore, it enhances the use of renewable energy and reduces dependence on traditional sources of energy. This showcases SCPS as an effective means of reaching energy access and sustainability in rural settings where grid extension may be impracticable or too expensive.

14.6.1 Lessons learned and best practices

These case studies provide insightful lessons and best practices for implementing SCPS-based smart grids. In urban settings, the roles of edge computing and IoT are critical in managing high data volumes and achieving low latency responses. Meanwhile, in rural settings, decentralizing energy management through microgrids and P2P trading highlights SCPS's effectiveness in achieving energy autonomy and sustainability. Blockchain and secure data-sharing protocols play crucial roles in both environments by fostering transparency, security, and trust in distributed energy transactions. Overall, these implementations stress the need for adaptable SCPS designs tailored to the specific demands and constraints of urban and rural contexts. Advanced technologies like AI, IoT, blockchain, and edge computing are essential for addressing unique environmental challenges, making SCPS a transformative tool for sustainable energy management and reliable grid operations across diverse settings.

14.7 CONCLUSION AND FUTURE DIRECTIONS

14.7.1 Key takeaways from SCPS applications in smart grids

The deployment of SCPS in smart grids has led to significant progress in energy management, decentralized energy systems, and sustainable energy practices. Integrating cyber technologies like IoT, machine learning (ML), and blockchain with physical power infrastructure transforms traditional grids into responsive, adaptive systems that handle the complexities of renewable energy sources, shifting demands, and decentralized energy generation more effectively. A notable takeaway is the effectiveness of edge computing and IoT devices in processing real-time data closer to where it is generated. This approach reduces latency and supports quick decision-making essential for load balancing, demand response, and fault detection, which are crucial to maintaining grid stability in variable conditions. As illustrated by the Amsterdam Smart City project, these technologies work together to optimize urban energy usage through real-time data collection and local processing [16].

Blockchain and secure data-sharing protocols have proven invaluable in facilitating decentralized energy trading, enabling transparent P2P energy markets that support renewable energy use and energy independence. These technologies also strengthen security by ensuring data integrity and transparency for all energy market stakeholders. The Rajasthan Smart Microgrid Project in India, which supports P2P trading through a blockchain-enabled platform, demonstrates the success of decentralized systems in rural settings, leveraging local renewable sources to ensure energy availability for isolated communities [17]. Another key takeaway is the role of ML and big data analytics in forecasting energy demands and optimizing grid resource allocation. ML models have been crucial in predicting variations in energy generation from renewable sources and adjusting load management strategies accordingly. This predictive capability is essential for maximizing renewable energy use while ensuring grid stability and efficiency.

14.7.2 Future research directions: integrating advanced AI and security protocols

As SCPS applications continue to develop, several research areas show promise for further enhancing smart grid capabilities. A primary area is the integration of more advanced AI and ML techniques to increase grid intelligence. Advances in deep learning, reinforcement learning, and federated learning can enhance demand forecasting models, enable real-time anomaly detection, and support adaptive, resilient demand response strategies. Future AI-powered control systems could autonomously optimize energy distribution in decentralized grids, further boosting efficiency and stability in real-time. Another critical research area is strengthening cybersecurity measures to counteract rising cyber threats. Given SCPS's reliance on interconnected digital systems, vulnerabilities in IoT devices, data transmission protocols, and control systems pose risks to grid stability. Building a robust blockchain infrastructure and developing hybrid security frameworks that combine cryptographic solutions with AI-based threat detection could create resilient security mechanisms for smart grids. Additionally, secure multiparty computation and homomorphic encryption can support privacy-preserving data-sharing among stakeholders without compromising on the security, a critical aspect for decentralized energy trading and sensitive data management [9,10].

Interoperability among SCPS components is another important research area, given the diversity of devices, protocols, and software systems used in modern grids. Developing standardized interfaces and protocols will help overcome integration challenges by allowing SCPS components from different vendors to work seamlessly together. This level of standardization will also make grid architecture more modular and scalable, allowing new technologies to be easily incorporated without major changes to the existing infrastructure.

14.7.3 SCPS as a foundation for sustainable, efficient, and resilient energy systems

Smart cyber-physical systems (SCPS) are poised to become the foundation of sustainable, efficient, and resilient energy infrastructure worldwide. SCPS's capabilities allow smart grids to manage renewable energy sources effectively, supporting a transition away from fossil fuel-based energy production. By integrating technologies like AI, IoT, and blockchain, SCPS enables a flexible and transparent energy market where generation, distribution, and consumption are optimized to meet sustainability goals. SCPS also enhances energy system resilience by enabling fault detection, adaptation, and recovery mechanisms. With self-organizing, self-healing, and predictive maintenance capabilities, SCPS-based grids maintain operational stability even under adverse conditions. For instance, ML algorithms can predict maintenance needs, allowing grid operators to address potential failures before they escalate, reducing downtime and repair costs.

Finally, SCPS's ability to facilitate decentralized energy systems will be crucial in expanding energy access and promoting energy independence, especially in remote or underserved areas. Microgrids, P2P energy trading, and distributed renewable sources can be effectively managed within SCPS frameworks, creating local, sustainable energy ecosystems with less reliance on centralized power grids. Continued investment in SCPS research and development will help stakeholders move closer to global energy sustainability goals, creating energy systems that are not only more efficient and secure but also more equitable and accessible.

REFERENCES

1. Baheti, R. and Gill, H. (2011). "Cyber-physical systems," *The Impact of Control Technology*, vol. 2011, pp. 161–166.
2. Yu, X. and Xue, Y. (2016). "Smart grids: A cyber–physical systems perspective," *Proceedings of the IEEE*, vol. 104, no. 5, pp. 1058–1070.
3. National Institute of Standards and Technology. (2020). *Framework for Cyber-Physical Systems: Volume* 1, Overview and Fundamentals. National Institute of Standards and Technology.
4. de Sépibus, J. (2017). *The integration of electricity from renewable energy sources in the European Union Electricity Market: The Case for 'Smart Grids.'* In T. Cottier & I. Espa (Eds.), *International Trade in Sustainable Electricity: Regulatory Challenges in International Economic Law* (pp. 262–286). Cambridge: Cambridge University Press.
5. Marburger, J.H., Kvamme, E.F., Scalise, G. and Reed, D.A. (2007). *Leadership under Challenge: Information Technology R&D in a Competitive World. An Assessment of the Federal Networking and Information Technology R&D Program.* Executive Office of the President Washington DC President's Council of Advisors on Science and Technology.

6. Rajhans, Akshay, Cheng, Shang-Wen, Schmerl, Bradley, Garlan, David, Krogh, Bruce, Agbi, Clarence and Bhave, Ajinkya. (2009). An Architectural Approach to the Design and Analysis of Cyber-Physical Systems. *ECEASST*, vol. 21. https://doi.org/10.14279/tuj.eceasst.21.286.
7. Graham, S., Baliga, G., and Kumar, P.R. (2009). "Abstractions, architecture, mechanism, and middleware for networked control," *IEEE Transactions on Automatic Control*, vol. 54, no. 7, pp. 1490–1503.
8. Insup Lee and Oleg Sokolsky. (2010). Medical cyber physical systems. In Proceedings of the 47th Design Automation Conference (DAC '10). Association for Computing Machinery, New York, NY, 743–748. https://doi.org/10.1145/1837274.1837463.
9. Yassine, Maleh, Shojafar, Mohammad, Haqiq, Abdelkrim and Darwish, Ashraf. (2019). Cybersecurity and Privacy in Cyber Physical Systems.
10. Gopstein, Avi, Nguyen, Cuong, O'Fallon, Cheyney, Hastings, Nelson and Wollman, David. (2021). NIST Framework and Roadmap for Smart Grid Interoperability Standards, Release 4.0. 10.6028/NIST.SP.1108r4.
11. Mackall, D., Nelson, S. and Schumann, Johann. (2002). Verification and Validation of Neural Networks of Aerospace Applications.
12. G. J. FitzPatrick and D. A. Wollman. (2010). "NIST interoperability framework and action plans," *IEEE PES General Meeting*, Minneapolis, MN, 2010, pp. 1–4, https://doi.org/10.1109/PES.2010.5589699.
13. K. Ghanem, R. Asif, S. Ugwuanyi and J. Irvine. (2020). "Bandwidth and Security Requirements for Smart Grid," *2020 IEEE PES Innovative Smart Grid Technologies Europe (ISGT-Europe)*, The Hague, Netherlands, 2020, pp. 36–40, https://doi.org/10.1109/ISGT-Europe47291.2020.9248932.
14. Cavus, Muhammed. (2025). "Advancing Power Systems with Renewable Energy and Intelligent Technologies: A Comprehensive Review on Grid Transformation and Integration," *Electronics*, vol. 14, no. 6, p. 1159. https://doi.org/10.3390/electronics14061159.
15. Marco Cook, Angelos Marnerides, Chris Johnson, and Dimitrios Pezaros. (2023). A Survey on Industrial Control System Digital Forensics: Challenges, Advances and Future Directions. *Communications Surveys and Tutorials*, vol. 25, no. 3 (thirdquarter 2023), pp. 1705–1747. https://doi.org/10.1109/COMST.2023.3264680.
16. Esfandi, Saeed, Safiyeh Tayebi, John Byrne, Job Taminiau, Golkou Giyahchi, and Seyed Ali Alavi. (2024). "Smart Cities and Urban Energy Planning: An Advanced Review of Promises and Challenges" *Smart Cities*, vol. 7, no. 1, pp. 414–444. https://doi.org/10.3390/smartcities7010016.
17. Singh, Dr and Pal, Nitai. (2020). Obstacles and Comparative Analysis in the Advancement of Photovoltaic Power Stations in India. *Sustainable Computing: Informatics and Systems,* vol. 25. https://doi.org/10.1016/j.suscom.2020.100372.
18. Xin, Chen, Shen, Jiachen, Cao, Zhenfu, and Dong, Xiaolei. (2020). A Blockchain-Based Privacy-Preserving Scheme for Smart Grids. 120–124. ICBCT '20: Proceedings of the 2020 2nd International Conference on Blockchain Technology. https://doi.org/10.1145/3390566.3391667.

Chapter 15

Investigating system vulnerabilities in digital environments

Aromma Singh, Ramesh Chandra Poonia, and Kamal Upreti

15.1 INTRODUCTION

This is a fast and fascinating world with technology advancements, which go hand in hand with the attacks to make the system vulnerable and nullify all the security measures that one provides. With ever-increasing leaks of data, functionality, and increased-factor attacks, attacks that exploit multiple vulnerabilities or attack vectors ensure that, as it may be an advanced cyber-physical system (CPS), as one proclaims, it becomes putty in the hands of the attackers [1]. The continuous regeneration of cyber threats behoves a visionary procedure for cyberneticists, where routine appraisals and updates become essential to defense mechanisms [2]. Furthermore, understanding the layout of digital vulnerabilities is easily explained by the solid scripted tools and techniques that one without cyber knowledge might need to learn [3,4]. This paper delves into understanding the intricacies of [5] system vulnerabilities, whether by exploitation through host discovery [6], system attacks, or malicious injections. It confirms all the defense mechanisms, suggesting insights into developing a resilient and self-knowledge cybersecurity stance that will withhold all the threats imposed.

15.2 BACKGROUND AND HISTORY

Cyber threats have become complex; thus, the need for network security has also been evolving. Starting from the 1980s where simple patterns like early viruses and straightforward attacks dwelt, advanced threats such as ransomware and DoS attacks have taken center stage. One of these is the Slowloris DoS attack, a certain consumption-oriented attack that was discovered in the year 2009. It brings web servers down by keeping http connections open for as long as it can to a point that the targeted web server runs out of resources.

In order to overcome some of these issues, Nmap and Wireshark have been very useful tools at work. Nmap is used to perform host discovery and scanning for exploits, as seen in this research, when it was used to find

DOI: 10.1201/9781003542513-15

out that Slowloris CVE-2007-6750 was on HTTP port 80. But Wireshark is used to analyze the network on the packet basis, and in this work, it showcases the effects of Slowloris attack using trend statistics on the traffic. Moreover, the Metasploit framework was also used for the purpose of attack simulation, with hackers in the making utilizing the Slowloris Module of the Framework for real-life simulation.

This arms race between the attackers and the defenders has also led to the use of artificial intelligence (AI) or machine learning techniques in cybersecurity. Once trained, such technology improves the already existing detection and response mechanisms by making them capable of recognizing and responding to patterns that cannot be detected by the traditional modes of operation. Building upon historical progress, this chapter describes an AI-Enhanced Vulnerability Management System to reinforce defense mechanisms against changing dangers.

15.3 LITERATURE REVIEW

The complexity of modern systems has increased the need for robust security measures to combat evolving cyber threats (Figure 15.1). Numerous vulnerabilities remain unnoticed but exploited by cyber-criminals. The availability of tools and resources has fuelled the rise of common cyberattacks. Specialized tools like Nessus, Nmap, Nexpose, and Wireshark, designed for vulnerability scanning and assessment, play a vital role in recognizing and mitigating potential security risks across various sectors, including educational institutes and governmental entities.

Cihan Cobanoglu [5] wrote that restaurants face increasing cyber threats from viruses and phishing attacks, often due to limited IT expertise and budget constraints. To mitigate these risks, robust security practices like firewall protection, password management, and employee training are essential. However, many restaurants lack basic network security tools like anti-virus software [2], physical locks, and firewalls and do not utilize intrusion detection systems or vulnerability assessment scans. Catalyzed by the study [1], for ensuring network security of education institutes' computer network, a proper and effective vulnerability management process (VMP) has become inevitable with the use of tools since it highlights the efficiency of integrating Nmap for information gathering and Nexpose scanning, providing a notable emphasis on its ability to prioritize vulnerabilities based on risk scores, thus allowing security teams to focus on high-risk areas.

Hence, Input the challenges faced by administrators in securing systems and networks and provided the details of the penetration testing process: reconnaissance, scanning, exploitation, post-exploitation, and reporting. The significance of automated tools such as [7, 3] Nmap, Metasploit, and Kali Linux in effectively scanning and testing for vulnerabilities is emphasized to periodically evaluate security, particularly considering the growing number

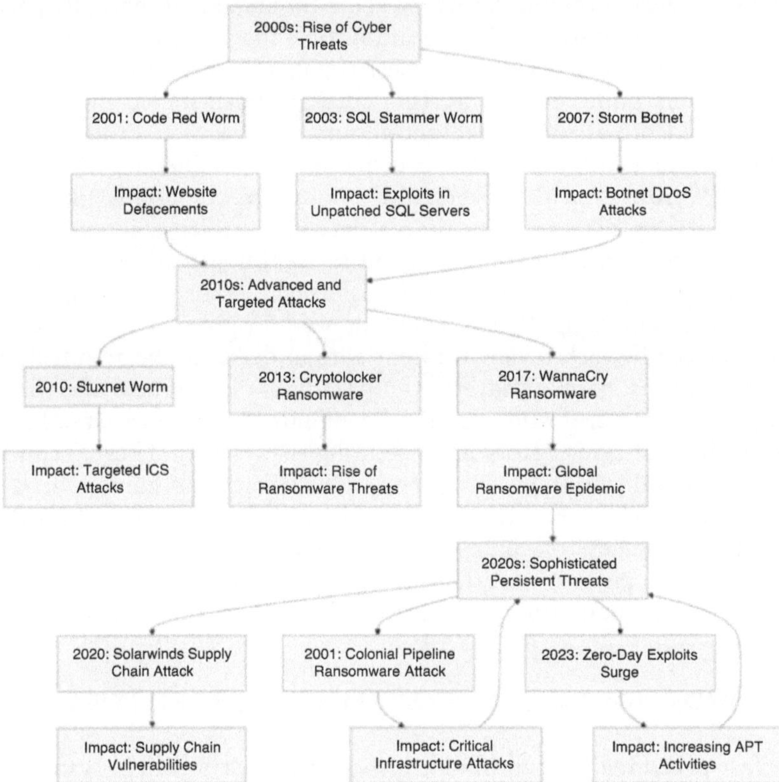

Figure 15.1 The evolution of cyber threats.

of linked devices. Automated solutions are becoming increasingly common and enable even small organizations to perform comprehensive penetration testing with limited resources, which is one of its main contributions.

Vega is a tool used for web application penetration testing [7, 8]. The comprehensive methodology of penetration testing is a crucial tool for identifying system vulnerabilities, ensuring financial loss prevention, industry compliance, customer and stakeholder protection, and defending corporate image. In a well-known system [4], the research methodology involved penetration testing on a Redhat Linux system using a network test bed to simulate attacks with tools like Nmap, Nikto, and Hydra. The study utilized the Snort intrusion detection system to identify malicious traffic and analyzed system logs to trace attack attempts, from which later, in conclusion, the Redhat system apt vulnerabilities and system failed to identify penetration testing [2, 1, 4] (white, black, and gray box) involves simulating cyberattacks to identify vulnerabilities in a system or organization follows several phases including reconnaissance, scanning and enumeration, vulnerability analysis, gaining and maintaining access, and reporting through steps like

IP discovery, port scanning, directory brute forcing, and eventually gaining root access. It demonstrates the use of [8] tools like Nmap, Burp Suite, and reverse shells. The situational awareness is palpable when the audience and service markets know about the risks created and installed by exploiting party or hired professionals like pen-testers and hackers. Ethical hacking is a crucial practice for identifying cyber-security vulnerabilities in organizations [9, 10]. Ethical hackers test common entry points like remote networks, dial-up networks, local area networks, stolen equipment, social engineering, and physical entry to identify weaknesses in defense mechanisms. This proactive security measure is crucial as the increasing number of victims of hacking highlights the urgent need for proactive measures. The limitation faced by each review is essential to understand the architecture of operating systems, network layers, and paradigm of weakness and risk factors. Cybersecurity professionals must be vigilant against a variety of tactics employed by malicious actors to breach digital systems. These techniques often exploit vulnerabilities in software, hardware, or human behavior. Understanding the diverse strategies used by attackers is essential for developing effective defense mechanisms. Attackers rely on specialized tools and strategies at each phase of an attack. These tools assist with reconnaissance, exploitation, and maintaining access to systems once compromised.

15.4 FUNDAMENTALS OF SYSTEM VULNERABILITIES

In the fast-paced realm of technology, there are always gaps in systems that may affect the level of security provided; that is, the existence of such gaps may mean effective security, while the absence of such means total collapse. Whether caused by a simple mistake in an application's code or a complex defect in its hardware, weaknesses usually have far-reaching results. Defining these weaknesses is necessary for any professional who works daily in cybersecurity, system administration, or even average technology consumption. This section will explore the different kinds of vulnerabilities systems face, their origins, and some widespread occurrences.

15.4.1 Types of vulnerabilities

Vulnerabilities can be seen in many forms, each with a distinct effect. For clarification, it is mainly classified into three groups: software vulnerabilities, hardware vulnerabilities, and network vulnerabilities.

15.4.1.1 Software vulnerabilities

Everyone is familiar with the concept of jigs in computer applications. Errors can be as few serious ones as gaping holes in the system security. Software vulnerabilities are inherent weaknesses in the programming code, which usually give the attackers an enormous scope of the system.

Such vulnerabilities can be caused by simple human inaccuracy in the system design. Let us go on and analyze some of these instances more closely.

- **Bugs in Code:** It is generally accepted that software development is a human activity, and hence, errors are bound to occur. Some code flaws are latent and can be taken advantage of by an attacker to break into a system and manipulate it in ways that were not intended. For example, a simple bug like buffer overflow can, unfortunately for the developer, be used by a hacker to load and run an arbitrary code at a given address.
- **Unpatched Software:** Software companies are well known for issuing new versions or upgrades of their software to offer improvements or corrections. This is especially true for security vulnerabilities; however, they are less effective where a number of systems persist with earlier versions, and thus, built-in protection mechanisms are compromised. In fact, some of the most devastating attacks in history, like the Ransomware attack that took place in the year 2017, called WannaCry, were due to systems that were not updated, causing panic all over the world.
- **Insecure APIs:** Application Programming Interfaces are a crucial component for different applications to be able to talk to each other. Even with these APIs being well protected by design, they can be an easy target for attackers. Attacks exploit weak authentication or poorly secured API endpoints.

15.4.1.2 Hardware vulnerabilities

We often think of vulnerabilities as software-related, but hardware vulnerabilities can be just as dangerous, if not more so. These vulnerabilities exploit flaws in physical devices, such as CPUs, memory, or other critical aspects of the system.

- **Spectre and Meltdown:** The vulnerabilities were discovered in 2018 and were found to exist on almost all modern processors. In order to breach those barriers, Spectre and Meltdown exploit speculative execution—a performance-enhancing feature employed within CPUs—by gleaning information from other processes. The consequences of such flaws were far-reaching since they touched many devices, from personal computers and servers to phones. Identifying these weaknesses resulted in a tremendous campaign across the entire sector to fix devices and implement measures to counter the risks.
- **Firmware Vulnerabilities:** Numerous peripheral devices, such as smart devices, wireless routers, and hard drives, contain software called firmware, which dictates the operations of a given hardware. If an attacker manages to compromise the firmware, he or she can take over the hardware, bypassing the other security defense mechanisms. For example, the well-known Stuxnet worm aimed at the firmware embedded within ICS resulted in actual damage to the Iranian nuclear enrichment plant.

15.4.1.3 Network vulnerabilities

The modern world is deeply interconnected, and as we rely more on networks to connect our devices, we expose ourselves to vulnerabilities in the communication channels. Network vulnerabilities can stem from weak encryption, poor configurations, or human error.

- **Open Ports:** Think of ports as doors to the system. If a port is left open without proper security, attackers can use it to access the system. Open ports are often a target in cyberattacks, where attackers can scan for vulnerabilities and exploit them to gain unauthorized access.
- **Unencrypted Communications:** Many systems rely on data traveling across networks. Attackers can intercept and read if the data are not properly encrypted. This is especially risky when transmitting sensitive information, like passwords or personal details. Think of it like sending a postcard in the mail—anyone along the way can read its contents.

15.4.2 Open ports and their vulnerabilities

Open ports are communication endpoints offering scope for one system to communicate with other devices within the network. They are useful for carrying on genuine communication; however, open ports that are not well secured may lead to threats as they make the system prone to possible attacks easily. Ports are divided into three main classes:

- **Well-known Ports (0–1023):** Reserved for specific services, like HTTP (Port 80), FTP (Port 21), and SSH (Port 22).
- **Registered Ports (1024–49151):** Assigned to particular applications but less universally recognized.
- **Dynamic or Private Ports (49152–65535):** Typically used for short-term client application connections.

Leaving open ports unmonitored or unsecured provides attackers with easy entry points. Here are some common vulnerabilities associated with widely used ports:

- **Port 80 (HTTP):** Used for web traffic, this port is often exploited when web servers (e.g., Apache and Nginx) have known vulnerabilities. Attackers might inject malicious scripts or gain unauthorized access to web applications.
- **Port 22 (SSH):** SSH enables remote system administration. Exposed SSH ports without strong access controls or multi-factor authentication are vulnerable to brute-force attacks, especially if weak or default credentials are used.
- **Port 23 (Telnet):** An outdated protocol transmitting data in plaintext, Telnet is susceptible to interception and credential theft. Replacing Telnet with SSH is generally advised for improved security.

- **Port 3389 (RDP):** RDP provides remote access to Windows systems, but it's a common target for attackers seeking to exploit unpatched vulnerabilities. Attackers can use brute-force methods or exploit known issues to gain control over the system.
- **Port 445 (SMB):** Although SMB was designed for file sharing and network communication on Windows Operating Systems, it has some known weaknesses. The assailants exploit such weaknesses, say in the instance of the WannaCry virus, to spread harmful software among a variety of systems and networks, while also collecting private data.

15.4.2.1 Practices for securing open ports

Efficient port administration is vital in mitigating the chances of the network being hindered. Primary ways are as follows:

1. **Regular Port Scanning:** Consistent monitoring of an open port aids in identifying unauthorized applications or a service being incorrectly configured, thereby facilitating quick correction.
2. **Access Control:** Access to users or applications, and even systems, is limited to what is strictly necessary, by the use of firewalls, IP filtering, and role-based access control and hence also increases security.
3. **Port Forwarding and VPNs:** When it is necessary to have open ports that are critical, the use of a VPN or a reverse proxy mitigates access from outside and adds an extra layer of security.

15.4.3 Sources of vulnerabilities

What are the origins of these vulnerabilities? What causes systems to be breached? This is not a question with a definitive answer; the breach of a system is usually due to a complex relationship between design oversights, reckless user behavior, and deliberate sabotage. Let us analyze some of the frequently recognized contributors to vulnerabilities:

15.4.3.1 Coding errors

One of the most significant sources of vulnerabilities in software is the human factor. It is also the case with highly skilled and seasoned developers that they can commit errors, which often leave an opening in the code structure that can be taken advantage of. Failing to secure code using safe coding practices such as proper input validation or memory management can cause serious security risks. Occasionally, it has been noted that security weaknesses exist for years, only to be discovered later by an attacker trying to find one.

For instance, Heartbleed, a widely known bug affecting millions of servers, was caused by a programming bug in the Open SSL library that lets a hacker gain access to the server's available memory. It was dormant in the code for at least 2 years until it was noticed, although journalist sources noted the outbreak of code misuse.

15.4.3.2 Configuration errors

Another primary source of vulnerabilities arises from misconfigurations. This happens when systems are set up incorrectly or default configurations are left unchanged. Administrators often leave unnecessary services running or must remember to implement proper security settings. This exposes systems to attacks that could have been easily prevented with a more secure configuration.

For example, a system might enable file-sharing services without proper access controls, or a firewall might be configured to allow all incoming traffic. These misconfigurations give attackers an easy entry point. The infamous "default password" vulnerabilities are another classic example, where attackers use default usernames and passwords that come with many devices, such as routers, cameras, and printers.

15.4.3.3 Insider threats

When discussing security, we often think of external attackers trying to break into a system. However, insider threats can be just as dangerous, if not more so. Insiders, such as employees or contractors, have access to sensitive systems and data, which gives them the ability to cause serious harm—either maliciously or negligently.

Insider threats can come in various forms, like an unhappy worker trying to abscond with trade secrets or, at the bottom of the ladder, a simple mistake like using the wrong address, causing inadvertent leakage of confidential information. In most instances, insider threat actors are more challenging to conceptualize and represent because the perpetrator of the malicious act has a legal right of access to the given system.

An illustrative case of an insider threat would be the publication of classified intelligence materials by Edward Snowden, who previously worked as a National Security Agency (NSA) contractor. Although this instance is very much on the far end of the spectrum, it is essential to point out the risk that insider abuse comes with, especially concerning information-sensitive retrieval and exposure.

15.5 SYSTEM ATTACKS AND THEIR TYPES

Systems are constantly subjected to some form of attack in this digital age. There are hackers, cybercriminals, and other malevolent forces that are always impatiently looking for weaknesses. Learning the categories of attack that the systems can suffer is the primary step in constructing thorough defense mechanisms. In this section, we will discuss a few attack vectors and their effects on systems.

15.5.1 Common types of system attacks

- **Malware Attacks:** Malware, an ever-increasing problem is a software designed to damage or harm any device, service, or network [8]. This encompasses anything from file-based viruses to adware that takes control of the device and holds the data in exchange for ransom.
  - i. **Viruses:** Virus, a type of malicious program, attaches itself to files that are uninfected with any malware so as to spread whenever execution of those files is done.
  - ii. **Worms:** These do not need a host file. They replicate across networks and can bring entire systems down.
  - iii. **Trojans:** These sneak in disguised as legitimate software but unleash havoc once they are installed.
  - iv. **Ransomware:** This type locks users out of their own data and demands payment to regain access.
- **Man-in-the-Middle (MitM) Attacks:** In an MITM attack, the attacker listens to the transmission between two parties and can also intercept and alter the transmission. This can take place in the form of emails, web interactions, or even telephonic conversations. When the communication channel is unprotected and unencrypted, hackers can access information, tamper with it, and relay it to the undesigned recipient without either of the parties being aware.
  - i. **Session Hijacking:** Attackers grab an active session to impersonate the user.
  - ii. **Eavesdropping:** Clear by the name, it is a silent intruder keeping tabs on the line of connection between two parties without having an actual authorization.
- **Privilege Escalation:** This is referred to as privilege escalation. Suppose an individual begins using a computer as a normal user, in that case, the user may exploit existing loopholes within the system to assume control of an administrator account, thereby gaining all administrative privileges.
- **SQL Injection:** SQL injection is a code-injection method that exploits a security vulnerability of an application's software by allowing the insertion of a malicious SQL query through inserted data from the client to the application's server. Attacks like these usually involve websites and applications whose security levels are not of a high standard.
- **Zero-Day Attacks:** Such attacks make use of the vulnerabilities that have never been exploited before. Such attacks are more or less lethal as there is no availability of a bug fix or patch. Whenever a new weakness is discovered, it is understood that an outburst of attacks will follow when appropriate protection mechanisms have not yet been put in.

- **Phishing and Social Engineering:** Phishing is one of the oldest tricks in the book, where attackers disguise themselves as trusted entities to steal information like login credentials [8]. It is often done through deceptive emails or websites that look legitimate.
  - **Pretexting:** The attacker pretends to be someone else to get users to share sensitive information [11].
  - **Baiting:** Offering something free or enticing (like free software) to trick users into downloading malware.
- **Denial-of-Service (DoS) Attacks:** DoS attacks aim to take down a website or service by overwhelming it with traffic [8, 4]. The idea is simple: flood the system until it cannot handle any more requests, causing it to crash.
  1. **Flood Attacks:** Attackers send so much traffic that the server runs out of resources [7].
  2. **Application Layer Attacks:** They are more perplexing and are focused on targeting specific applications that functionalize on the server by consuming resources including memory and CPU cycles.

15.5.2 The impact of system attacks

The consequences faced in these instances can be very destructive, varying from grips of financial constraints to lasting damage of a company's standing. Few of the familiar occurrences are enumerated below:

15.5.2.1 Private sector effect

Perhaps, system attacks are the most threatening to private sectors [5, 2]. Earning opportunistic strategies, criminals can potentially compromise systems and place significant hurdles in the processes of carrying out business or even extorting the companies. Loss of data can translate to loss of revenue and damage in terms of public relations or even law suits. Ransomware attacks can cripple companies and thus force a lot of money into operators' pockets to recover normal functioning of their systems. On top of all that, it can also slow down growth of businesses by robbing them of their patents.

15.5.2.2 Teenage impact

Teenagers are also at great risk of system attacks on account of their high online activity coupled with often limited computer skills [2]. They are likely to suffer emotionally from activities like cyber bullying, abuse, and being subjected to distasteful content. Young people's abilities are also exploited toward such unfashionable elements as predation. These experiences make them liable to loss of their identity as well as con-man scams even when they do not know any better.

15.5.2.3 Emotional perspective

System attacks can have emotional aspects in the individuals who come across them. Victims of cyber offenses, for instance, are likely to feel anxious, stressed, or depressed [2]. It can be stressful, especially regarding the negative implications of losing one's identity or funds or losing one's good name. Also, in addition to fear of such tangible attacks, people have to contend with the invisible yet threatening cyberattacks.

15.5.2.4 Impact on adolescents

Youth are significantly affected by system intrusion, especially in the educational and information technology aspects [2]. When cyber instabilities affect schools, it often leads to reduced education and growth. Furthermore, the non-uniform availability of technology and education on how to stay safe online enhances existing disparities among the youth. Most youths are also at risk of online con games, harassment, and other forms of witchcraft because they do not have computer knowledge.

15.5.3 General challenges of system attacks

15.5.3.1 Weak authentication

Insufficient authentication approaches pose a significant challenge to information security. These weaknesses can be exploited by attackers in different ways to access systems and data without permission. The authentication procedure is liable to gaining the verification and validated data's access to safeguard the aforesaid data from unnecessary and most importantly unsupervised access [2]. The access to accounts, profiles, and pages is easily gained either by close friends, family, members, or someone who wants to potentially terminate the owner's access. In such a digital world, it is no secret that the footprint of the user's access of any services is saved, be it in the form of cookies or other [1]. For attackers and scammers, gaining such crucial credentials becomes easy because of weak password policies, dependence on single-factor authentication without any alternative, or the non-application of encryption to protect stored credentials and their transmission. These deficiencies are even more conspicuous in networks where computerized intruders use feeble authentication procedures to access networks, escalate privileges, and move laterally within the networks [9].

15.5.3.2 Password attacks

Some types of attacks which are aimed at passwords include brute force attacks, which employ a computer and software and try to crack a password by running all possible combinations of numbers, characters, letters, and symbols in it and play Guess the Password. This attack is useful regardless of the password length or complexity. Another common method is the Dictionary Attack, wherein outlined attackers take a large list of frequently used passwords and try them as a guess for the user's password.

15.5.4 How to defend against system attacks

Luckily, however, few things can be done to reduce the chances of falling prey to such attacks:

- **Update Software Regularly:** Always install the latest security patches. Many attacks exploit known vulnerabilities that could have been fixed with an update.
- **Use Firewalls and Intrusion Detection Systems (IDS):** A good firewall blocks unwanted traffic, while an IDS alerts you if something suspicious is happening in the system.
- **Encrypt Data:** Encryption helps protect sensitive information, especially when it's in transit. Even if hackers intercept it, they won't be able to read it.
- **Train Users:** Make sure everyone in your organization knows the basics of security, from recognizing phishing attempts to using strong, unique passwords.
- **Enable Multi-Factor Authentication** (MFA): This suggests providing a layered protection to data by implementing an additional security measure from determined threats from accessing sensitive information even with the correct password.

15.6 TOOLS AND TECHNIQUES FOR VULNERABILITY DETECTION

Identifying weaknesses is essential for securing a given system. A number of tools and techniques are deployed in a bid to find flaws in network, system, and application environments. Here is a summary of the main types of instruments used and their comparative analysis of functions performed.

15.6.1 Overview of key tools

15.6.1.1 Nmap (network mapper)

Nmap also known as Network Mapper, is a utility for unmatched open-source network scanning which focuses on host and service discovery on a network [7]. It enables the determination of active ports, operating systems, and active services on the targeted computers [3]. It provides many scanning methods including tailored TCP and UDP port scans, OS

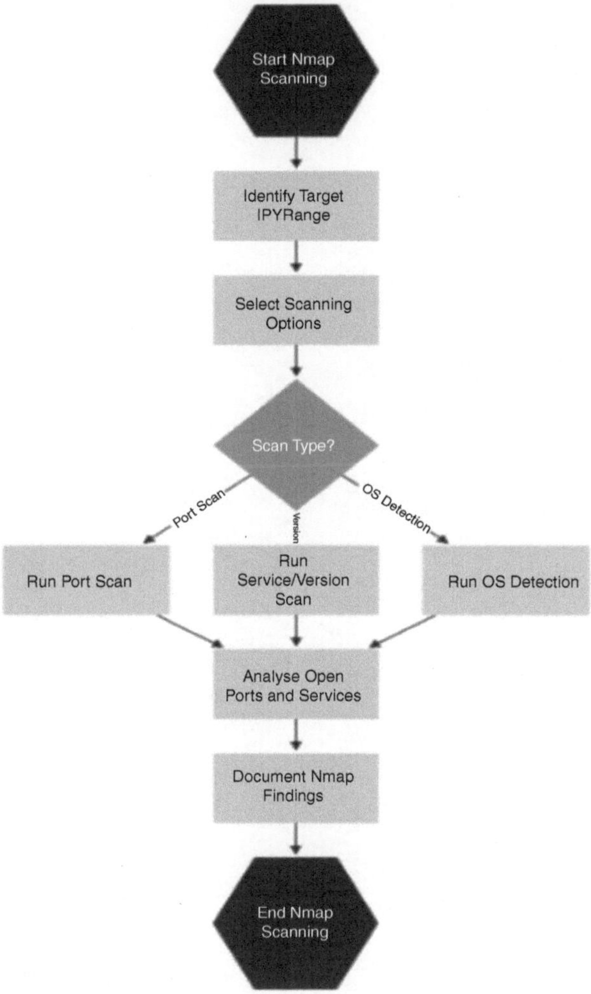

Figure 15.2 Working of Nmap.

detection, service versioning, and even scanning for weaknesses. Nmap can be employed in arranging complex networks, detecting possible components of network security threats, and even diagnosing network issues [9]. As such, the convenience and adaptability of the tool are reasons why it is essential to network admins and security experts alike (Figure 15.2).

Use case: Network mapping and identifying exposed services.

15.6.1.2 Metasploit

Due to its penetration testing capabilities, Metasploit can be described as a versatile and useful tool for all ethical hackers who want to perform real

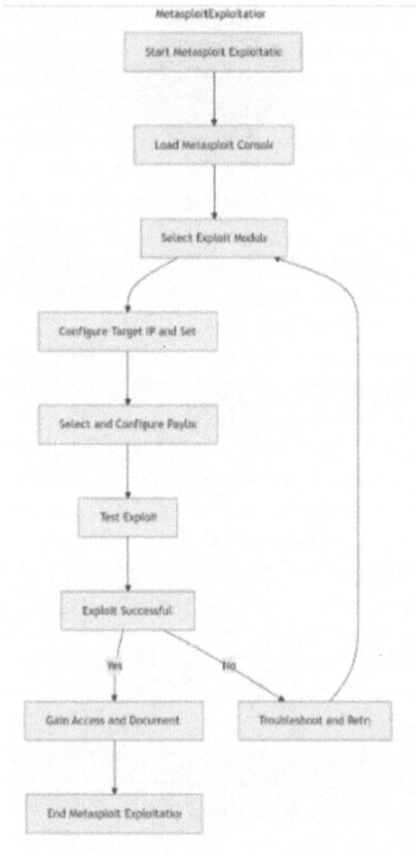

Figure 15.3 Working of Metasploit.

attack simulation and scan for the weaknesses that exist in systems and networks [7, 8]. It offers a diverse range of utilities and corresponding modules tailored toward several offensive approach patterns such as exploitation, post-exploitation, and privilege escalation modules [3, 4, 9]. Metasploit is more than just an exploit development framework; it also provides a means for securing web applications, networks, and operating systems for taking organizations' security assessment a notch higher (Figure 15.3).

Use case: Penetration testing and vulnerability exploitation.

15.6.1.3 Wireshark

Wireshark serves as a network protocol dissecting and capturing programs. It helps observe and study packets, assess the performance of the given network, diagnose difficulties, and scan for potential threats. The program is

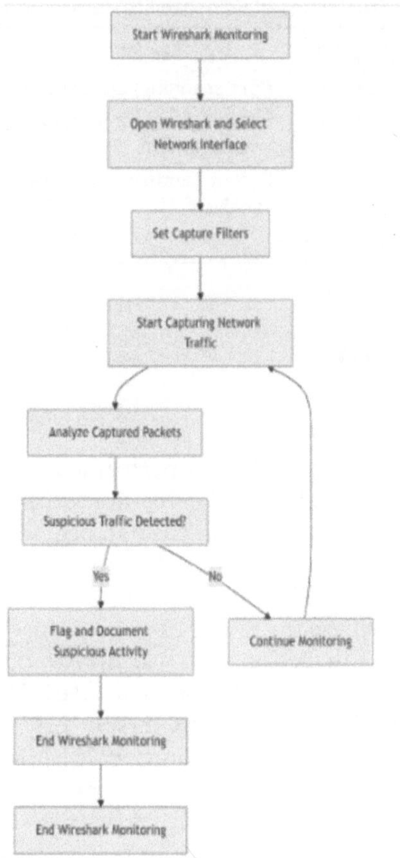

Figure 15.4 Working of Wireshark

capable of a number of protocols including TCP/IP and other higher layers such as HTTP, HTTPS, file transfer protocol, and domain name system, among others [12]. It has a Graphic User Interface coupled with high level and efficient search and filtering options, allowing the user to swiftly look for and obtain any desired packet. Wireshark, in other words, is a key tool for every network administrator, security investigator, and indeed anyone who must deal with networks (Figure 15.4).

Use case: Real-time network traffic analysis and detecting abnormal network behavior.

15.6.2 Comparative analysis

See Table 15.1.

Table 15.1 Features and capabilities of vulnerability detection tools

Tools	*Primary purpose*	*Key features*	*Use case*
Nmap	Network scanning	Port scanning, OS detection	Mapping network
Metasploit	Penetration testing	Exploit development, payload management	Exploitation testing
Wireshark	Traffic analysis	Packet capture, protocol analysis	Anomaly detection
Packet capture and protocol analysis			

15.7 METHODOLOGY

15.7.1 Nmap

In network security investigation, the performed Nmap scans show how critical host enumeration and service detection techniques can be. The first simple command access "nmap scanme.nmap.org" reconstructs basic port discovery attempts on the target address (45.33.32.156), which enumerated four open ports: SSH-22, HTTP-80, nping-echo-9929, and Elite-31337 [3]. In the following scan, where the -sV option was added ("nmap -sV scanme.nmap.org"), this version detection scan was able to provide specific help in service identification, revealing OpenSSH 6.6.1p1 Ubuntu as the service running on port SSH 22, Apache httpd version 2.4.7 running on port 80, and Nping echo service listening on port 9929. It also classified the OS architecture as Linux, adding more insights into the target's environment. This refers to the infamous scan depth versus the scan time, essential in security assessment and research techniques.

These scans were simply being performed on scanme.nmap.org, the official test site of Nmap; hence, ethical research was observed during the network reconnaissance strategies. Within the domain of vulnerability assessment research, Nmap scans show the ability to perform advanced security testing using the command ("sudo nmap –A –script vuln scanme.nmap.org"). This scan is considered a complete vulnerability assessment that uncovered several shocking aspects. The most shocking aspect was that the assessment found a Slowloris DOS attack (CVE-2007-6750) on the host [1, 9]. This is an important finding since it exposes the target to DOS attacks from any user. The scan found various vulnerabilities, including CSRF (Cross-Site et al.), in two search forms under the search path'/search/' with form IDs of "not-head-search" and "not-foot-search." More importantly, it reported the presence of CVE-2008-2928, a Slowloris DOS attack posing a high risk to web application port 80 [7]. The second output identifies large-scale vulnerability enumeration concerning the OpenSSH service, which operates through port 22, as there are many CVE entries, some with a severity score of 10.0 being critical to 0.0, which is informational.

Table 15.2 Nmap detailed scanning insights

Nmap function	*Purpose*	*Results*
nmap scanme.nmap.org	Basic port discovery	Enumerated four open ports: SSH (22), HTTP (80), Nping Echo (9929), and Elite (31337).
nmap -sV scanme.nmap.org	Version detection	Identified services and versions: OpenSSH 6.6.1p1 Ubuntu (port 22), Apache httpd 2.4.7 (port 80), and Nping echo (port 9929). OS classified as Linux.
sudo nmap -A –script vuln scanme.nmap.org	Advanced scan with vulnerability scripts	Detected vulnerabilities: Slowloris DOS attack (CVE-2008-2928), CSRF vulnerabilities under "/search/," critical OpenSSH vulnerabilities (e.g., CVE-2023-38408).

The identified critical vulnerabilities include CVE-2023-38408 and numerous possibilities for exploitation designated by labels "EXPLOIT."

For ethical reasons, these scans are performed on scanme.nmap.org, a test server run by nmap.org, to present enhanced security evaluation techniques on research work (Table 15.2).

15.7.2 Metasploit

Invoking Metasploit Shell saying "msfconsole" in the Kali terminal [8]. This command line invocation initiates the Metasploit console, a command interface that matches the framework. We began by looking for vulnerabilities in Metasploit exploits by using the command: "search slow loris" from the [3] Nmap scan since this vulnerability showed up in open port 80. With this command, Slowloris attack modules from the Metasploit framework are searched, and a couple of modes of attack were performed using the Slowloris module, and the target's HTTP service was subjected to DoS. Ultimately, a port scan of the target was performed using Nmap, which clearly confirmed that port 80 (HTTP) was open. Slowloris attacks constitute [4] Denial-of-Service (DoS) attacks that seek to bring down FTP servers by bombarding them with empty HTTP requests. After selecting the Exploit Module with the command, we give the command "use auxiliary/dos/HTTP/slowloris." This command selects the Slowloris module to be used for the attack.

Setting the target with the command "set RHOST 45.33.32.156" [13]. This command sets the target IP address of the vulnerable system. We ensure the attacker machine (192.168.206.140) can overflood the HTTP servers by setting the socket count to the desired number. We enabled the framework to recognize the port to attack with "set REPORT 80," which attacks https at port 80; this kind of configuration indicates an effort to perform a DoS attack on the target site. Configuring additional options (Optional) showed all the available options that can be configured for the selected module. We can change options such as the remote telephone number (RPORT), the

Table 15.3 Metasploit Penetration Detailed Insights

Command	*Purpose*	*Results*
msfconsole	Launch Metasploit console	Opens Metasploit framework's command-line interface.
search slow loris	Search for Slowloris attack modules	Displays available Slowloris modules in the Metasploit framework.
use auxiliary/dos/ http/slowloris	Select the Slowloris exploit module	Prepares the Slowloris module for attack configuration and execution.
set RHOST 45.33.32.156	Set the target's IP address	Targets the specific vulnerable server (HTTP service on port 80).
set RPORT 80	Set the target's port	Specifies port 80 (HTTP) for the attack.
set SOCKETS 150	Configure the number of sockets	Increases connection load to overwhelm the target with partial HTTP requests.

use of SSL (SSL), the directory structure (URIPATH), and so forth so that specific weaknesses may be targeted while increasing the socket to 150–200.

The attack employed 150 or more threads aimed at opening connections by sending partial HTTP requests so that the target server could not serve its clients due to a high number of open connections. Then execute the attack with "run." This starts the attack on the target server (45.33.32.156) from the attacker machine (192.168.206.140- Kali system), and in this case, many slow half-HTTP requests (GET/HTTP/1.1) were deployed to the target server, which may lead it to the DoS—Slowloris attack on the IP address 45.33.32.156 targeting port 80 (HTTP). In the said attack, 150 to 200 sockets are opened to send half-opened HTTP requests to the target server and flood it with many connections, resulting in a denial of service to the server and the number (count) of payload iterations dispatched to exploit [13]. Flooding the target with half-open HTTP requests to render its service unavailable is also occurring. The assault has been executed for a specific duration since the number of active sockets has been increasing (Table 15.3).

15.7.3 Wireshark

Wireshark justifies its inclusion in the study on the impact of a Slowloris DoS attack on HTTP traffic, primarily due to its excellent filters that target specific packet aspects of the attack [14]. To begin with, there was an [14] HTTP filter applied on port 80 (HTTP or tcp.port == 80) so that the capture could center on any HTTP packets, while other communications were omitted so that the focus would be on analyzing the server's interaction with external agents. Within the Slowloris siege, Wireshark has shown that numerous HTTP requests remained sideways as in their in-progress state with open connections, whereby each was without a reply from the server. After using the follow TCP stream option, the data exchange patterns between the attacker and the server during the attack were studied.

Table 15.4 Wireshark detailed analysis

Wireshark filter/option	*Purpose*	*Results*
tcp.port == 80	Filter HTTP traffic on port 80	Isolated HTTP packets to analyze server interaction during the Slowloris attack.
Follow TCP Stream	Analyze data exchange patterns	Revealed incomplete HTTP requests, showing the server's open connections being exploited.
tcp.flags.syn == 1 or tcp.flags.ack == 1	Observe SYN and ACK packets	Detected a high number of synchronization and acknowledgment packets, indicating server stress.
TCP.analysis. retransmission	Identify retransmitted packets	Showed lost or delayed packets, highlighting server overload during the attack.
tcp.flags.reset == 1	Filter reset packets	Displayed server resets, showing attempts to purge open connections to relieve resource strain.

It was noted how step by step, every underserved request, which the server accepted but not completed, kept the connection alive, gradually consuming the server's resources. This cut in the data stream also showed how these requests kept on suspending and contributed to more and more resources being used up and using the Wireshark to specify the specific port being exploited (tcp port == 80).

To complicate the analysis of the stress on the server, Wireshark advanced its SYN and ACK filters (tcp.flags.syn == 1 or tcp.flags.ack == 1) as a means of observing synchronization and acknowledgment packets attack, noting that such packets were disproportionately high [12]. This shows the server's stresses as it tried to handle far too many open connections. Moreover, during the attack, Wireshark applied the retransmission filter (TCP.analysis. retransmission) to identify the transmission of some packets that were lost in the process or others whose packets arrived late to the sender's computer, indicating the overload of the server.

The reset packet filter (tcp.flags.reset == 1) depicted many resets as the server tried purging these connections to alleviate the pressure on its resources. Consequently, over time, patterns formed, showing further effects of the Slowloris on the server and evidence of the server response becoming lower and lower (Table 15.4).

15.8 RESULTS AND DISCUSSION

A meticulous Nmap scan was carried out to examine the target machine in further detail, exposing services and software versions that could potentially be targeted. This imposed scanning of specific vulnerabilities, that included determining port and associated threats, which aided in finding vulnerabilities that simple scans cannot reveal. A DoS attack was then carried out following the identification of the exploit by Metasploit's Slowloris attack, which initiated multiple [15] HTTP requests to the server but in incomplete form to make it busy. Network traffic was analyzed in real time using Wireshark, which also incorporated the motion of the attack and the state of the server during the attack.

As the attack continued, attention was directed to the HQE request packet scrutiny to get rid of the Slowloris module's traffic. This enabled us to see the half-open requests and the SYN ACK packets, which were obtained from the stalled connections. As time progressed and with increasing attack, the throughput of incoming requests caused the server to slow down due to several incomplete connections. Analysis of the Wireshark tool recorded the effects of the attack on the server, where it was seen that there were more cases of TCP retransmissions and RST packets, which were all indications of the infection by the Slowloris-related resource consuming DOS. Other results included lost packets and slow service requests, revealing the effectiveness of the attack while also showing vulnerabilities to the network.

15.8.1 Vulnerability findings

To systematically identify and analyze the vulnerabilities of the target system, a deep analysis consisting of several stages utilizing Nmap, Metasploit, and Wireshark was carried out. First of all, Nmap was put to use to carry out an extensive exploration scan with emphasis on version discovery of open and filtered ports. This high-level scanning did not only concern with port mapping but also detailed the processes and software in use in the mapped ports and the related threats presented by their weak points. These scans revealed a number of flaws, especially those associated with open and poorly secured ports. Open ports allowed access to critical services, and coupled with out-of-date service software, they provided a wide attack vector. For example, several ports were running services with known vulnerabilities, making those ports easily exploitable.

Slowly, we also incorporated a more complex layer of the model of a vulnerability scan that acted out specific functions again on each port, as previously defined. However, this added depth of scanning was to make sure that no security gaps were missed as most simple scans would have done. Some of these flaws stemmed from services which were not properly locked down, thus making it easier for an attacker to breach. Thereafter, once these flaws were recognized, the entire evaluation moved to the next stage exploitation testing with the use of Metasploit.

15.8.2 Exploitation insights

See Table 15.5.

Table 15.5 Impact of exploited vulnerabilities

Vulnerability	*Exploited service*	*Exploitation method*	*Severity*	*Potential impact*
Open port 22 (SSH)	SSH	Remote code execution	High	Unauthorized remote access, potential data compromise
Outdated web server	Web server	Exploit known CVE for outdated software	Medium	Data breach, potential loss of sensitive information
Unauthenticated port 80	HTTP	Denial-of-Service (DoS) via Slowloris	High	Server resource exhaustion, server unresponsiveness
Misconfigured database	MySQL	SQL injection	High	Data manipulation, unauthorized access to sensitive data
Outdated openSSL version	OpenSSL	Known CVE exploit	High	Man-in-the-middle attacks, data interception

15.9 CONCLUSION AND FUTURE SCOPE

15.9.1 Conclusion

The analysis of system weaknesses within the digital realm reiterates the need for a proactive and relentless approach to security. In this chapter, an in-depth analysis of vulnerabilities was given, tracing its origin, typologies, methods of analysis, and eradication of vulnerabilities. With advanced technologies such as Nmap, Metasploit, and Wireshark, organizations are capable of detecting weaknesses and countering possible threats promptly. Most importantly, case studies have shown real-life incidences of attacks and how the vulnerabilities were taken advantage of, which are useful in building digital infrastructures.

These findings are a clear testament of the importance of automated vulnerability assessment frameworks, as well as the training requirements to tackle the soft component of the cybersecurity risk. In 2 years, it will be possible to extend the systems of vulnerability detection and remediation due to the introduction of AI with ML-enhanced systems. Cyber threats, if not well managed, can turn out to be very expensive and even lead to closure of most organizations; hence, it is important for organizations to exercise the practices suggested and adjust to the emerging threats. Securing digital spaces is an objective that has to be reached through a conglomeration of technology, tactics, and enhancing the security in place on a relentless basis.

15.9.2 Future research directions

Combining AI, ML, and tools such as NMAP, Metasploit, and Wireshark also seems to be a viable option in as far as enhancing the safety of a given network is concerned. For example, risk assessment and anomalous network activity detection can go on without the aid of human beings. The next logical step would be for institutes to build systems that will respond accordingly to the possible threats posed.

Moreover, for such systems to be incorporated even in the existing emergency response strategy, threats may be eliminated in the least time. Concerning auto-incident response scenarios, these tools would prove to be helpful because they reduce response measures as well as the influence of the attack. Further research might also seek to explore the extensiveness of the tools in other new technologies, especially cloud computing and IoT. With every innovation that is introduced into the marketplace, new risks are introduced, which calls for modification of existing security measures or introduction of better ones to cover the new threats.

To a great extent, there are observations of systemic deficiencies, design vulnerabilities and faults with Nmap, Metasploit, and Wireshark usage. Next, principal use of the technologies will include tools such as AI and ML, which will make the video security management system more detectable, preventive, and responsive to cyber threats.

ACKNOWLEDGMENT

First, I would like to thank Dr. Ramesh Chandra Poonia for his constant and unwavering support during this research work. His expertise and encouragement greatly contributed to this work. I would also like to thank my colleagues, sister, and faculties who contributed with discussions, technical help, and support. Their assistance helped me a lot in this field. In the end, my family, and friends deserve a very special thank you for the love, support, and understanding which has enabled me to complete this research work despite the odds.

REFERENCES

[1] Prasad, M. R., & Manjula, B. (2014). Ethical hacking tools: A situational awareness. *International Journal of Emerging Technologies in Computer Science & Electronics*, *11*, 33–38.

[2] Das, S., & Nayak, T. (2013). Impact of cybercrime: Issues and challenges. *International Journal of Engineering Sciences & Emerging Technologies*, *6*(2), 142–153.

[3] Kaushik, S., Bhutto, A., & Pandey, B. (2019). Efficient information gathering using NMAP and NBTSCAN: Case study on 172.19.19.0 IP address. *Indian Journal of Science and Technology*, *12*(28), 1–13.
[4] Almaarif, A., & Lubis, M. (2020). Vulnerability assessment and penetration testing (VAPT) framework: Case study of government's website. *International Journal on Advanced Science, Engineering and Information Technology*, *10*(5), 1874–1880.
[5] Cobanoglu, C. (2007). A critical look at restaurant network security: Attacks, prevention tools, and practices. *Journal of Foodservice Business Research*, *10*(1), 31–50.
[6] Hussain, M. Z., Hasan, M. Z., Taimoor, M., & Chughtai, A. (2017). Penetration testing in system administration. *International Journal of Scientific & Technology Research*, *6*(6), 275–278.
[7] Stiawan, D., Idris, M. Y., & Abdullah, A. H. (2015). Penetration testing and network auditing: Linux. *Journal of Information Processing Systems*, *11*(1), 104–115. Korea Information Processing Society.
[8] Kettani, H., & Wainwright, P. (2019, March). On the top threats to cyber systems. In *2019 IEEE Second International Conference on Information and Computer Technologies (ICICT)*, Kahului, HI, USA (pp. 175–179). IEEE.
[9] Chhillar, K., & Shrivastava, S. (2021). University computer network vulnerability management using Nmap and Nexpose. *International Journal*, *10*(6), 3084–3090.
[10] Mohan, A., Swaminathan, G. A., Kubsa, A. D., Tolasa, F. T., Tesfayohanis, M., & Bakala, G. K. (2021). Security implementation using penetration testing. *Design Engineering*, *2021*(8) 5408–5421.
[11] Salama, R., Al-Turjman, F., Bhatla, S., & Yadav, S. P. (2023, April). Social engineering attack types and prevention techniques – A survey. In *2023 International Conference on Computational Intelligence, Communication Technology and Networking (CICTN)*, Ghaziabad, India (pp. 817–820). IEEE.
[12] Mabsali, N. A., Jassim, H., & Mani, J. (2023, January). Effectiveness of wireshark tool for detecting attacks and vulnerabilities in network traffic. In *First International Conference on Innovation in Information Technology and Business (ICIITB 2022)*, Muscat, Sultanate of Oman (pp. 114–135). Atlantis Press.
[13] Katkar, V., Zinjade, A., Dalvi, S., Bafna, T., & Mahajan, R. (2015, February). Detection of DoS/DDoS attack against HTTP servers using naive Bayesian. In *2015 International Conference on Computing Communication Control and Automation*, Pune, India (pp. 280–285). IEEE.
[14] Soepeno, R. A. A. P. (2023). Wireshark: An effective tool for network analysis. University of Arizona.
[15] Argeshwara, D. K., Hadi, M. S., Sendari, S., & Irvan, M. (2024). Comparative analysis of application layer protocols in EV charging stations: Evaluating HTTP, MQTT, and Websocket Performance Metrics. *Bulletin of Social Informatics Theory and Application*, *8*(1), 86–96.

Chapter 16

Adaptive multimodal human–machine interaction in cyber-physical systems for enhanced decision-making

Ansu Samuel, Maligi Bindu, Rajat Dang, Aman Singh, Sonali Samdarshini, and Ramesh Chandra Poonia

16.1 INTRODUCTION

Due to the rapid advancements in technology, cyber-physical systems (CPS) are becoming a vital component of several sectors, including but not limited to manufacturing and health care as well as smart cities. Embedded computers use feedback loops to perceive the outer physical environment and act upon it. Such systems include very close physical, networking, and computational activities. Increased and more complex systems demand optimization of the decision-making process and system performance through the seamless integration of both human and machine inputs. It is important to take into consideration the role of HMI systems in CPS because therein lies the efficiency of interacting, managing, and making use of the system's intelligence by operators. Traditionally, the more straightforward and easier HMI elements have been limited to the use of images and text with very few interaction modes. Nevertheless, the evolution of real-time processing capabilities combined with artificial intelligence (AI) technologies allows for the possibility of systems that incorporate user input other than the standard speech and text formatting, such as the use of voice commands, moving one's hands, and even touching. In this manner, the system will become more flexible and easier to use.

There is evidence that the benefits of machine and robot-based production are mainly in the automated and repetitive areas [1], whereas the intelligent production still involves many human beings. People's cognitive skills and tacit knowledge are irreplaceable when the management of production is unpredictable, ambiguous, and adaptable. In this context, what will the work environment of humans look like in the future with technology that is at the human–machine interface. Therefore, the US National Science Foundation came up with a project in 2016 [2]. This research shows that the cooperation between humans and technology is the key to interdependent

DOI: 10.1201/9781003542513-16

human–technology relationships. Talking about the Japanese philosophy, Society 5.0 is a holistic nationwide strategy to implement human-centered design in the context of a super-smart society [3,4]. A recent industrial transformation has occurred with the European Union's formulation of Industry 5.0 [5–8], in which the focus is on shifting to intelligent production that is people-oriented.

The development and inheritance of Industry 4.0 are Industry 5.0. Industry 5.0, in contrast, empowers people, using CPS tools aiming to aid human proficiency rather than replace human workers with the latest technology, as Industry 4.0 uses CPS [9,10] to boost production by employing digital, networked, and intelligent capabilities. Hence, the basis of Industry 5.0 is the fusion of the physical, cyber, and human worlds (HCPS) [11–15].

Designing user interfaces that are not only responsive to user inputs but also adapt dynamically to task-specific requirements, user preferences, and contextual data is a difficult exercise. Multimodal interaction systems provide a promising solution that makes it possible for users to interface with CPS through various means, e.g., speech, touch, gestures, and visual feedback. This elasticity of the system enhances the ability to back-up decision-making processes, especially in routings where it is vital to provide timely and precise responses, such as big-scale manufacturing operations or healthcare establishments.

This study deals with the development and examination of CPS's adaptive multimodal HMI systems which can modulate human interaction with them. First, this study continuously adapts their interface and interaction models with the use of AI-driven insights and real-time data by these systems, which in turn makes human decision-making easier. The study is basically concerned with such domains where different advantages are gained due to the combined pathways of machine precision and human judgment, for achieving efficiency, safety, and correctness. These areas are healthcare, manufacturing, and smart cities. The primary objectives of this research work are as follows:

1. The first objective is to examine how human–machine interaction in CPS may be improved via adaptive multimodal interaction systems.
2. The second objective is to provide an adaptable multimodal HMI framework for CPS applications that incorporates several interaction modalities (such as speech, gestures, haptics, and visual feedback).
3. The third objective is to assess how real-time AI-driven insights affect human decision-making and system performance in intricate settings including smart cities, manufacturing, and healthcare sectors.
4. The fourth objective is to evaluate the efficacy of dynamic and personalized interface modifications inside CPS according to the task difficulty, user preferences, and real-time feedback.

The first section of the research discusses the introduction part, which describes how good HMI has been for CPS and also the research objectives. The second section discusses existing research solutions and problem statements. The third section discusses the research methodology and the algorithms employed for initial implementation. The fourth section discusses results and discussion. Finally, the conclusion, limitations, research implications, and key findings of the proposed work are provided.

16.2 LITERATURE REVIEW

16.2.1 Existing research and solutions

An essential component of "dynamic energy benchmark" development is the correct adjustment of the energy consumption values that appear to be abnormal owing to their malfunctioning operation. Thus, the energy patterns that are made up of the main structure and its substructures can serve as the basis for the whole composed structure definition. A power standard is a statistical characteristic showing the "typical power" of the model when the TUMs are running in normal conditions. Consequently, the method in the thesis applies to the evaluation of the energy benchmark day average of the year [16].

A key component of the HitL paradigm, human–robot interaction is a new trend in Industry 5.0. Despite being widely used in Industry 4.0, industrial robots are kept apart from operators and gated in [17]. A fixed mode like this lacks the flexibility needed to adapt to the frequent changes brought forth by the mass personalization of Industry 5.0. Physical limitations are overcome by human–robot interaction in the unit-level production cell, which promotes flexibility for quick reconfiguration to satisfy a range of manufacturing requirements [18].

Real-time safety distance assessment and collision avoidance are the key topics of this field's research [19,20]. Robot kinematics and dynamics for collaboration modes are controlled by speed/separation monitoring, which takes into account the relative velocities and minimum separation distances between the operator and the robot [21]. To guarantee safety in cooperative modes, hand directing and power/force limiting systems are used. Through physical touch, the former enables people to directly control a robot while maintaining a mild engagement force. By adjusting the force and strength of touch, the latter prevents possible harm to humans [22]. Additionally, by anticipating human movements and body motion trajectories, preplanned robot trajectories or task breakpoints may be adjusted to guarantee safety [23].

Table 16.1 shows the progress made in multimodal interaction, security, decision support systems, and personalization in the current study of HMI

Table 16.1 Comparative analysis of existing studies

Existing (R)	*Year*	*Title*	*Focus area*	*Key findings*
[24]	2024	"Human-machine interaction towards Industry 5.0: Human-centric smart manufacturing"	Manufacturing HMI	Proposed AI-driven adaptive interfaces for real-time decision-making in manufacturing. Showed enhanced operator performance in complex production environments.
[25]	2023	"Cyber-Physical System Security Based on Human Activity Recognition through IoT Cloud Computing"	Multimodal interaction	Explored voice, gesture, and haptic inputs for seamless user interaction in CPS. Demonstrated how multimodal HMI improves usability in healthcare.
[26]	2024	"AI-Based Decision Support Systems in Industry 4.0"	Decision support	Analyzed how AI-powered insights improve decision-making in CPS, particularly in smart city traffic management. Demonstrated improved system efficiency and user satisfaction.
[27]	2023	"Cognitive Load Management"	Cognitive load	Investigated how adaptive HMIs can adjust interface complexity based on user cognitive load in healthcare applications, leading to more accurate decision-making under stress.
[28]	2024	"Towards Attack Detection in Multimodal Cyber-Physical Systems"	Security and HMI	Examined the security vulnerabilities introduced by multimodal HMI systems and proposed adaptive security mechanisms. Found that real-time adjustments reduce the risk of cyber threats in smart city infrastructures.
[29]	2024	"Software and Architecture Orchestration for Process Control"	Personalization	Developed a personalized HMI framework that adjusts based on operator skills and preferences in industrial environments. Results showed improved task completion time and error reduction.

within CPS. The research named "Human-Machine Interaction Towards Industry 5.0: Human-Centric Smart Manufacturing" was directed to the HMI in manufacturing in an Industry 5.0 environment. The research described the situation of synchronized AI-based adaptive user interfaces in a proper time allowing the effective management of the highly complicated production processes. The conclusions showed that such instruments are helpful for operators who need to offset cognitive load and perform tasks when units have divergent demands in a human–machine environment.

In a 2023 research study titled "Cyber-Physical System Security Based on Human Activity Recognition through IoT Cloud Computing," the researchers explored multimodal interaction by analyzing the CPS's sensory, gesture, and voice inputs. The study further demonstrated that multimodal HMI has gained considerable popularity and adoption, especially in situations where the smoothness of communication is the critical component of successful operation. The role of AI in CPS, specifically for decision support in smart city applications such as traffic and pedestrian management, was the main point of a 2024 investigation under the title "AI-Based Decision Support Systems in Industry 4.0." The study states that AI-inducing information that improves the decision-making process is provided; thereby, the system and customer satisfaction are significantly enhanced.

Recent research labeled "Cognitive Load Management" focused on the healthcare sector and investigated cognitive load management and correspondingly unfolded the concepts of adaptive HMIs, ensuring a more precise decision-making process, and at the same time, decreasing the complexity of the interface for the user if he/she is overwhelmed with information. The chart also emphasizes the safety factor of multimodal HMI, which was the main focus of a 2024 paper under entitled "Towards Attack Detection in Multi-Modal Cyber-Physical Systems." This research study investigated the limitations of multimodal systems, especially in the city's smart infrastructures, and a backup measure which adapted to each situation and still provided real-time protection was implemented, thus decreasing the likelihood of cyber-attacks. The choice of HMI customization was the main move of the 2024 study named "Software and Architecture Orchestration for Process Control." This research came out with a customized framework for HMI, aligned with the operators' skill level and preferences in industrial situations, thus shortening the task execution duration and error decrease. All these studies, therefore, provide very important insights into how multimodal, user-centric, and adjustable interfaces for HMI contribute to the enhancement of CPS across various industry types.

Another perspective integrates stakeholders along the industrial value chains to highlight the impact of social networks on CPS. A cyber–physical–social system (CPSS) [30] may bring together the interactions between manufacturers, designers, customers, and other service providers to enhance product lifecycle management. It seeks to achieve Industry 5.0's system-of-systems level interconnections. In order to analyze the cyber organization utilizing social networks and social media with the aid of social computing, CPSS emphasizes social systems [31,32]. It makes it easier for different manufacturers and their stakeholders to communicate with one another.

16.2.2 Problem statement

The need for efficient HMI is growing as key industries like healthcare, manufacturing, and smart cities depend more and more on CPS. Existing HMIs, however, frequently rely on constrained and inflexible interaction models that do not take into consideration the many demands, preferences, and real-time difficulties that users encounter in dynamic and data-intensive situations. Traditional HMIs are not enough to enable optimal decision-making and system performance as CPS grows increasingly autonomous and data-driven, particularly in settings where human judgment and adaptation are essential. Furthermore, the majority of existing systems are unable to dynamically adjust to several user input modalities, including speech, gestures, and haptics, or to change according to the cognitive load, task difficulty, or ambient conditions of the user. Decreased user satisfaction, increased error rate, and far-from-ideal decision-making are the results of this. Moreover, given that many input streams might be a source of vulnerabilities, the connection and the operation of the multimodal systems bring extra security problems. We deal with the mentioned problems by developing a flexible HMI framework, implementing AI and real-time data in CPS contexts to manufacture user-centric interfaces in the last line. The primary purpose is to optimize the system function and human cognitive activities, while still ensuring security and usability in a variety of application domains.

16.3 RESEARCH METHODOLOGY

The study begins with a comprehensive review of literature for the purpose of identifying existing gaps in HMI designs, specifically in the realm of CPS, in order to explore them. An adaptable multimodal HMI structure that includes multiple input designs, such as gestures, speech, haptic feedback, and visual signals, serves as a vehicle for this research. The characteristic of adaptability, which is the outcome of the development process powered by AI-based decision-making support tools, is therefore a major constituent part of the system. In machine learning, the models are used to analyze the data in real-time obtained from CPS, and thus they are able to give the users the necessary perspectives which will help them take more informed decisions.

It is most notably interesting about this system that it can autonomously modify the interface based on user preferences, level of cognitive engagement, and the situational context. It can detect cognitive load using physiological information and tailor its level to the user's mental state. Security is also an important aspect of the system, and a unified security policy prevents all modalities from being hijacked. In CPS environments such as manufacturing and healthcare, even at the stage of prototyping, the aspects of the system will be user-tested with the purpose of evaluating its accuracy, versatility, and usability. The parameters of the system are analyzed according to the following: task accomplishment duration, decision-making precision, and feedback provided by the user. The AI models and interaction mechanisms are progressively updated by users to enhance these characteristics and individualization. To ensure security which is a crucial part of the system, it is well-tested.

16.3.1 Research workflow

This study's approach, which includes both design and experimental phases, is organized to meet each of the stated goals. It includes the subsequent steps:

- **Step 1. Literature Review and Analysis:** Examine the CPS HMI systems in depth, focusing on multimodal interfaces (voice, gestures, and haptics) and AI-driven decision support systems. Identify any issues with the functionality and designs of the HMIs utilized in various sectors, including manufacturing, healthcare, and smart cities.
- **Step 2. Design and Development:** Construct a flexible HMI architecture that integrates speech, gesture, visual, and haptic inputs. In order to assist users in making decisions, employ AI algorithms to analyze CPS data in real-time and generate context-sensitive suggestions. Build systems that can adapt their interface dynamically to user preferences, task complexity, and ambient conditions. This entails incorporating biometric sensors for cognitive load monitoring in order to modify system complexity.
- **Step 3. Initial Implementation:** HMI systems integrated with CPS configurations (specific industries and applications for example, health care) should be prototyped. Use the appropriate hardware (speech input—microphones and non-touch-based interaction—gesture) and software (real-time applications like machine learning for feedback) to design mechanisms for each modality. Make use of machine learning algorithms which contain functions that actively process the incoming information and provide the user with appropriate feedback to increase the efficiency and correctness of his/her decision. These algorithms are expected to be trained from the older datasets and be able to adapt themselves to changes in data, which have not been seen before by the algorithm.

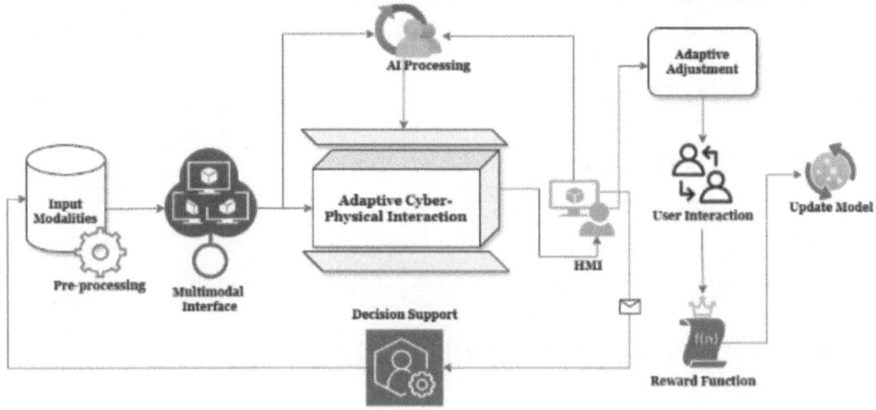

Figure 16.1 Architectural design for multimodal HMI in CPS. (This architecture is crafted using design software.)

- **Step 4. Security Integration:** A layer of security is designed to protect multimodal input pathways to keep an eye on and reduce any security risks during human–machine interactions, using encryption protocols and anomaly detection methods.
- **Step 5. Optimization and Refinement:** Enhance the AI-powered interaction models and decision support algorithms by using system performance metrics and user input. Iteratively improve the system's responsiveness and adaptability while focusing on reducing cognitive load and increasing user pleasure.

Figure 16.1 depicts the architectural design for multimodal HMI in CPS. The first component of the architecture is the database with input modalities. This component is then preprocessed with modalities. The second component is the multimodal interface which acts as the base interface for all the models. Then, the third component is adaptive cyber-physical interaction, which gets into contact with HMI, AI processing, and multimodal interface. This streamlines to adaptive adjustment, user interaction, and reward function, which leads to the updation of the model. In case there is some issue in the final output, the HMI leads to decision support.

16.3.2 Algorithm employed

Algorithm 1: Adaptive multimodal HMI for CPS

STEP 1. INITIALIZE VARIABLES

- Define input modalities M (voice, gesture, haptics, and visual)
- Define the AI decision support function D

- Initialize real-time data input X_t from CPS
- Initialize user preferences P, cognitive load C, and environmental factors E
- Set security threshold S_{th}

STEP 2. MULTIMODAL INTERACTION

For each interaction modality, $m \in M$:

- Capture input $m(t)$ at time t
- Applying preprocessing to convert $m(t)$ into actionable data $I_m(t)$

STEP 3. REAL-TIME DATA PROCESSING

- Capture real-time CPS data $X(t)$
- Feed $X(t)$ and $I_m(t)$ into the AI model D
- Output $O(t)$ represents AI-driven recommendations or actions for decision-making

STEP 4. ADAPTIVE ADJUSTMENT

- Measure cognitive load $C(t)$ using biometric sensors
- If $C(t)$ exceeds the threshold C_{th}, adjust interface complexity:

$$\text{Adjust}(I_m) \leftarrow \text{Simplify}(I_m)$$

- Adjust the interaction based on user preferences and environmental conditions:

$$I_m(t) \leftarrow \text{Personalize}(I_m, P, E)$$

STEP 5. REWARD FUNCTION

- Monitor system interactions for anomalies
- If anomaly $A(t)$ detected:

$$A(t) > S_{\text{th}} \Rightarrow \text{Trigger Security Protocol}$$

- Ensure data integrity across all modalities

STEP 6. UPDATED MODEL

- Continuously update AI decision support function D based on new data $X(t)$ and user feedback F:

$$D \leftarrow D + \Delta D\left(X(t), F\right)$$

16.3.3 Algorithm explanation

The adaptive multimodal HMI system algorithm in CPS incorporates a variety of input modalities, including voice, gestures, haptics, and visual signals, to facilitate human–machine contact. The system initially captures user input in real time using various modalities. Each modality, including voice and gesture, is examined and transformed into information that the system may utilize to inform its choices. At the same time, the CPS generates real-time data streams that show the current state of the physical environment. A decision-making model that uses AI processes this information and user input. The AI analyzes the data and gives the user helpful feedback. One of the key features of this system is its ability to adapt. It uses special sensors to measure how much mental effort the user is putting in and changes the interface to match. If the user is feeling very stressed, the system makes the interface easier by reducing the number of ways to input information or by showing information more congenially.

There are six major steps in deriving the algorithm for CPS-led multimodal HMI. The first step comprises variable initialization mainly defining the input modalities and AI decision support function. To set the security threshold, at this step, the initialization of cognitive load, environment factors, user preferences, and real-time input to CPS is done. The second step includes a multimodal interaction for capturing input and applying preprocessing to convert $m(t)$ into actionable data $I_m(t)$. The third step is real-time data processing for capturing real-time CPS data $X(t)$ to feed $X(t)$ and $I_m(t)$ into the AI model D in order to represent AI-driven recommendations or actions for decision-making as Output $O(t)$. The fourth step is an adaptive adjustment to measure the cognitive load $C(t)$ using biometric sensors. There is a condition that if $C(t)$ exceeds threshold C_{th}, the interface complexity must be adjusted. Otherwise, interaction must be adjusted based on user preferences and environmental conditions. The fifth step is the reward function; in this, the model monitors the system interactions for anomalies. There is a condition in this that if an anomaly $A(t)$ is detected, only then ensure that across all modalities there is data integrity. The final step of this algorithm is updating the model by continuously updating the AI decision support function D based on new data $X(t)$ and user feedback F.

16.4 RESULTS AND DISCUSSION

16.4.1 Experimental setup

The results of the study focus on the performance of the adaptive multimodal HMI system in CPS across several industries, including manufacturing, healthcare, and smart cities. The tests were carried out in real-world scenarios where decision-making, work completion, and security were essential.

16.4.2 Simulation analysis

The metrics of usability contained in the adaptive HMI system for the speech, gesture, haptic, and visual modes are visualized in Table 16.2 (all levels). Task completion time is the duration of a task from the perspective of each of the users referring to the different modalities. The most recent completion times were 42 seconds for haptics and 25 seconds for visual inputs. The error rate refers to the ratio of the tasks that users did not accomplish correctly. Haptics, with a 3% error rate, had the lowest due to its accuracy, whereas gestures were the main culprit, with a 6% error rate, which can be attributed to the difficulties in the precise gesture identification. The user satisfaction levels were measured on a scale of 1–10. The user satisfaction score for visual inputs, probably, because of their accuracy and user-friendliness, was the highest (9.1). Haptics had the second highest user satisfaction score of 8.8, indicating that the users were likely to experience the tactile sensation. The higher error rates and the learning curve of gestures might have been the reasons for the lowest satisfaction score of 7.9 for gesture inputs.

Discussion: Figure 16.2 displays the Usability Metrics for Various Input Modalities. The visual modality shows the fastest work completion time with a relatively low error rate, even though voice and haptic feedback were evaluated as having higher user satisfaction. Because real-time interpretation in dynamic conditions is challenging, gesture-based inputs exhibited a somewhat higher error rate while being relatively quick.

Table 16.3 presents a survey of the improvement in decision accuracy in the three applications (healthcare, manufacturing, and smart cities) that have AI assistance and those that do not. The No AI column, on the other hand, shows the base decision accuracy without AI support. Healthcare, by far, had the largest baseline decision accuracy without AI at 72%, and manufacturing had the smallest at 68%. The "With AI support" column, on the other hand, documents how decision accuracy was improved by AI's insightful analysis. Accuracy had the same increase in all applications, with healthcare increasing to 90%, manufacturing to 88%, and smart cities to 92%. The accuracy enhancement column shows the percentage of improvement in decision accuracy. Out of the remaining, that is, manufacturing apps, the smartest cities had

Table 16.2 Usability metrics for different input modalities

Modality	*Average task completion time (s)*	*Error rate (%)*	*User satisfaction (1–10)*
Voice	35	5	8.5
Gesture	28	6	7.9
Haptics	42	3	8.8
Visual	25	4	9.1

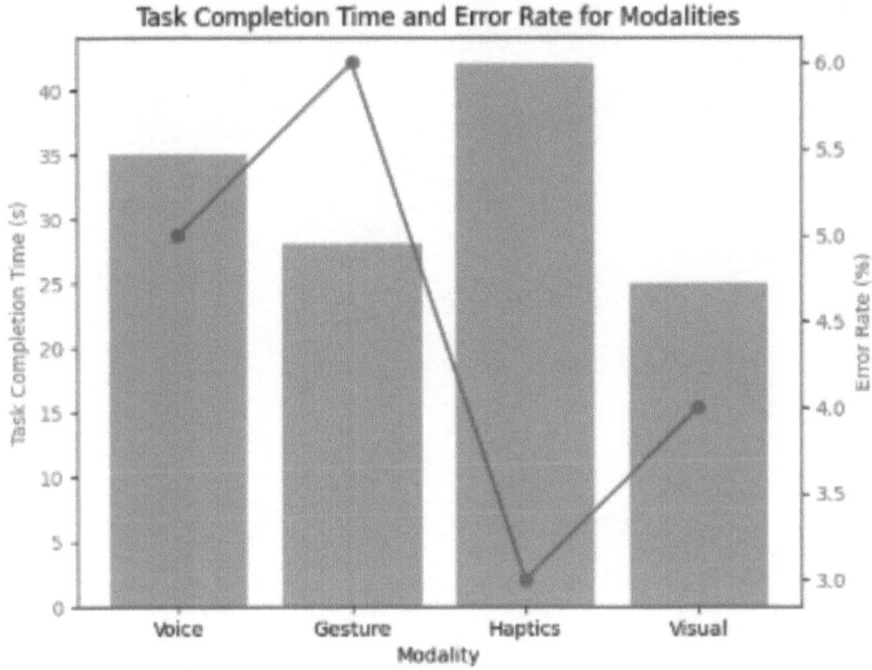

Figure 16.2 Usability metrics for different input modalities.

Table 16.3 AI-driven decision support impact on decision accuracy

Application	*Without AI support (%)*	*With AI support (%)*	*Accuracy improvement (%)*
Healthcare	72	90	18
Manufacturing	68	88	20
Smart cities	75	92	17

a decision accuracy of 17% and healthcare got the least from performance, 18%. It thus shows that embedding AI in the decision-making process contributes to such outstanding results when it comes to this entangled performance.

Discussion: AI-driven decision support significantly improved decision accuracy in all studied scenarios. The largest benefits (20%) were seen in the industrial sector, indicating that AI insights might help solve complex problems, particularly when dealing with large CPS data sets (Figure 16.3).

Table 16.4 investigates the influence of cognitive load (CL) and interface complexity (IC) on the time of task completion and user input for interface

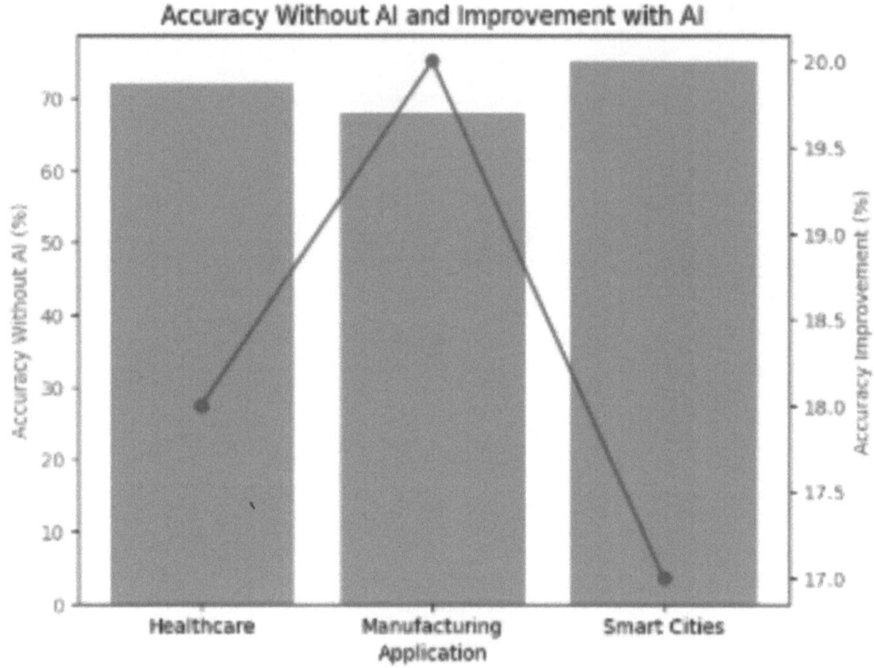

Figure 16.3 AI-driven decision support impact on decision accuracy.

Table 16.4 Cognitive load vs interface complexity adaptation

Cognitive load (CL)	*Interface complexity (IC)*	*Task completion time (s)*	*User feedback on complexity (1–10)*
Low (CL < 30%)	High	32	7.5
Moderate (30% ≤ CL < 60%)	Medium	27	8.6
High (CL ≥ 60%)	Low	24	9.2

complexity. The completion time was 32 seconds mainly due to low cognitive load (less than 30%) and over-controlling complexity. The range of correctness of the complexity of the interactive adaptation system is the value of the algorithm to the degree 7.5. This suggests that peoples' continuing response is still being able to cope with more complex interfaces, even though they might not be so complex users, but at the same time, users' memory hardly gets involved in the operation. The cognitive load was from moderate to high or 30%–60%, and the interface complexity was scaled down to the medium level. The test completion time improved

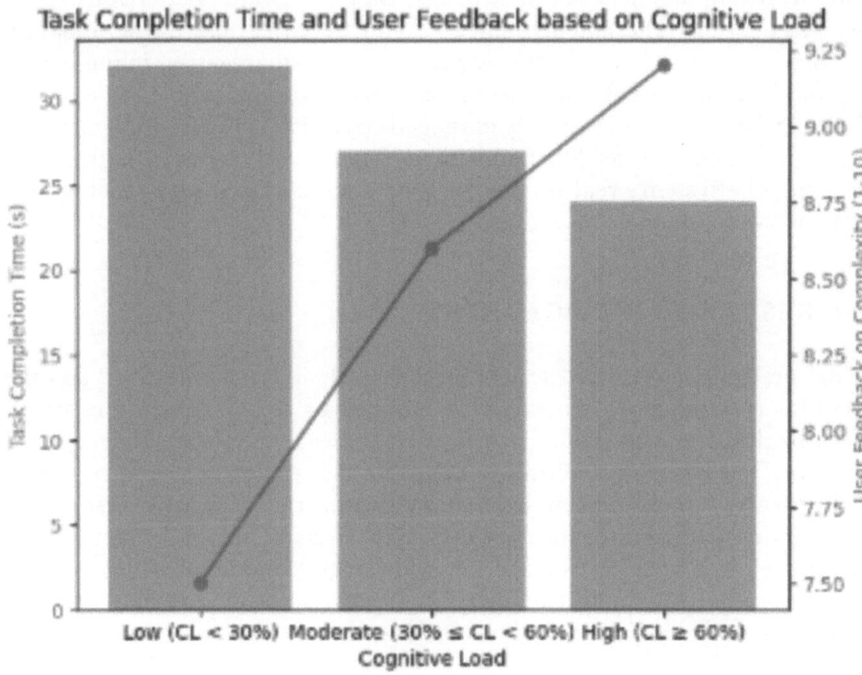

Figure 16.4 Cognitive load vs interface complexity adaptation.

to 27 seconds, and user feedback on complexity increased to 8.6, indicating that reducing complexity at higher cognitive loads improves user satisfaction and performance. When the cognitive load was substantial (more than 60%), the interface complexity was further decreased to low. Task completion time decreased to 24 seconds, while user ratings on complexity peaked at 9.2. This suggests that users are less overwhelmed by information when the interface is made simpler while they are under a lot of cognitive loads, which significantly improves user comfort and productivity.

Discussion: The adaptive interface complexity effectively reduced cognitive overload during tasks (Figure 16.4). Better user feedback and faster work completion times were seen when interface complexity was reduced at high cognitive load levels, indicating that the adaptive system was effective in making changes in real time.

16.5 KEY FINDINGS

Multimodal interaction efficiency, which combines speed and accuracy, is one of the study's main conclusions. Of the modalities examined, haptic and visual were the most effective. However, because voice interfaces are easier to use in less demanding settings, they were rated better for overall user satisfaction. The employment of AI in decision-making significantly improved accuracy in applications pertaining to smart cities, manufacturing, and healthcare, with increases in accuracy ranging from 17% to 20%. The adaptive system successfully managed cognitive load by adjusting interface complexity. User feedback indicated that when there was a high cognitive load, the system's real-time adjustments were crucial for reducing task complexity.

16.6 RESEARCH IMPLICATIONS

AI-driven decision support in conjunction with adaptive interfaces has the potential to completely transform a number of sectors, including smart cities, healthcare, and manufacturing, by increasing choice efficiency and accuracy. The interface's ability to adapt to user preferences and cognitive load provides a more personalized and efficient user experience, especially in situations with significant data and high levels of strain. In sectors like healthcare, data integrity is essential, and the multimodal HMI framework provides a novel method of maintaining security across a variety of input modalities.

16.7 LIMITATIONS

The system has been highly successful in manufacturing, healthcare, and smart cities, however, it could be a little bit less flexible depending on the industry, especially if certain input modalities are needed. Along with the addition of many input modalities and real-time adaptation techniques, the technical difficulties may particularly be in the case where there is insufficient hardware support or infrastructure. The efficiency of AI-driven decision assistance is largely associated with the amount and quality of real-time data from CPS. Some flaws or loudness in the system might decrease its effectiveness.

16.8 CONCLUSION

The study analyzed the effectiveness of adaptable interactive human–machine systems in CPS to the processes of decision-making and the function of the whole system. Applying the AI techniques and the actual data from sensors, the methods that the authors proposed can lead to the productivity, user satisfaction, and accuracy in making decisions in difficult areas such as smart city infrastructure, manufacturing, and health care to considerably be improved. This simulative application allows for changes to user preferences, cognitive load, and environmental conditions, which

6. J. Leng, W. Sha, B. Wang, P. Zheng, C. Zhuang, Q. Liu, T. Wuest, D. Mourtzis, and L. Wang, "Industry 5.0: Prospect and retrospect," *Journal of Manufacturing Systems*, vol. 65, pp. 279–295, Oct. 2022, doi: 10.1016/j.jmsy.2022.09.017.
7. European Commission, Directorate General for Research and Innovation, Industry 5.0: towards a sustainable, human-centric, resilient European industry. LU: Publications Office, 2021. Accessed: Sep. 4, 2022. [Online]. Available: https://data.europa.eu/doi/10.2777/308407.
8. P. K. R. Maddikunta, Q.-V. Pham, B. Prabadevi, N. Deepa, K. Dev, T. R. Gadekallu, R. Ruby, and M. Liyanage, "Industry 5.0: A survey on enabling technologies and potential applications," *Journal of Industrial Information Integration*, vol. 26, p. 100257, Mar. 2022, doi: 10.1016/j.jii.2021.100257.
9. C. Hildebrandt, A. Köcher, C. Küstner, C.-M. López-Enríquez, A. W. Müller, and B. Caesar, "Ontology building for cyber-physical systems: Application in the manufacturing domain," *IEEE Transactions on Automation Science and Engineering*, vol. 17, no. 3, pp. 1266–1282, Jul. 2020, doi: 10.1109/TASE.2020.2991777.
10. W. Hao, T. Yang, and Q. Yang, "Hybrid statistical-machine learning for real-time anomaly detection in industrial cyber-physical systems," *IEEE Transactions on Automation Science and Engineering*, vol. 20, no. 1, pp. 1–15, 2021, doi: 10.1109/TASE.2021.3073396.
11. B. Wang, P. Zheng, Y. Yin, A. Shih, and L. Wang, "Toward human-centric smart manufacturing: A human-cyber-physical systems (HCPS) perspective," *Journal of Manufacturing Systems*, vol. 63, pp. 471–490, Apr. 2022, doi: 10.1016/j.jmsy.2022.05.005.
12. X. Li, J. He, P. Vijayakumar, X. Zhang, and V. Chang, "A verifiable privacy-preserving machine learning prediction scheme for edge-enhanced HCPSs," *IEEE Transactions on Industrial Informatics*, vol. 18, no. 8, pp. 5494–5503, Aug. 2022, doi: 10.1109/TII.2021.3110808.
13. S. Lou, Y. Zhang, R. Tan, and C. Lv, "A human-cyber-physical system enabled sequential disassembly planning approach for a human-robot collaboration cell in Industry 5.0," *Robotics and Computer-Integrated Manufacturing*, vol. 87, p. 102706, Jun. 2024, doi: 10.1016/j.rcim.2023.102706.
14. Y. Feng, X. Jiang, Z. Hong, Z. Li, H. Si, B. Hu, Z. Zhang, H. Qiu, and J. Tan, "Human-cyber-physical system for operation in nuclear reactor possessing asymmetric multi-task learning-based predicting framework," *Journal of Manufacturing Systems*, vol. 64, pp. 443–453, Jul. 2022, doi: 10.1016/j.jmsy.2022.07.008.
15. L. Zong, F. H. Memon, X. Li, H. Wang, and K. Dev, "End-to-end transmission control for cross-regional industrial internet of things in industry 5.0," *IEEE Transactions on Industrial Informatics*, vol. 18, no. 6, pp. 4215– 4223, Jun. 2022, doi: 10.1109/TII.2021.3133885.
16. D. Singh, M. Arshad, B. Tyagi, and G. Kalia, "Predictive maintenance strategies for HVAC systems: Leveraging MPC, dynamic energy performance analysis, and ML classification models," *Iconic Research and Engineering Journals*, vol. 7, no. 4, pp. 98–108, 2023.

17. A. K. Inkulu, M. V. A. R. Bahubalendruni, A. Dara, and S. K., "Challenges and opportunities in human-robot collaboration context of Industry 4.0- a state-of-the-art review," *Industrial Robot: The International Journal of Robotics Research and Application*, vol. 49, no. 2, pp. 226–239, Feb. 2022, doi: 10.1108/IR-04–2021–0077.
18. S. Proia, R. Carli, G. Cavone, and M. Dotoli, "Control techniques for safe, ergonomic, and efficient human-robot collaboration in the digital industry: A survey," *IEEE Transactions on Automation Science and Engineering*, vol. 19, no. 3, pp. 1798–1819, Jul. 2022, doi: 10.1109/TASE.2021.3131011.
19. R. Jahanmahin, S. Masoud, J. Rickli, and A. Djuric, "Human-robot interactions in manufacturing: A survey of human behavior modeling," *Robotics and Computer-Integrated Manufacturing*, vol. 78, p. 102404, Dec. 2022, doi: 10.1016/j.rcim.2022.102404.
20. S. Kumar, C. Savur, and F. Sahin, "Survey of human-robot collaboration in industrial settings: Awareness, intelligence, and compliance," *IEEE Transactions on Systems, Man, and Cybernetics: Systems*, vol. 51, no. 1, pp. 280–297, Jan. 2021, doi: 10.1109/TSMC.2020.3041231.
21. L. Gualtieri, E. Rauch, and R. Vidoni, "Emerging research fields in safety and ergonomics in industrial collaborative robotics: A systematic literature review," *Robotics and Computer-Integrated Manufacturing*, vol. 67, p. 101998, Feb. 2021, doi: 10.1016/j.rcim.2020.101998.
22. N. Nikolakis, V. Maratos, and S. Makris, "A cyber-physical system (CPS) approach for safe human-robot collaboration in a shared workplace," *Robotics and Computer-Integrated Manufacturing*, vol. 56, pp. 233–243, Apr. 2019, doi: 10.1016/j.rcim.2018.10.003.
23. A. Pereira, M. Baumann, J. Gerstner, and M. Althoff, "Improving efficiency of human-robot coexistence while guaranteeing safety: Theory and user study," *IEEE Transactions on Automation Science and Engineering*, vol. 20, no. 4, pp. 2706–2719, 2022, doi: 10.1109/TASE.2022.3201970.
24. J. Yang, Y. Liu, and P. L. Morgan, "Human-machine interaction towards Industry 5.0: Human-centric smart manufacturing," *Digital Engineering*, vol. 2, p. 100013, 2024, ISSN 2950-550X, doi: 10.1016/j.dte.2024.100013.
25. S. Achar, N. Faruqui, M. Whaiduzzaman, A. Awajan, and M. Alazab, "Cyber-physical system security based on human activity recognition through IoT cloud computing," *Electronics*, vol. 12, no. 8, p. 1892, 2023, doi: 10.3390/electronics12081892.
26. M. Soori, F. K. G. Jough, R. Dastres, and B. Arezoo, "AI-based decision support systems in industry 4.0, a review," *Journal of Economy and Technology*, 2024, ISSN 2949–9488, doi: 10.1016/j.ject.2024.08.005.
27. E. E. Howie, H. Dharanikota, E. Gunn, O. Ambler, R. Dias, S. J. Wigmore, R. J. E. Skipworth, and S. Yule, "Cognitive load management: An invaluable tool for safe and effective surgical training," *Journal of Surgical Education*, vol. 80, no. 3, pp. 311–322, Mar. 2023, doi: 10.1016/j.jsurg.2022.12.010. Epub 2023 Jan 19. PMID: 36669990.

28. A. E. Hong, P. P. Malinovsky, and S. K. Damodaran, "Towards attack detection in multimodal cyber-physical systems with sticky HDP-HMM based time series analysis," *Digital Threats*, vol. 5, no. 1, Article 5, p. 21, Mar. 2024, doi: 10.1145/3604434.
29. C. Serôdio, P. Mestre, J. Cabral, M. Gomes, and F. Branco, "Software and architecture orchestration for process control in industry 4.0 enabled by cyber-physical systems technologies," *Applied Sciences*, vol. 14, no. 5, p. 2160, 2024, doi: 10.3390/app14052160.
30. J. J. Zhang, F.-Y. Wang, X. Wang, G. Xiong, F. Zhu, and Y. Lv, "Cyber-physical-social systems: The state of the art and perspectives," *IEEE Transactions on Computational Social Systems*, vol. 5, no. 3, pp. 829–840, Sep. 2018, doi: 10.1109/TCSS.2018.2861224.
31. G. Xiong, T. S. Tamir, Z. Shen, X. Shang, H. Wu, and F.-Y. Wang, "A survey on social manufacturing: A paradigm shift for smart prosumers," *IEEE Transactions on Computational Social Systems*, vol. 10, no. 5, pp. 1–19, 2022, doi: 10.1109/TCSS.2022.3180201.
32. K. Wang, J. An, M. Zhou, Z. Shi, X. Shi, and Q. Kang, "Minority-weighted graph neural network for imbalanced node classification in social networks of internet of people," *IEEE Internet of Things Journal*, vol. 10, no. 1, pp. 330–340, 2022, doi: 10.1109/JIOT.2022.3200964.

Index

leads to a reduction in cognitive load and an increase in the synergy of user participation. The existence of dependency on data and the scope of the technologies" application to other sectors are factors in the discussion of the fact that AI-driven technologies are positive, yet decision-making, via multimodal means, is the main point of issue. Multimodal interaction efficiency, which combines speed and accuracy, is one of the study's main conclusions. Of the modalities examined, haptic and visual were the most effective. However, because voice interfaces are easier to use in less demanding settings, they were rated as better for overall user satisfaction. The employment of AI in decision-making significantly improved the accuracy in applications pertaining to smart cities, manufacturing, and healthcare, with increases ranging from 17% to 20%. The adaptive system successfully managed cognitive load by adjusting interface complexity. User feedback indicated that when there was a high cognitive load, the system's real-time adjustments were crucial for reducing task complexity. Further studies must be carried out to increase the system's durability in white noise and the lack of data interference, as well as to improve the adaptability of the model in the occurrence of novel scenarios. In a word, this research provides a platform for more advanced HMI systems, enabling access to more secure, smarter, and user-centered interactions in the rapidly developing field of CPS.

REFERENCES

1. Z. Liu, H. Wang, H. Wei, M. Liu, and Y.-H. Liu, "Prediction, planning, and coordination of thousand-warehousing-robot networks with motion and communication uncertainties," *IEEE Transactions on Automation Science and Engineering*, vol. 18, no. 4, pp. 1705–1717, Oct. 2021, doi: 10.1109/TASE.2020.3015110.
2. K. S. Welfare, M. R. Hallowell, J. A. Shah, and L. D. Riek, "Consider the human work experience when integrating robotics in the workplace," in *2019 14th ACM/IEEE International Conference on Human-Robot Interaction (HRI)*, Daegu, Korea (South): IEEE, Mar. 2019, pp. 75–84. doi: 10.1109/HRI.2019.8673139.
3. X. Yao, N. Ma, J. Zhang, K. Wang, E. Yang, and M. Faccio, "Enhancing wisdom manufacturing as industrial metaverse for industry and society 5.0," *Journal of Intelligent Manufacturing*, vol. 35, pp. 235–255, Nov. 2022, doi: 10.1007/s10845-022-02027-7.
4. S. Huang, B. Wang, X. Li, P. Zheng, D. Mourtzis, and L. Wang, "Industry 5.0 and society 5.0—Comparison, complementation and co-evolution," *Journal of Manufacturing Systems*, vol. 64, pp. 424–428, Jul. 2022, doi: 10.1016/j.jmsy.2022.07.010.
5. A. Akundi, D. Euresti, S. Luna, W. Ankobiah, A. Lopes, and I. Edinbarough, "State of industry 5.0—Analysis and identification of current research trends," *Applied System Innovation*, vol. 5, no. 1, p. 27, Feb. 2022, doi: 10.3390/asi5010027.